Universitext

T0234363

Universitext

Universitext is a series of textbooks that presents material from a wide variety of mathematical disciplines at master's level and beyond. The books, often well class-tested by their author, may have an informal, personal even experimental approach to their subject matter. Some of the most successful and established books in the series have evolved through several editions, always following the evolution of teaching curricula, to very polished texts.

Thus as research topics trickle down into graduate-level teaching, first textbooks written for new, cutting-edge courses may make their way into *Universitext*.

For further volumes:
http://www.springer.com/series/223

Sergei Ovchinnikov

Graphs and Cubes

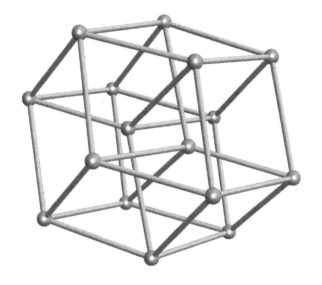

 Springer

Sergei Ovchinnikov
Department of Mathematics
San Francisco State University
San Francisco, CA 94132
USA
sergei@sfsu.edu

ISSN 0172-5939   e-ISSN 2191-6675
ISBN 978-1-4614-0796-6   e-ISBN 978-1-4614-0797-3
DOI 10.1007/978-1-4614-0797-3
Springer New York Dordrecht Heidelberg London

Library of Congress Control Number: 2011935732

Mathematics Subject Classification (2010): 05CXX, 68R10

Printed on acid-free paper

Springer is part of Springer Science+Business Media (www.springer.com)

To my dear wife, Galya

Preface

This book is an introductory text in graph theory, focusing on partial cubes, that is, graphs that are isometrically embeddable into hypercubes of an arbitrary dimension. This branch of graph theory has developed rapidly during the past three decades, producing exciting results and establishing links to other branches of mathematics. Because of their rich structural properties, partial cubes have found applications in theoretical computer science, coding theory, data transmission, genetics, and even the political and social sciences. However, this research area has previously failed to trickle down into graduate-level teaching of graph theory. In fact, even the term "partial cube" can't be found in standard textbooks on graph theory. In this book, I attempt to remedy this situation.

Exercising a concrete approach to graph theory, this book focuses on three classes of graphs: bipartite graphs, cubical graphs, and partial cubes (introduced in Chapters 2, 4, and 5, respectively). Cubical graphs are graphs that are embeddable into hypercubes; if they are isometrically embeddable into hypercubes, then they are called partial cubes.

Cubical graph theory is a branch of graph theory that is reasonably small, yet deep enough to demonstrate the power and tools of the general theory. It can serve as a launching pad for studies of other topics in graph theory and their applications.

The book is organized into eight chapters. The first four provide the concepts and tools needed to understand Chapter 5 (Partial Cubes), and the remaining three chapters build on this understanding with examples, applications, and a foundation for further exploration. The specific topics addressed in each chapter are as follows.

Chapter 1 introduces several basic concepts of graph theory that are used throughout the book. I adopt the terminology and notations used by J. A. Bondy and U. S. R. Murty in their influential text *Graph Theory with Applications* (Bondy and Murty, 1976) and in their recent book *Graph Theory* (Bondy and Murty, 2008). Many other fundamental concepts of graph theory are introduced gradually in the rest of the book, as needed.

Because all cubical graphs are bipartite, Chapter 2 presents an elementary theory of bipartite graphs. This chapter establishes various characterizations of bipartite graphs and discusses their structural properties. In the discussion, I emphasize the importance of geometric structures of betweenness and convexity, concepts that recur frequently in the treatment of cubical graphs later in the book.

Chapter 3 focuses on the beautiful geometric and combinatorial objects known as hypercubes (or cubes, as I generally call them for the sake of simplicity). The first five sections of Chapter 3 consider various instances of cubes in geometry, algebra, and graph theory. Section 3.6 explores the subtleties of how a cube's dimensionality affects its classification as a Cartesian product (finite-dimensional cubes are Cartesian products, but infinite-dimensional cubes are not; they are weak Cartesian products), whereas later sections address the nature of cubes as highly symmetrical objects. Section 3.8 describes symmetry groups, and Section 3.10 characterizes finite cubes.

Chapter 4 presents practically everything that is known about cubical graphs (and we do not know much). Unlike with bipartite graphs (Chapter 2) and partial cubes (Chapter 5), there is no effective characterization of cubical graphs in general. However, a criterion based on c-valuations (Section 4.2) is a useful tool for establishing properties of some special classes of cubical graphs, such as dichotomic trees (Section 4.3).

Chapter 5, the central chapter of the book, presents the concept of the partial cube as a cubical graph that admits an isometric embedding into a cube. This chapter deals mainly with structural properties of partial cubes, using techniques introduced in Chapters 2 through 4. One goal of this chapter is to establish several characterizations of partial cubes. Another goal is to demonstrate how general mathematical techniques of constructing new objects from old ones work for partial cubes.

Chapter 6 expands the understanding of partial cubes by defining them as graphs isometrically embeddable into integer lattices (grids), which are themselves definable as partial cubes. This chapter is devoted entirely to lattices and their isometric subgraphs.

Chapter 7 moves on to a particularly beautiful geometric example of partial cubes: region graphs of hyperplane arrangements. In this chapter, I prove that such graphs are indeed partial cubes, and I present their algebraic and geometric applications.

Finally, Chapter 8 lays down a mathematical foundation for cubical graph applications. Two kinds of token systems—cubical systems and media—are presented and defined axiomatically by imposing compelling independent conditions on the systems. The last section provides a stochastic model for system evolution.

There are few prerequisites for the main text. It is assumed that the reader is familiar with basic mathematical concepts and methods on the level of undergraduate courses in discrete mathematics, linear algebra, group theory, and topology of Euclidean spaces. Although the book is intended for lower-division graduate students, I believe that it will find readership in a much wider audience.

I have chosen a very geometric mode of presentation for this book, in accordance with its topic. The reader will find many drawings illustrating concepts, proofs, and the exercises included (along with historical notes) at the end of every chapter. I encourage readers to explore the exercises fully, and even to use them as the basis for research projects.

I want to thank my Springer editor Kaitlin Leach for her support throughout the preparation of this book.

Berkeley, California   *Sergei Ovchinnikov*
June 2011

Contents

1

Graphs

The main goal of this chapter is to introduce some very basic concepts and constructions of graph theory. We consciously made this chapter short and proved only a few simple facts. Combined with exercises at the end of the chapter this material can be used as a concise introduction to graph theory.

1.1 Graphs and Drawings

A *graph* G is an ordered pair (V, E) consisting of a nonempty set of *vertices* V and a set of *edges* E. Each edge in E is an unordered pair uv of distinct vertices of G. A graph is *finite* if its vertex set is finite, and *infinite* otherwise.

When dealing with several graphs $G, H, \ldots$ at the time, we distinguish their vertex and edge sets by writing $V(G), E(G), V(H), E(H)$, and so on.

Example 1.1. Let $G = (V(G), E(G))$, where

$$V(G) = \{u, v, x, y\}$$
$$E(G) = \{uv, ux, uy, vx, vy, xy\}.$$

This graph is known as the complete graph on four vertices K_4.

Example 1.2. Let $H = (V(H), E(H))$ where

$$V(H) = \{u, v, w, x, y, z\}$$
$$E(G) = \{ux, uy, uz, vx, vy, vz, wx, wy, wz\}$$

The graph H is called the complete bipartite graph $K_{3,3}$.

The terms "vertex" and "edge" have a distinct geometric flavor: their prototypes are vertices and edges of 3-dimensional polytopes. Vertices of a polytope are points in the 3-dimensional space and edges are straight line segments connecting vertices (see the tetrahedron in Figure 1.1 and the cube in Figure 1.6).

We draw a finite graph in the plane $\mathbf{R}^2$ or in the 3-dimensional space $\mathbf{R}^3$ by selecting points representing the vertices of the graph and depicting each edge by a line joining the points representing its ends. To make vertices clearly visible we represent them by small circles.

There are many ways to draw a graph. The choice of points representing vertices and the shapes of lines representing edges has no significance. A drawing of a graph merely represents the incidence relation holding between its vertices and edges. For instance, the graph depicted in Figure 1.1 is a drawing of the complete graph K_4 from Example 1.1.

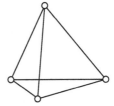

Figure 1.1. The tetrahedron and its graph K_4.

Many concepts of graph theory are suggested by graph drawings. If $e = uv$ is an edge of a graph G, then e is said to *join* vertices u and v that are the *ends* of e. In this case, the vertices u and v are said to be *adjacent* in G and *incident* to the edge e. Two edges are *incident* or *adjacent* if they have a common end. Two adjacent vertices are *neighbors*; the set of all neighbors of a vertex v is denoted by $N_G(v)$ (or $N(v)$ if no ambiguity arises from the notation).

The number of neighbors of a vertex v in a finite graph $G = (V, E)$ is called the *degree* of v and denoted $\deg_G(v)$ (or simply $\deg(v)$). The following result sometimes is called "the first theorem of graph theory."

$$\sum_{v \in V} \deg_G(v) = 2|E|. \tag{1.1}$$

This formula is obtained by "counting in two ways" the number m of pairs (v, e) such that $v \in V$ is an end of $e \in E$. Each vertex v has $\deg_G(v)$ neighbors, so $m = \sum_{v \in V} \deg_G(v)$. On the other hand, each edge e has two ends, so $m = 2|E|$.

The lines representing two edges in a drawing of a graph may intersect at a point that does not represent a vertex (see Figure 1.1, right). A graph is called *planar* if it can be drawn in the plane in such a way that edges meet only at vertices that are their common ends, and such a drawing is called a *planar embedding* of the graph. Two such embeddings of the graph K_4 are shown in Figure 1.2.

The three drawings of K_4 depicted in Figures 1.1 and 1.2 do not tell us which points in the plane are assigned to particular vertices of K_4 in Example 1.1. We can do it by labeling vertices and edges of the drawing by elements of the vertex and edge sets, respectively; see Figure 1.3. A labeling of vertices of $K_{3,3}$ from Example 1.2 is shown in the same figure. It is clear that vertices of K_4 can be labeled in any possible way; the edges must be labeled accordingly. This is not true for the graph $K_{3,3}$, as the reader should verify.

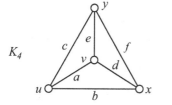

Figure 1.2. Two planar embeddings of the complete graph K_4.

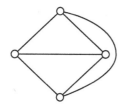
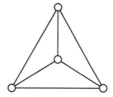

Figure 1.3. Labeled graphs K_4 and $K_{3,3}$. Edges of K_4 are labeled by $a = uv$, $b = ux$, $c = uy$, $d = vx$, $e = vy$, and $f = xy$. Edges of $K_{3,3}$ are not labeled.

Not every graph is planar. Examples of nonplanar graphs are shown in Figure 1.4. (Compare the two drawings of $K_{3,3}$ in Figures 1.3 and 1.4 by labeling vertices of the graph in Figure 1.4.)

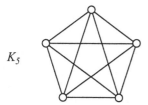
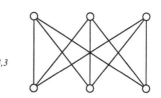

Figure 1.4. Nonplanar graphs K_5 and $K_{3,3}$.

The 1-skeleton[1] of the tetrahedron in Figure 1.1 together with the vertices of the tetrahedron can be considered as a 3-dimensional drawing of the graph K_4. In fact, any finite graph can be represented in $\mathbf{R}^3$ with edges that are straight line segments intersecting only at vertices (cf. Exercise 1.7).

1.2 Isomorphisms and Automorphisms

The drawing of a graph G is itself a graph, say H. The vertices of H are points in the plane and its edges are lines connecting some of these points. The graphs G and H are not identical but have the same structure defined by the adjacency relation. To formalize this idea, we introduce two important concepts.

Two graphs G and H are *identical*, written $G = H$, if $V(G) = V(H)$ and $E(G) = E(H)$. Two graphs are *isomorphic*, written $G \cong H$, if there is a bijection $\varphi : V(G) \to V(H)$ preserving the adjacency relation; that is,

$$uv \in E(G) \quad \text{if and only if} \quad \varphi(u)\varphi(v) \in E(H).$$

The map φ is called an *isomorphism* between G and H.

By definition of a drawing of a graph G, the graph defined by the drawing is isomorphic to the graph G. We do not normally distinguish between isomorphic graphs and we denote them by the same symbol. For instance, every graph in Figures 1.1 and 1.2 is the graph K_4 from Example 1.1. We assign labels to vertices and edges (cf. Figure 1.3) mainly for reference purposes as needed in proofs and examples. Moreover, we often write $G = H$ instead of $G \cong H$.

An *automorphism* of a graph is an isomorphism of the graph onto itself. Thus, an automorphism is a permutation of the vertex set. The set of all automorphisms of a graph G forms a group (cf. Exercise 1.10). We denote this group by $\mathrm{Aut}(G)$.

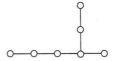

Figure 1.5. A graph with the trivial automorphism group.

The size of the group $\mathrm{Aut}(G)$ can be regarded as a "degree" of symmetry of the graph G. For instance, it can be easily seen that any permutation of the five vertices of the graph K_5 in Figure 1.4 defines an automorphism of K_5. Thus, $\mathrm{Aut}(K_5) = S_5$, the symmetric group of all permutations of a 5-element set. There are $5! = 120$ automorphisms of K_5. On the other hand, there are no nontrivial automorphisms of the graph in Figure 1.5 (cf. Exercise 1.11).

---

[1] The 1-skeleton of a polytope is the union of its edges.

1.3 Walks, Paths, and Cycles

A *walk* in a graph G is a sequence

$$W = v_0 e_1 v_1 e_2 \cdots v_{i-1} e_i v_i \cdots v_{k-1} e_k v_k,$$

where $v_0, v_1, \ldots, v_k$ are vertices of G, $e_1, e_2, \ldots, e_k$ are edges of G, and v_{i-1} and v_i are the ends of e_i for $1 \le i \le k$. We say that W is a walk from v_0 to v_k, or a $v_0 v_k$-walk, and refer to v_0 and v_k as *initial* and *terminal vertices* of W, respectively. They are also called the *ends* of W. All other vertices (if any) are *internal* vertices of W. Because an edge is a pair of its ends, the walk W is determined by the sequence of its vertices and we usually write it as $W = v_0 v_1 \cdots v_k$. The integer k (the number of edges in W) is the *length* of the walk W. A walk is *even* or *odd* as its length is even or odd. It is convenient to define a *trivial walk* as a single vertex; it has zero length.

By reversing the order of terms in a $v_0 v_k$-walk W, we obtain a $v_k v_0$-walk denoted by W^{-1}:

$$W = v_0 v_1 \cdots v_{k-1} v_k \quad \xrightarrow[\text{the order of vertices}]{\text{reversing}} \quad v_k v_{k-1} \cdots v_1 v_0 = W^{-1}.$$

A $v_0 v_k$-walk W and a $v_k v_n$-walk W' can be concatenated at v_k resulting in the $v_0 v_n$-walk denoted WW':

$$W = v_0 \cdots v_{k-1} v_k \quad W' = v_k v_{k+1} \cdots v_n$$

$$\text{concatenating}$$

$$WW' = v_0 \cdots v_{k-1} v_k v_{k+1} \cdots v_n.$$

We note that there may be repetitions of vertices and edges in a walk. For instance, if $e = uv$ is an edge of a graph G, then any finite sequence in the form $ueve \cdots uev$ is a uv-walk in G.

Let $W = v_0 v_1 \cdots v_k$ be a nontrivial walk in G. For any $1 \le i < j \le k$, the $v_i v_j$-walk $v_i \cdots v_j$ is called a *segment* (or more precisely the $v_i v_j$-*segment*) of W. We say that W is *closed* if $v_0 = v_k$ and *open* otherwise ($v_0 \ne v_k$). In a closed walk W, any two vertices v_i and v_j with $i < j$ define two sequences: $v_i \cdots v_j$ and $v_j \cdots v_0 \cdots v_i$. We call both sequences *segments* of W.

A *path* P in G is a walk in which all vertices (and therefore all edges) are distinct. If u and v are initial and terminal vertices of P, respectively, we say that P is an uv-*path* or that P *connects* u and v. Clearly, every edge $e = uv$ can be regarded either as a uv- or vu-path. A trivial path consists of a single vertex. A *cycle* is a closed walk $v_0 v_1 \ldots v_{k-1} v_0$ of length $k \ge 3$ in which all vertices $v_0, v_1, \cdots, v_{k-1}$ are distinct.

Example 1.3. The cube and its graph Q_3 are depicted in Figure 1.6. In the graph Q_3,

$$W = uadcbadx$$

is a ux-walk of length 7 which is not a path,

$$P = uadcbvwx$$

is a path from u to x of length 7, and

$$C = uvbcdxu$$

is a cycle of length 6. The path P and the cycle C are shown in Figure 1.7.

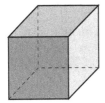 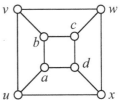

Figure 1.6. The cube and its graph Q_3.

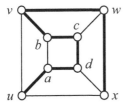 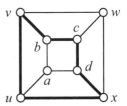

Figure 1.7. The path P and the cycle C from Example 1.3.

We say that a walk W contains a walk W' (equivalently, W' is contained in W) if the sequence W' is a subsequence of the sequence W. A segment of a walk is contained in the walk, but the converse is not always true. For instance, the walk $uadcbadx$ contains the walk $uadx$ in Figure 1.6, but the latter is not a segment of the former.

Theorem 1.4. *Any open uv-walk W contains a path from u to v.*

Proof. If $W = v_0v_1v_2\cdots v_k$ contains no repeated vertices, then it is itself a path. Suppose that $v_i = v_j$ for some $i < j$. Let us remove vertices $v_{i+1}, \ldots, v_j$ from W. The resulting walk has fewer vertices. By repeating (if necessary) this procedure, we obtain a path. □

By Theorem 1.4, a walk from a vertex u to another vertex v in a given graph G exists if and only if there is a path connecting these two vertices.

A closed walk does not necessarily contain a cycle. Consider, for instance, the walk $W = uvuvu$ in the graph Q_3 of Figure 1.6. However, the following result holds.

Theorem 1.5. *Every closed odd walk W contains an odd cycle.*

Proof. We use the same method as in the proof of Theorem 1.4. Given an odd closed walk W that is not a cycle, we construct an odd closed walk of lesser length and repeat the procedure until an odd cycle is obtained.

Suppose that $v_i = v_j$ ($i < j$) is a repeated vertex of an odd closed walk $W = v_0 v_1 \cdots v_{k-1} v_0$. One of the two closed walks $W' = v_0 \cdots v_i v_{j+1} \cdots v_0$ and $W'' = v_i \cdots v_j v_i$ must be odd, because the sum of their lengths is the length of W, an odd number. □

Note that the proofs of Theorems 1.4 and 1.5 are essentially by induction (cf. Exercise 1.18).

The concepts of path and cycle are very geometric (cf. Figure 1.7). Another important geometric property of a graph is its connectedness.

Two vertices u and v of a graph $G = (V, E)$ are said to be *connected* if there is a uv-walk in G. Each vertex is connected to itself by a trivial walk. Thus connectedness is a reflexive relation on V. It is symmetric, because we can reverse the order of terms of a uv-walk resulting in a vu-walk. By concatenating a uv-walk with a vw-walk we obtain a uw-walk. Hence, "connected to" is a transitive relation. It follows that connectedness is an equivalence relation on the set of vertices V. Equivalence classes of this relation are *components* of the graph G. If G has exactly one component, it is called *connected*; otherwise it is *disconnected*. Bearing in mind the result of Theorem 1.4, we can say that a graph G is connected if any two distinct vertices of G can be connected by a path.

All graphs in Figures 1.1–1.7 are connected. An extreme case of a disconnected graph is an *empty graph* on more than one vertex, that is, a graph $G = (V, E)$ such that $|V| > 1$ and $E = \varnothing$ (a graph with no edges; cf. Section 1.5).

Let G be a graph. For connected vertices u and v in G we define the *distance* $d_G(u, v)$ (or simply $d(u, v)$) between u and v to be the length of a shortest uv-path. Otherwise, we set $d_G(u, v) = \infty$. Thus defined *distance function* (or *metric*) makes the vertex set V of a connected graph G a *metric space*; that is, $d : V \times V \to \mathbf{R}$ satisfies the following conditions (cf. Exercise 1.19) for all $u, v, w \in V$.

(i) $d(u, v) \geq 0$, and $d(u, v) = 0$ if and only if $u = v$.
(ii) $d(u, v) = d(v, u)$.
(iii) $d(u, v) + d(v, w) \geq d(u, w)$ (the *triangle inequality*).

1.4 Subgraphs and Embeddings

A graph H is said to be a *subgraph* of a graph G if $V(H) \subseteq V(G)$ and $E(H) \subseteq E(G)$. We then say that H is *contained* in G or G *contains* H, and write $H \subseteq G$ or $G \supseteq H$, respectively. If $H \neq G$, then H is said to be a *proper subgraph* of G (denoted $H \subset G$).

Vertices and edges of a path or a cycle in a graph G define subgraphs of G (cf. Figure 1.7). A walk in G is not necessarily a subgraph of G. The empty graph $(V(G), \varnothing)$ is a subgraph of G as is the *trivial graph* on a vertex of G (i.e., the empty graph on a single vertex).

An *embedding* of a graph H in a graph G is a one-to-one mapping $\varphi : V(H) \to V(G)$ that maps the edges of H to edges of G; that is,

$$uv \in E(H) \quad \text{implies} \quad \varphi(u)\varphi(v) \in E(G).$$

The *image* $\varphi(H)$ of the graph H under embedding φ is the subgraph $(\varphi(V), \varphi(E))$ of G where $\varphi(E) = \{\varphi(u)\varphi(v) : uv \in E\}$. The graphs H and $\varphi(H)$ are isomorphic.

Let X be a nonempty subset of the vertex set V of a graph G. The subgraph of G whose vertex set is X and whose edges are the edges of G with ends in X is said to be *induced* by X and denoted by $G[X]$. The graph $G[X]$ is obtained from G by deleting all vertices from the set $V \setminus X$ together with all edges incident with these vertices. This operation is known as *vertex deletion*. If $V \setminus X$ consists of a single vertex v, the resulting subgraph is denoted by $G - v$.

Similarly, we may obtain a subgraph of G by *edge deletion*, that is, by deleting some of the edges of G but leaving the vertices intact. If a single edge e is deleted from the graph G, the resulting graph is denoted by $G - e$. A *spanning subgraph* of a graph G is a subgraph obtained by edge deleting only.

In Figure 1.7, the path P is obtained from the graph Q_3 by deleting edges uv, ab, cw, and xd. It is not an induced subgraph. On the other hand, the cycle C in the same figure is an induced subgraph.

A path in a subgraph H of a graph G is also a path in G, therefore we have $d_H(u, v) \geq d_G(u, v)$ for $u, v \in V(H)$. If $d_H(u, v) = d_G(u, v)$ on $V(H)$, we say that H is an *isometric subgraph* of G. In Figure 1.7, the cycle C is an isometric subgraph of Q_3 whereas the path P is not. An isometric subgraph of a graph is induced, but the converse does not hold (cf. Exercise 1.25).

In a more general setting, for arbitrary graphs H and G, a mapping $\varphi : V(H) \to V(G)$ is an *isometric embedding* if

$$d_H(u, v) = d_G(\varphi(u), \varphi(v))$$

for any $u, v \in V(H)$ (cf. Exercise 1.26). The image of H under isometric embedding is an isometric subgraph of G.

1.5 Special Families of Graphs

There are quite a few special types of graphs that are important in graph theory. Some of them have already been introduced in our examples.

The *complete graph* K_X on a nonempty set X has X as a set of vertices; edges of K_X are the unordered pairs of distinct elements of X. If X is a finite n-element set, the complete graph on X is denoted by K_n (see graphs K_4 and K_5 in Figures 1.1–1.4). The *empty graph* on X is the graph $(X, \varnothing)$. Standing alone is the *trivial graph* K_1, that is, the empty graph on a single vertex. To avoid cluttering the text with nontriviality conditions, we often implicitly assume that graphs under consideration are nontrivial.

A graph G is *bipartite* if its vertex set V can be partitioned into two subsets X and Y (thus, $X \neq \varnothing$, $Y \neq \varnothing$, $X \cap Y = \varnothing$, and $X \cup Y = V$) so that every edge has one end in X and one in Y. The pair (X, Y) is called a *bipartition* of V and the sets X and Y are called *parts* of the bipartition (X, Y). We denote a bipartite graph G with bipartition (X, Y) by $G[X, Y]$. If every vertex in X is joined to every vertex in Y, then $G[X, Y]$ is called a *complete bipartite graph* on (X, Y) and is denoted by $K_{X,Y}$. We use symbol $K_{m,n}$ if X and Y are finite sets of cardinalities m and n, respectively, (see Figures 1.4 and 1.8). A *star* is a complete bipartite graph $K_{X,Y}$ with $|X| = 1$ or $|Y| = 1$. Bipartite graphs are characterized and their properties are investigated in Chapter 2.

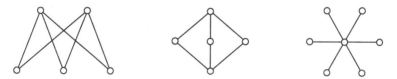

Figure 1.8. Two drawings of the graph $K_{2,3}$ and a drawing of the star $K_{1,6}$.

A *path* P_n is a graph on $n > 1$ distinct vertices $v_1, \ldots, v_n$ such that every vertex v_i is adjacent to v_{i+1} for $1 \leq i < n$. Likewise, a *cycle* C_n (or n-cycle) is a graph on $n \geq 3$ distinct vertices $v_1, \ldots, v_n$ such that every vertex v_i is adjacent to v_{i+1} for $1 \leq i < n$ and v_n is adjacent to v_1. Paths and cycles were already introduced as subgraphs in Section 1.3.

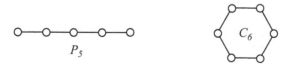

Figure 1.9. A path and a cycle.

The graphs P_5 and C_6 are depicted in Figure 1.9. As before (see Section 1.3), the length of a path or a cycle is the number of its edges. The cycles C_3, C_4, C_5, and C_6 are often called *triangle*, *quadrilateral*, *pentagon*, and *hexagon*, respectively.

A graph without cycles is called a *forest*. A *tree* is a connected component of a forest. A path and a star are trees. Any tree (and a forest) is a bipartite graph. We establish this result together with other properties of trees in Chapter 2.

A graph G in which every vertex has the same degree is called *regular*. If $\deg(v) = k$ for every $v \in V(G)$, then G is *k-regular*. Three-regular graphs are called *cubic* (see Q_3 in Figure 1.6). Unlike k-regular graphs with $k < 3$ (cf. Exercise 1.29), cubic graphs can be rather complex (see the Petersen graph in Figure 1.19).

The graph Q_3 of Figure 1.6 is also a member of a family of graphs usually called *hypercubes*. We simply call them *cubes* and study these graphs in Chapter 3.

1.6 New Graphs From Old Ones

Graphs are ordered pairs of sets, thus the set operations of union and intersections can be extended to graphs.

The *union* $G \cup H$ of two graphs G and H is the graph with vertex set $V(G) \cup V(H)$ and edge set $E(G) \cup E(H)$. If $V(G) \cap V(H) \neq \varnothing$, we define the *intersection* $G \cap H$ as a graph that has vertex set $V(G) \cap V(H)$ and edge set $E(G) \cap E(H)$. Otherwise, we say that the graphs G and H are *disjoint*. For instance, the components of a disconnected graph are disjoint connected graphs. Note that both G and H are subgraphs of the union $G \cup H$, and the intersection $G \cap H$ is a subgraph of both G and H, provided of course that G and H are not disjoint.

A powerful way of forming a new graph from two graphs is the operation of Cartesian product. The *Cartesian product* of graphs G and H is the graph $G \square H$ whose vertex set is the Cartesian product $V(G) \times V(H)$ and whose edge set is the set of pairs $(u, v)(x, y)$ such that either $u = x$ and $vy \in E(H)$ or $ux \in E(G)$ and $v = y$ (cf. Figure 1.10). The graphs G and H are called *factors* of the product $G \square H$.

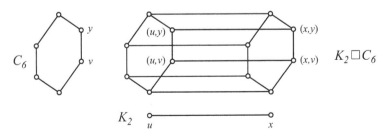

Figure 1.10. The Cartesian product $K_2 \square C_6$.

The Cartesian product $P_n \,\square\, P_m$ of two paths is called the $(n \times m)$-grid. The (5×3)- and (3×5)-grids are shown in Figure 1.11.

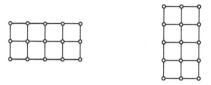

Figure 1.11. The (5×3)- and (3×5)-grids. These graphs are isomorphic.

If $G \neq H$, the graphs $G \,\square\, H$ and $H \,\square\, G$ are not equal. However, they are isomorphic (cf. Exercise 1.30) as illustrated in Figure 1.11. In this sense the Cartesian product is commutative. In the same sense it is also an associative operation (cf. Exercise 1.30), so we can write

$$\prod_{k=1}^{n} G_k = G_1 \,\square\, G_2 \,\square\, \cdots \,\square\, G_n$$

for the product of a finite family of graphs $\{G_1, G_2, \ldots, G_n\}$ $(n \geq 2)$.

Let $G \,\square\, H$ be the Cartesian product of graphs G and H. The mappings

$$p_1 : (u, v) \mapsto u \quad \text{and} \quad p_2 : (u, v) \mapsto v$$

are *projections* of the vertex set $V(G \,\square\, H)$ onto $V(G)$ and $V(H)$, respectively. It follows immediately from the definition of the product that the image of an edge of $G \,\square\, H$ under one of the projections p_1 and p_2 is a vertex; then it is an edge under the other projection (cf. Exercise 1.31). This observation is instrumental for the proof of Theorem 1.7(ii).

Lemma 1.6. *Let P be a path in $G \,\square\, H$. The image $p_1(P)$ of P under projection p_1 is a path in G. Likewise, $p_2(P)$ is a path in H.*

The proof of Lemma 1.6 is straightforward and left to the reader (cf. Exercise 1.32).

Theorem 1.7. *Let G and H be connected graphs. Then*

(i) *The graph $G \,\square\, H$ is connected.*
(ii) $d_{G \,\square\, H}((u, v), (x, y)) = d_G(u, x) + d_H(v, y).$

Proof. (i) Let (u, v) and (x, y) be arbitrary vertices of $G \,\square\, H$. Because G and H are connected graphs, there is a path $u = u_1, \ldots, u_k = x$ in G and a path $v = v_1, \ldots, v_m = y$ in H. Then $(u, v), (u_2, v), \ldots, (x, v), (x, v_2), \ldots (x, y)$ is a path connecting (u, v) and (x, y) in $G \,\square\, H$.

(ii) We may assume that the two paths from part (i) are shortest paths in graphs G and H, respectively. Then the length of the path in $G \,\square\, H$ constructed in part (i) is $d_G(u, x) + d_H(v, y)$. Therefore,

$$d_{G\square H}((u, v), (x, y)) \leq d_G(u, x) + d_H(v, y). \tag{1.2}$$

Let P be a path connecting vertices (u, v) and (x, y) of the graph $G \,\square\, H$. By Lemma 1.6, $P_1 = p_1(P)$ and $P_2 = p_2(P)$ are paths in G and H, respectively. We have

$$d_G(u, x) + d_H(v, y) \leq |E(P_1)| + |E(P_2)| \leq |E(P)|,$$

where the last inequality holds because every edge of P is projected either into an edge of P_1 or into an edge of P_2. Suppose now that P is a shortest path in $G \,\square\, H$. Then

$$|E(P)| = d_{G\square H}((u, v), (x, y)).$$

Therefore,

$$d_G(u, x) + d_H(v, y) \leq d_{G\square H}((u, v), (x, y))$$

and the result follows from (1.2). $\square$

Other properties of Cartesian products and the concept of an infinite product are discussed in Chapter 3.

1.7 Other Notions of Graphs

We now introduce two other notions of graphs that appear less frequently in the book.

In a graph two vertices are joined by at most one edge and no vertex is joined to itself. In applications (cf. Chapter 8) and in graph theory itself, there is need for a more general concept that allows for multiple edges joining two vertices and for vertices that are joined to themselves.

A *multigraph* G is an ordered pair $(V(G), E(G))$ consisting of a nonempty set $V(G)$ of *vertices* and a set $E(G)$, disjoint from $V(G)$, of *edges* together with an *incidence function* ψ_G that associates with each edge of G an unordered pair of (not necessarily distinct[2]) vertices of G. An edge e such that $\psi_G(e) = \{u, v\}$ is said to *join* u and v, and the vertices u and v are called the *ends* of e.

An edge with distinct ends is called a *link*, and an edge with identical ends is called a *loop* (cf. Figure 1.12). Two links with the same ends are said to be *parallel edges*.

We retain the terminology of graph theory with some notable differences. First, a walk (path, circle) is not, in general, defined by the sequence of its vertices; edges of the walk must also be listed. Second, each loop at a vertex

---

[2] As usual, we identify a pair $\{x, x\}$ with the singleton $\{x\}$.

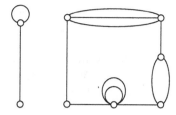

Figure 1.12. A multigraph with eight vertices, ten links, and three loops.

makes the vertex its own neighbor. Accordingly, it contributes 2 to the degree of the vertex. This convention saves equation (1.1). Finally, a multigraph may have cycles of length 1 or 2; they are loops and "double" edges (cf. Figure 1.12).

A multigraph can be "oriented" by assigning a "direction" to each of its edges. More formally, a *directed graph* (or *digraph*) G is an ordered pair $(V(G), E(G))$ consisting of a nonempty set $V(G)$ of *vertices* and a set $E(G)$, disjoint from $V(G)$, of *arcs*, together with an *incidence function*

$$\psi_G : E(G) \to V(G) \times V(G).$$

If a is an arc and $\psi_G(a) = (u, v)$, then a is said to *join u to v*. The vertex u is the *tail* of a, and the vertex v is its *head*.

Any digraph can be obtained from a multigraph (called *underlying graph*) by replacing every link by one of the two arcs with the same ends. Such a digraph is called an *orientation* of the underlying graph (see Figure 1.13 where arrows are attached to the heads of arcs).

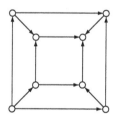

Figure 1.13. An orientation of the cube Q_3.

We often use the name "graph" for multigraphs and digraphs. It is clear from the context what kind of graph is under consideration.

1.8 Infinite Graphs

There are quite a few instances of infinite graphs in the book. Here we merely give some examples of infinite graphs. All graphs in the book have at most countable sets of vertices and edges. Moreover, we usually assume that graphs under consideration are *locally finite*; that is, every vertex has a finite number of neighbors.

In our definition of the complete graph K_X, we do not make assumptions about the cardinality of the vertex set X except that $|X| \neq 0$. For an infinite set X, the graph K_X is an example of an infinite graph that is not locally finite.

The notion of a path has two infinite analogues. A *ray* is a graph (V, E) with
$$V = \{x_0, x_1, x_2, \ldots\} \quad \text{and} \quad E = \{x_0 x_1, x_1 x_2, x_2 x_3, \ldots\},$$
and a *double ray* has the vertex and edge sets of the form
$$V = \{\ldots x_{-1}, x_0, x_1, \ldots\} \quad \text{and} \quad E = \{\ldots x_{-1} x_0, x_0 x_1, x_1 x_2, \ldots\}.$$

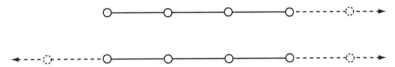

Figure 1.14. The ray and the double ray $\mathcal{Z}$.

These two graphs are shown in Figure 1.14. The set **Z** of integers is a natural choice for a vertex set of the double ray. We denote the resulting graph by $\mathcal{Z}$ and sometimes call it the 1-dimensional *integer lattice*.

The Cartesian product $\mathcal{Z}^2 = \mathcal{Z} \square \mathcal{Z}$ is called the 2-dimensional *integer lattice* (also known as the *square lattice*). Figure 1.15 depicts the graph $\mathcal{Z}^2$ and the *hexagonal lattice*.

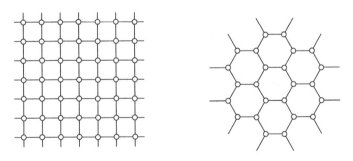

Figure 1.15. The square and hexagonal lattices.

The two rays and lattices depicted in Figures 1.14 and 1.15 are examples of locally finite infinite graphs. In geometry, infinite graphs arise naturally from locally finite hyperplane arrangements (see Chapter 7).

1.9 Matchings

A *matching* M in a graph G is a set of pairwise nonadjacent edges. The two ends of each edge of M are said to be *matched* under M, and each vertex incident with an edge of M is said to be *covered* (or *saturated*) by M. If a matching covers every vertex of G, then it is called a *perfect matching*. A *maximum matching* in a finite graph is one that covers as many vertices as possible. In other words, M is a maximum matching if $|M| \geq |M'|$ for any other matching M' in G. A matching in a finite graph is *maximal* if it cannot be extended to a larger matching. Equivalently, it is one that can be obtained iteratively by selecting an edge disjoint from previously selected edges.

It is clear that a perfect matching is maximum and a maximum matching is maximal. The matching in Figure 1.16a is maximum but not perfect; the one in Figure 1.16b is maximal but not maximum. The perfect matching in the path P_4 is depicted in Figure 1.16c.

Figure 1.16. Matchings in graphs $K_{1,3}$ and P_4.

There are mathematical problems and questions of practical interest that, in terms of graph theory, amount to finding a maximum matching in a graph. The notions of alternating and augmenting paths with respect to a given matching play an essential role in studies of maximum matchings in graphs.

Let M be a matching in a graph $G = (V, E)$. An *M-alternating path* in G is a path whose edges are alternately in M and $E \setminus M$. An *M-augmenting path* is an M-alternating path whose ends are not covered by M.

The following observation is instrumental. Let P be an M-augmenting path in G. Let us replace $M \cap E(P)$ with $E(P) \setminus M$ (see Figure 1.17).

Figure 1.17. Two matchings in P_8.

It is clear that $|E(P) \setminus M| = |M \cap E(P)| + 1$. Therefore, for the matching

$$M' = M \triangle E(P),$$

where $M \triangle E(P) = (M \setminus E(P)) \cup (E(P) \setminus M)$ is the *symmetric difference* of the sets M and $E(P)$, we have $|M'| = |M| + 1$. We proved the following assertion.

Lemma 1.8. *Let M be a matching in G. If G contains an M-augmenting path, then M is not maximum.*

It turns out that the converse assertion also holds.

Lemma 1.9. *If M is not a maximum matching in G, then G contains an M-augmenting path.*

Proof. Let M' be a maximum matching in G, so we have $|M'| > |M|$. Let H be a subgraph of G that has the same vertex set as G and whose edges are those edges of G that appear in exactly one of M and M'. Thus, $V(H) = V(G)$ and $E(H) = M \triangle M'$. Because each vertex v of H is an end of at most one edge of M and at most one edge of M', it must be that $\deg_H(v) \leq 2$. This implies that each connected component of H is either a path (perhaps trivial) or a cycle (cf. Exercise 2.36). Because $|M'| > |M|$, there must be a component, say H', with more edges from M' than from M. The component H' cannot be a cycle, inasmuch as edges from M and M' are alternately in H', so they occur in H' an equal number of times. Hence, H' is an M-alternating path with more edges from M' than from M. It follows that H' is an M-augmenting path. □

By combining the results of Lemmas 1.8 and 1.9, we obtain the following theorem.

Theorem 1.10. *A matching M in a graph G is a maximum matching in G if and only if G has no M-augmenting path.*

It is often desirable to find a matching in a bipartite graph $G[X, Y]$ that covers every vertex in X. If such a matching M exists, then it is clear that $|S| \leq |N(S)|$ for every subset S of X, where $N(S)$ denotes the set of neighbors of vertices in S. In particular, we have $|X| \leq |Y|$.

Theorem 1.11. *A bipartite graph $G = G[X, Y]$ has a matching that covers every vertex of X if and only if*

$$|N(S)| \geq |S| \qquad \text{for all } S \subseteq X. \tag{1.3}$$

Proof. Necessity was established above. For sufficiency, suppose that (1.3) holds, and consider a maximum matching M. Suppose that M does not cover X. To obtain a contradiction, we find a set S that violates (1.3).

Let u be a vertex in X that is not covered by M, and let S and T be the sets of vertices in X and Y, respectively, that are reachable from u by M-alternating paths (see Figure 1.18). Note that $u \in S$ and, by (1.3), $N(u) \neq \varnothing$. Because M is a maximum matching, it follows from Theorem 1.10 that M covers nonempty sets T and $S \setminus \{u\}$. Clearly the vertices of $S \setminus \{u\}$ are matched under M with the vertices of T. Hence, $|T| = |S| - 1$. By the same Theorem 1.10 and the definition of T, we have $N(S) = T$. But then

$$|N(S)| = |T| = |S| - 1 < |S|,$$

and this is the desired contradiction.   □

Theorem 1.11 provides a criterion for a bipartite graph to have a perfect matching.

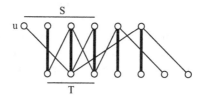

Figure 1.18. Proof of Theorem 1.11.

Corollary 1.12. *A bipartite graph $G[X, Y]$ has a perfect matching if and only if $|X| = |Y|$ and $|N(S)| \geq |S|$ for all $S \subseteq X$.*

Notes

The reader can find an attractive account of the first 200 years (up to 1936) of graph theory in Biggs et al. (1986).

The "first theorem of graph theory", equation (1.1), was established by Leonhard Euler in 1736 (see Paragraph 16 in the translation of Euler's original paper in Biggs et al. (1986)).

A careful study of planar embeddings has raised some topological questions that found answers about a century ago. The key result is the *Jordan Curve Theorem*. The theorem claims that a closed simple curve in the plane divides the plane into two open connected regions having that curve as a common boundary. Camille Jordan formulated this result as a theorem in 1887, but his proof was flawed according to most classical sources. These sources attribute the first correct proof to Oswald Veblen who published it in 1905. (For a defense and updated presentation of Jordan's proof see Hales, 2007). Standard proofs are found in topology textbooks; see, for instance, Munkres (2000).

A planar graph admits a planar representation in which all edges are represented by straight line segments. This result was established independently by K. Wagner in 1936 and I. Fáry in 1948 and is known as Fáry's Theorem. A proof is found in Behzad and Chartrand (1971); a more accessible reference is West (2001).

In 1930 Kazimir Kuratowski solved the long-standing problem of plane graph characterization. He proved that a graph is planar if and only if it contains no subdivision[3] of the graphs K_5 and $K_{3,3}$ (see Figure 1.4). These two graphs are known as the *Kuratowski graphs*. For a coherent introduction to planar (and other) embeddings see Mohar and Thomassen (2001).

As in many other branches of mathematics, the notions of isomorphism and automorphism are fundamental concepts in graph theory. A weaker version of isomorphism is the concept of a "homomorphism". A *homomorphism* of a graph G into another graph H is a mapping $\varphi : V(G) \to V(H)$ such that $\varphi(u)\varphi(v) \in E(H)$ for all $uv \in E(G)$. Any embedding (Section 1.4) is a homomorphism. The converse does not hold (cf. Exercise 1.37).

Deciding whether two finite graphs G and H on n vertices are isomorphic is a difficult problem. In fact, no efficient general algorithm for testing isomorphism is known. Of course one can use the "brute force" approach and test all $n!$ bijections between $V(G)$ and $V(H)$ in turn, and check if one of them is an isomorphism. Clearly, this approach is impractical even for moderately small graphs.

Testing isomorphism and other algorithmic problems of graph theory requires algebraic rather than geometric methods of specifying graphs. A graph can be efficiently represented by listing for every vertex the set of its neighbors in some order. A list of these lists is called an *adjacency list* of the graph. Adjacency lists are usually used to store graphs in computers. Two other ways of specifying a finite graph $G = (V, E)$ employ the incidence and adjacency matrices of G. Let $n = |V|$ and $m = |E|$. The *incidence matrix* of G is the $n \times m$ matrix $\mathbf{M}_G = (m_{ve})$, where m_{ve} is the number of times that vertex v and edge e are incident. The *adjacency matrix* of G is the square $n \times n$ matrix $\mathbf{A}_G = (a_{uv})$, where a_{uv} is the number of edges joining vertices u and v, each loop counting as two edges. A good introduction to algebraic methods in graph theory is found in Biggs (1993).

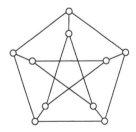 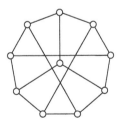

Figure 1.19. Two drawings of the Petersen graph.

In Section 1.5 we described only a few special families of graphs. Other families are introduced later in the book. Of course there are many more. Even a single graph may form an important family. An example is given by

---

[3] A subdivision of a graph is obtained by replacing some of its edges by paths of length greater than one.

the *Petersen graph* depicted in Figure 1.19. The entire book by Holton and Sheehan (1993) is devoted to this graph.

In addition to the Cartesian product there are other ways of forming from two graphs a new graph whose vertex set is the Cartesian product of their vertex sets. A comprehensive presentation of these products is found in Imrich and Klavžar (2000).

The mere purpose of Section 1.8 was presenting a few basic examples of infinite graphs. There are more examples of infinite graphs in subsequent chapters. Moreover, a significant part of the results presented in the book is concerned specifically with infinite graphs. For the theory of infinite graphs per se see Chapter 8 in Diestel (2005).

Theorem 1.10 is known in graph theory as Berge's Theorem. The result was established by Claude Berge (1957) who is considered to be one of the founders of modern graph theory. The result of Theorem 1.11 was established by Philip Hall (1935) and often bears his name. Hall's Theorem is also known as the "Marriage Theorem" because it can be restated as follows. If every group of girls in a village collectively likes at least as many boys as there are girls in the group, then each girl can marry a boy she likes. In combinatorics, Hall's Theorem is often reformulated in terms of "systems of distinct representatives". Let $\{A_i\}_{i \in I}$ be a finite family of nonempty subsets of a finite set A. A *system of distinct representatives* (SDR) for the family A is a set $\{a_i\}_{i \in I}$ of distinct elements of A such that $a_i \in A_i$ for all $i \in I$. In this language, Hall's Theorem tells us that A has an SDR if and only if $|\cup_{i \in J} A_i| \geq |J|$ for all subsets J of I. In this form it appeared in Hall's original paper (1935) where it was used to answer a question in group theory. It is hard to underestimate the importance of matching theory in both pure and applied mathematics. For a good account of the history and methods of the theory the reader is referred to Lovász and Plummer (1986).

Exercises

1.1. Show that in a finite graph the number of vertices of odd degree is even.

1.2. Is there a cubic graph with 9 vertices?

1.3. Prove that every finite nontrivial graph has at least two vertices of equal degree.

1.4. Describe graphs G such that $N(u) \cup N(v) = V(G)$ for every pair of distinct vertices u and v.

1.5. How many edges does the complete graph K_n have?

1.6. Let G be a graph with n vertices and m edges.

a) Does there exist a graph G with $n = 4$ and $m = 7$?

b) Prove that

$$0 \le m \le \binom{n}{2} = \frac{n(n-1)}{2}.$$

c) Show that for any two positive integers n and m satisfying inequalities from part b) there exists a graph G with $n = |V(G)|$ and $m = E(G)|$.

1.7. Let C be the curve in $\mathbf{R}^3$ defined by the equations

$$x = t, \quad y = t^2, \quad z = t^3, \quad t \in \mathbf{R}.$$

Show that no four distinct points on C lie in the same plane. Deduce that a finite graph can be drawn in $\mathbf{R}^3$ with all edges straight.

1.8. Show that the complete graph K_5 is not planar. You may assume that edges in drawings of planar graphs are straight line segments.

1.9. Obtain a planar embedding of the graph in Figure 1.20 in which every edge is a straight line segment. This is the graph of a polytope. Construct this polytope.

1.10. Show that $\mathrm{Aut}(G)$ is a group under the operation of composition.

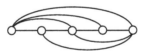

Figure 1.20. Exercise 1.9.

1.11. A nontrivial graph G is said to be *asymmetric* if the group $\mathrm{Aut}(G)$ is trivial.

a) Show that the graph in Figure 1.5 is asymmetric.
b) For each $n \ge 6$, find an asymmetric graph on n vertices.
c) Show that there is no asymmetric graph on $n \le 5$ vertices.

1.12. Show that, for $n \ge 2$, $\mathrm{Aut}(P_n)$ is the symmetric group S_2.

1.13. Describe $\mathrm{Aut}(K_{3,3})$, where $K_{3,3}$ is the complete bipartite graph from Example 1.2.

1.14. Prove that if a graph G is bipartite and $H \cong G$ then H is also bipartite.

1.15. Let φ be an embedding of graph G into graph H. Show that the graphs G and $\varphi(G)$ are isomorphic.

1.16. Show that the two graphs in Figure 1.19 are isomorphic.

Figure 1.21. Exercise 1.17.

1.17. Show that the two graphs in Figure 1.21 are not isomorphic.

1.18. Prove Theorems 1.4 and 1.5 by induction.

1.19. Prove that the distance function $d_G(u, v)$ on a connected graph G satisfies properties (i)–(iv) at the end of Section 1.3.

1.20. Let G be a finite graph such that every vertex has a degree at least two. Prove that G contains a cycle.

1.21. Let G be a connected graph. Show that for any two disjoint nonempty sets $S, T \subseteq V(G)$ there is a path with ends in S and T, respectively, that does not contain other vertices from $S \cup T$.

1.22. Show that two paths of maximum length in a finite connected graph have a common vertex.

1.23. Prove that if a graph has exactly two vertices of odd degrees then they are connected by a path.

1.24. Describe all graphs G such that every proper subgraph of G is trivial.

1.25. Show that an isometric subgraph is induced. Give an example of an induced subgraph that is not isometric.

1.26. Justify the term "isometric embedding" by showing that an isometric embedding is a one-to-one mapping.

1.27. Show that:

a) Every path is bipartite.
b) A cycle is bipartite if and only if its length is even.

1.28. Let G be a bipartite graph with bipartition (X, Y).

a) Show that
$$\sum_{u \in X} \deg(u) = \sum_{u \in Y} \deg(u).$$

b) Deduce that if G is k-regular, with $k > 0$, then $|X| = |Y|$.

1.29. Characterize k-regular graphs for $k = 0, 1, 2$.

1.30. Prove that the Cartesian product is a commutative and associative operation up to graph isomorphism; that is,

$$G \square H \cong H \square G \quad \text{and} \quad (G \square H) \square F \cong G \square (H \square F).$$

1.31. Show that one of the projections p_1 and p_2 sends an edge of $G \square H$ into a vertex and the other one sends it into an edge.

1.32. Prove Lemma 1.6.

1.33. Prove that the Cartesian product of two graphs is connected if and only if both factors are connected (cf. Theorem 1.7(i)).

1.34. Let $G = \prod_{k=1}^{n} G_k$, where the graphs G_k are connected. Show that

$$d_G((u_1, \ldots, u_n), (v_1, \ldots, v_n)) = \sum_{k=1}^{n} d_{G_k}(u_k, v_k)$$

(cf. Theorem 1.7(ii)).

1.35. Let G be a graph. Show that two vertices $(u_1, \ldots, u_n)$ and $(v_1, \ldots, v_n)$ are adjacent in
$$G^n = \underbrace{G \square \cdots \square G}_{n \text{ factors}}$$
if and only if there is $1 \leq j \leq n$ such that $u_i = v_i$ for $i \neq j$, and u_j and v_j are adjacent in the graph G.

1.36. Let $\varphi : G_1 \to H_1$ and $\psi : G_2 \to H_2$ be two graph embeddings. Show that $G_1 \square G_2$ is isomorphic to the subgraph $\varphi(G_1) \square \psi(G_2)$ of the graph $H_1 \square H_2$.

1.37. Give an example of a homomorphism (see the definition in the Notes section) that is not an embedding.

1.38. Find the number of maximum matchings in the complete bipartite graph $K_{m,n}$.

1.39. Show that there are $(2n)!/(2^n n!)$ perfect matchings in the complete graph K_{2n}.

1.40. Find the minimum size of a maximal matching in the cycle C_n.

1.41. Prove that every tree has at most one perfect matching.

1.42. Let M, M' be matchings in a bipartite graph $G[X, Y]$. Suppose that $S \subseteq X$ is covered by M and that $T \subseteq Y$ is covered by M'. Show that G contains a matching that covers $S \cup T$.

2

Bipartite Graphs

In this book, we deal mostly with bipartite graphs. The main thrust of this chapter is to characterize bipartite graphs using geometric and algebraic structures defined by the graph distance function. Fundamental sets and the two theta relations introduced in Section 2.3 play a crucial role in our studies of partial cubes in Chapter 5.

2.1 Standard Characterizations

We begin with a simple but important property of bipartite graphs that the reader should verify (cf. Exercise 2.1).

A walk in a bipartite graph $G[X, Y]$ is even if and only if its ends belong to the same part of the bipartition (X, Y). Equivalently, a walk is odd if and only if its ends belong to different parts of (X, Y).

Examples of even and odd paths are shown in Figure 2.1.

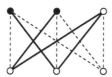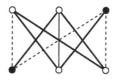

Figure 2.1. Even and odd paths in the graph $K_{3,3}$.

It follows that a closed walk in a bipartite graph must be of even length. This gives us a necessary condition for a graph to be bipartite. In fact this condition is also sufficient.

Theorem 2.1. *A graph $G = (V, E)$ is bipartite if and only if it contains no closed walk of odd length.*

Proof. We need to prove sufficiency only and may assume that G is connected (cf. Exercise 2.2). For a fixed vertex $v \in V$ we define subsets X and Y of the vertex set V by

$$X = \{u \in V : d(u, v) \text{ is even}\},$$
$$Y = \{u \in V : d(u, v) \text{ is odd}\}.$$

Note that $v \in X$. Because G is nontrivial (this is our usual assumption) and connected, (X, Y) is a bipartition. To prove that G is bipartite it suffices to show that there is no edge with ends in the same part. We assume that there is an edge uw with ends in, say, X, and derive a contradiction. Let P be a shortest vu-path and Q be a shortest wv-path. Both paths are even. By concatenating the path P, the edge uw, and the path Q, we obtain a closed walk of odd length. This contradicts the absence of odd walks. □

Theorem 2.2. *A graph G is bipartite if and only if it does not contain an odd cycle.*

Proof. By Theorem 1.5, a graph does not contain an odd walk if and only if it does not contain an odd cycle. The result follows from Theorem 2.1. □

The next characterization uses the graph distance function.

Theorem 2.3. *A connected graph G is bipartite if and only if for every vertex v there is no edge uw with $d(v, u) = d(v, w)$.*

Proof. (Necessity.) Recall that the distance between two vertices is the length of a shortest path connecting these vertices. The ends of an edge belong to different parts of the vertex set, therefore the numbers $d(v, u)$ and (v, w) have opposite parity.[1] Thus, in a bipartite graph, we must have $d(v, u) \neq d(v, w)$ for any choice of vertex v and edge uw.

(Sufficiency.) We need to show that if $d(v, u) \neq d(v, w)$ for all $v \in V(G)$ and $uw \in E(G)$, then G is a bipartite graph.

For a given vertex v, let (X, Y) be the same bipartition as in the proof of Theorem 2.1. Let uw be an edge of G. By the triangle inequality, we have

$$|d(v, u) - d(v, w)| \leq d(u, w) = 1.$$

Because $d(v, u) \neq d(v, w)$, we must have

$$|d(v, u) - d(v, w)| = 1. \tag{2.1}$$

Therefore the integers $d(v, u)$ and $d(v, w)$ have opposite parity. From this we deduce that the edge uw has ends in different parts of the bipartition (X, Y). It follows that G is a bipartite graph. □

---

[1] Two integers have the same parity if they are either both even or both odd. Otherwise they have opposite parity.

One can say that bipartite graphs are characterized by the lack of isosceles triangles with a base of length one. Here, by a "triangle" we mean just a configuration of three distinct vertices; it does not have "real" sides unless we have a cycle of length three.

From equation (2.1) we immediately obtain the following result.

Corollary 2.4. *Let uw be an edge of a connected bipartite graph G. Then for any vertex v of G we have either*

$$d(v, u) = d(v, w) + 1$$

or

$$d(v, w) = d(v, u) + 1$$

2.2 Betweenness and Convexity in Graphs

Throughout the book we widely use "geometric" concepts that have their roots in the usual Euclidean geometry (cf. the paragraph following the proof of Theorem 2.3). However, the reader should be warned that the analogy with classical geometry is rather superficial and must be used with care.

In order to utilize the distance function on a graph, we assume that the graphs under consideration are connected. The next two lemmas establish some basic properties of shortest paths in a graph.

Lemma 2.5. *Let $P = u_0, u_1, \ldots, u_n$ be a shortest path in a graph G. Then any $u_i u_j$-segment of P is a shortest path in G. Accordingly,*

$$d_G(u_i, u_j) = |j - i| \text{ for } 1 \leq i, j \leq n.$$

Proof. Suppose that there is a $u_i u_j$-path Q in G of length less than the length of the $u_i u_j$-segment of P. By concatenating the $u_0 u_i$-segment of P, the path Q, and the $u_j u_n$-segment of P, we obtain a walk in G of length less than the length of P. This contradicts our assumption that P is a shortest path.

Because the $u_i u_j$-segment of P is a shortest path in G, we have, for $i \leq j$,

$$j - i = d_P(u_i, u_j) = d_G(u_i, u_j).$$

Hence $d_G(u_i, u_j) = |j - i|$ for all $1 \leq i, j \leq n$. $\square$

Lemma 2.6. *Let u and v be two vertices of a graph $G = (V, E)$. A vertex w belongs to a shortest path connecting u and v if and only if*

$$d(u, w) + d(w, v) = d(u, v). \tag{2.2}$$

Proof. (Necessity.) Let P be a shortest uv-path, and let P_1 and P_2 be the uw- and wv-segments of P, respectively. Because P_1 and P_2 are shortest paths (by Lemma 2.5) and the length of a path is the number of its edges, we have (2.2).

(Sufficiency.) Let w be a vertex satisfying equation (2.2). By concatenating two arbitrary shortest uw- and wv-paths, we obtain a uv-walk P in G of length $d(u, w) + d(w, v)$. By (2.2), the walk P is a shortest path. Clearly, $w \in P$. □

We say that a vertex w *lies between* vertices u and v if w belongs to a shortest path with ends u and v, or (equivalently, by Lemma 2.6) if equation (2.2) holds for w. The *interval* $I(u, v)$ between two vertices of a graph is the set of all vertices that lie between u and v:

$$I(u, v) = \{w \in V : d(u, w) + d(w, v) = d(u, v)\}.$$

It is clear that $I(u, u) = \{u\}$, and $I(u, v) = \{u, v\}$ if uv is an edge. The reader should distinguish between paths and intervals: a path is a subgraph whereas an interval is a set of vertices. Also note that $u, v \in I(u, v)$ and $I(u, v) = I(v, u)$.

Example 2.7. Let $K_{X,Y}$ be a complete bipartite graph with $|X| > 1$, $|Y| > 1$ (cf. Exercise 2.5 for the case of a star). As we noted above, the edges of G are intervals. Suppose that a pair $\{u, v\}$ is not an edge. Then both u and v belong to the same part, say X, of the bipartition (X, Y). Every vertex of the part Y lies between u and v, and no other vertex of X lies between vertices u and v. Thus, $I(u, v) = \{u, v\} \cup Y$ (see Figure 2.2 where $X = \{u, v, w\}$ and $Y = \{x, y\}$).

Figure 2.2. Interval $I(u, v) = \{u, v, x, y\}$ in the complete bipartite graph $K_{2,3}$.

A set S of vertices of a graph is said to be *convex* if $I(u, v) \subseteq S$ for every two vertices u and v in S. In words: a set of vertices is convex if for every two vertices in the set it contains all vertices that lie between them.

Intervals need not be convex sets as can be seen from the graph $K_{2,3}$ shown in Figure 2.2. Indeed, the interval $I(u, v) = \{u, v, x, y\}$ is not a convex set because the vertex w lies between vertices x and y but does not belong to the interval (cf. Example 2.7).

A connected subgraph H of a graph G is said to be *convex* if any shortest path in G connecting two vertices of H is already in H. Clearly, the graph G itself as well as its vertex set are convex.

If H is a convex subgraph of a graph G, then $V(H)$ is a convex set of vertices (cf. Exercise 2.8). The converse is not true; see Figure 2.3 where edges of the subgraph H of the cycle C_4 are bold line segments.

However, we have the following result that relates the two notions of convexity. The proof is straightforward and left to the reader (cf. Exercise 2.8).

Theorem 2.8. *A subgraph $H \subseteq G$ is convex if and only if it is induced by a convex set of vertices of G.*

Figure 2.3. The subgraph H is not convex. Its vertex set is convex.

It is clear that not every metric d on a nonempty set V is the distance function on a graph G with the vertex set V. One clear restriction is that d must take values in the set of nonnegative integers. A characterization of metric spaces that are arising from connected graphs is the result of the next theorem. (The proof is left to the reader as Exercise 2.10.)

Theorem 2.9. *A metric d on a set V is the distance function on some graph $G = (V, E)$ if and only if it satisfies conditions:*

(i) $d(u, v)$ *is an integer;*
(ii) *If $d(u, v) > 1$, then there is $w \in V$, distinct from both u and v, such that $d(u, v) = d(u, w) + d(w, v)$;*

for all $u, v \in V$.

One can say that an integral metric on a set V is a distance function on some graph if, for any two elements of V on a distance greater than one from each other, there is an element of V distinct from these two elements and lying between them.

2.3 Fundamental Sets and Relations

It is evident that two distinct edges of a graph either belong to a shortest path in the graph or there is no shortest path containing both edges. This simple observation is a basis for the graph concepts introduced in this section. First, we consider two examples.

Example 2.10. The edges $e = xy$ and $f = uv$ of the cube Q_3 in Figure 2.4 belong to the shortest xv-path $xyuv$. Likewise, the edges f and h belong to the shortest path uvw. On the other hand, the edges e and h do not belong to a shortest path in Q_3. These edges are parallel in the usual 3-dimensional drawing of the cube. It is clear that any pair of distinct edges of Q_3 in Figure 2.4 either belong to a shortest path or are parallel. Note that any connected subgraph of Q_3 has the same geometric property.

Figure 2.4. The cube Q_3 and the complete graph K_4.

Example 2.11. The situation is quite different in the case of the complete graph K_4 shown in Figure 2.4. Clearly, there are no pairs of distinct edges of K_4 that belong to the same shortest path. In particular, the vertices of two adjacent edges xy and yv of K_4 in Figure 2.4 do not form a shortest path. On the other hand, in a bipartite graph (note that K_4 is not bipartite), the vertices of any two adjacent edges form a shortest path (cf. Exercise 2.9).

Let $G = (V, E)$ be a connected graph. Suppose that two distinct edges $e = xy$ and $f = uv$ belong to a shortest path (see Figure 2.5).

Figure 2.5. Two edges in a shortest path.

For the labeling of these two edges shown in Figure 2.5, we have

$$d(x, v) = d(x, u) + 1 \text{ and } d(y, v) = d(y, u) + 1,$$

by Lemma 2.6. (Recall that a segment of a shortest path is a shortest path itself; see Lemma 2.5.) Therefore,

$$d(x, u) + d(y, v) = d(x, v) + d(y, u). \tag{2.3}$$

The three remaining possible labelings of e and f are obtained by permutations $x \leftrightarrow y$ and $u \leftrightarrow v$. It is clear that equation (2.3) is invariant under these permutations. Therefore, condition (2.3) is necessary for two edges to lie on the same shortest path. It is not a sufficient condition, because the edges xy and uv of the graph K_4 in Figure 2.4 satisfy equation (2.3) but do not belong to a shortest path. However, for bipartite graphs we have the following result.

Theorem 2.12. *Let $e = xy$ and $f = uv$ be two distinct edges of a connected bipartite graph G. Then e and f belong to a shortest path in G if and only if their ends satisfy condition (2.3).*

Proof. We need to prove sufficiency only. Without loss of generality, we may assume that the number $d(y, u)$ is the least number among the four distances in (2.3). By Theorem 2.3, we have $d(y, u) < d(y, v)$ and $d(y, u) < d(x, u)$. By Corollary 2.4, we must have

$$d(y, v) = d(y, u) + 1 \text{ and } d(x, u) = d(y, u) + 1.$$

Therefore, by (2.3),

$$d(x, v) = d(y, u) + 2 = d(x, y) + d(y, u) + d(u, v)$$

(see Figure 2.5). The length of any xv-walk is greater than or equal to $d(x, v)$, therefore the xv-walk obtained by concatenating the edge xy, a shortest yu-path, and the edge uv is a shortest xv-path containing edges e and f.   □

Pairs of edges that do not satisfy condition (2.3) are of considerable interest later in the book.

Definition 2.13. *Let $e = xy$ and $f = uv$ be two edges of a connected graph $G = (V, E)$. The edge e is in* relation Θ *to the edge f if*

$$d(x, u) + d(y, v) \neq d(x, v) + d(y, u). \tag{2.4}$$

More formally,

$$\Theta = \{(xy, uv) \in E \times E : d(x, u) + d(y, v) \neq d(x, v) + d(y, u)\}$$

The relation Θ is a binary relation on the edge set E. We often write $e \Theta f$ to indicate that $(e, f) \in \Theta$.

It is clear that the relation Θ is reflexive (i.e., $e \Theta e$ for all $e \in V$) and symmetric (i.e., $e \Theta f$ implies $f \Theta e$ for all $e, f \in V$). This relation need not be transitive as examples of the complete bipartite graph $K_{2,3}$ and the odd cycle C_5 demonstrate. Indeed, for both graphs depicted in Figure 2.6, we have $a \Theta b$ and $b \Theta c$, but $(a, c) \notin \Theta$.

Let us recall that for the ends of any two distinct edges $e = xy$ and $f = uv$ in a shortest path, equation (2.3) holds. It follows that two distinct edges e and f with $e \Theta f$ cannot belong to the same shortest path. Moreover, we have the following result for bipartite graphs.

Figure 2.6. Intransitivity of the relation Θ.

Theorem 2.14. *Let G be a connected bipartite graph and $e = xy$, $f = uv$ be two edges of G with $e \Theta f$. There are two possible cases:*

(i) $d(x, v) = d(x, u) + 1 = d(y, v) + 1 = d(y, u)$.
(ii) $d(x, u) = d(x, v) + 1 = d(y, u) + 1 = d(y, v)$.

Proof. By Corollary 2.4, we have either $d(x, v) = d(x, u) + 1$ or $d(x, u) = d(x, v) + 1$. It suffices to establish case (i) of the theorem because the other case is obtained from (i) by exchanging variables u and v.

By the same Corollary 2.4, we have either $d(y, v) = d(y, u) + 1$ or $d(y, u) = d(y, v) + 1$. If $d(y, v) = d(y, u) + 1$, then we have

$$d(x, u) + d(y, v) = d(x, v) - 1 + d(y, u) + 1 = d(x, v) + d(y, u)$$

contradicting condition (2.4). Therefore, $d(y, u) = d(y, v) + 1$ (the last equality in (i)).

By the triangle inequality,

$$d(x, u) + 1 = d(x, v) \leq d(x, y) + d(y, v) = 1 + d(y, v).$$

Hence, $d(x, u) \leq d(y, v)$. Likewise,

$$d(y, v) + 1 = d(y, u) \leq d(y, x) + d(x, u) = 1 + d(x, u),$$

yielding $d(y, v) \leq d(x, u)$. It follows that $d(x, u) = d(y, v)$. $\square$

The two cases of Theorem 2.14 are illustrated by the diagrams in Figure 2.7 where dotted lines designate shortest paths between vertices.

Figure 2.7. Two cases of Theorem 2.14.

Theorem 2.15. *Suppose that a walk W in a graph G connects ends of an edge e but does not contain it. Then W contains an edge f with $f \Theta e$.*

Proof. Let $W = u_0 u_1 \cdots u_n$ be a walk in G with $e = u_0 u_n$ and let $e_i = u_{i-1} u_i$ for $1 \le i \le n$. We need to show that there is i such that

$$d(u_0, u_{i-1}) + d(u_n, u_i) \ne d(u_0, u_i) + d(u_n, u_{i-1}).$$

The proof is by contradiction. Suppose that

$$d(u_0, u_{i-1}) + d(u_n, u_i) = d(u_0, u_i) + d(u_n, u_{i-1}),$$

for all $1 \le i \le n$. Let us add together all the above equations:

$$\sum_{i=1}^{n} d(u_0, u_{i-1}) + \sum_{i=1}^{n} d(u_n, u_i) = \sum_{i=1}^{n} d(u_0, u_i) + \sum_{i=1}^{n} d(u_n, u_{i-1}).$$

By cancelling common terms on both sides of this equation, we have

$$0 = d(u_0, u_n) + d(u_n, u_0) = 2,$$

a contradiction. Therefore, there is i such that

$$d(u_0, u_{i-1}) + d(u_n, u_i) \ne d(u_0, u_i) + d(u_n, u_{i-1}),$$

that is, $e_i \ominus e$. □

Corollary 2.16. *Let C be a cycle of a graph G. For any edge e of C there is another edge f of C with $f \ominus e$.*

Let ab be an edge of the cube Q_3. The edges that are parallel to the edge ab in the drawing in Figure 2.8 are shown in bold. These and only these edges stand in the relation $\ominus$ to ab. They join two disjoint subgraphs W_{ab} and W_{ba} of Q_3 that we call "semicubes".

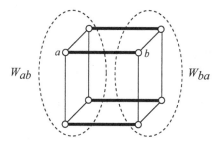

Figure 2.8. Semicubes in the cube Q_3.

Definition 2.17. Let $G = (V, E)$ be a connected graph. For any two adjacent vertices $a, b \in V$, let W_{ab} be the set of vertices that are closer to a than to b:

$$W_{ab} = \{x \in V : d(x, a) < d(x, b)\}. \tag{2.5}$$

The sets W_{ab} and subgraphs induced by these sets are called *semicubes* of the graph G. The semicubes W_{ab} and W_{ba} are called *opposite semicubes*.

Note that the subscript ab in W_{ab} stands for an ordered pair of vertices. Examples of pairs of opposite semicubes are displayed in Figure 2.9.

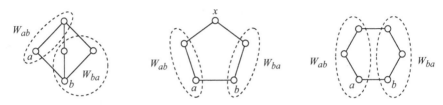

Figure 2.9. Semicubes in $K_{2,3}$, C_5, and C_6.

It is clear that a semicube W_{ab} is a nonempty set (because $a \in W_{ab}$) and that two opposite semicubes are disjoint (we cannot have both $d(x, a) < d(x, b)$ and $d(x, b) < d(x, a)$). Moreover, for any edge $ab \in E$,

$$d(x, a) = d(x, b) \text{ if and only if } x \notin W_{ab} \cup W_{ba}.$$

Thus, by Theorem 2.3, we have the following characterization of bipartite graphs (cf. Exercise 2.14).

Theorem 2.18. *A connected graph $G = (V, E)$ is bipartite if and only if the semicubes W_{ab} and W_{ba} form a partition of the vertex set V for every edge $ab \in E$, that is,*

$$W_{ab} \cap W_{ba} = \varnothing \text{ and } W_{ab} \cup W_{ba} = V,$$

for all $ab \in E$.

This theorem is illustrated by three drawings in Figure 2.9 where the graphs $K_{2,3}$ and C_6 are bipartite and C_5 is not.

By definition, $d(x, a) < d(x, b)$ for $x \in W_{ab}$. In fact, one can say more about the relation between these two distances. By Corollary 2.4, we have the following result.

Lemma 2.19. *Let $x \in W_{ab}$ for some edge ab of a connected graph. Then*

$$d(x, b) = d(x, a) + 1.$$

Accordingly,

$$W_{ab} = \{x \in V : d(x, b) = d(x, a) + 1\}.$$

Lemma 2.19 reveals an interrelationship existing between the relation Θ and semicubes in a connected graph. Suppose that an edge $f = xy$ has both ends in the semicube W_{ab} (the left diagram in Figure 2.10). By Lemma 2.19, we have

$$d(x, b) = d(x, a) + 1 \text{ and } d(y, b) = d(y, a) + 1.$$

By adding the opposite sides of these two equations, we obtain an equation

$$d(x, a) + d(y, b) = d(x, b) + d(y, a)$$

implying that the edges $e = ab$ and $f = xy$ are not in relation Θ to each other.

On the other hand, an edge $f = xy$ with ends in opposite semicubes W_{ab} and W_{ba} (the right diagram in Figure 2.10) is in relation Θ to the edge ab, because

$$d(x, b) + d(y, a) = d(x, a) + 1 + d(y, b) + 1 \neq d(x, a) + d(y, b),$$

by Lemma 2.19. The converse does not hold as the graph C_5 in Figure 2.9 demonstrates. Indeed, each of two edges incident with vertex x is in the relation Θ to the edge ab of C_5 but does not join opposite semicubes W_{ab} and W_{ba}.

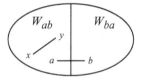 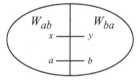

Figure 2.10. Relative positions of an edge $f = xy$ with respect to the edge $e = ab$.

Definition 2.20. Let $G = (V, E)$ be a connected graph and $e = ab$ and $f = xy$ be two edges of G. The edge f is in *relation* θ to the edge e if f joins a vertex in W_{ab} with a vertex in W_{ba}. The notation can be chosen such that $x \in W_{ab}$ and $y \in W_{ba}$.

It is clear that θ is a reflexive binary relation. It is not immediately evident that the relation θ is symmetric.

Lemma 2.21. *The relation θ is a symmetric relation on the edge set E.*

Proof. Let $e = ab$ and $f = xy$ be two edges in E. Suppose that $e\,\theta\,f$ with $x \in W_{ab}$ and $y \in W_{ba}$. By Lemma 2.19 and the triangle inequality, we have

$$d(x, a) = d(x, b) - 1 \leq d(x, y) + d(y, b) - 1 = d(y, b)$$
$$= d(y, a) - 1 \leq d(y, x) + d(x, a) - 1 = d(x, a).$$

Hence,

$$d(a, y) = d(a, x) + 1 \text{ and } d(b, x) = d(b, y) + 1.$$

Therefore, $a \in W_{xy}$ and $b \in W_{yx}$. It follows that $(f, e) \in \theta$.   □

As in the above proof, let $e = ab$ and $f = xy$ be two edges with $x \in W_{ab}$ and $y \in W_{ba}$, that is, $(e, f) \in \theta$. By Lemma 2.19, we have

$$d(a, x) + d(b, y) = d(a, y) - 1 + d(b, x) - 1 \neq d(a, y) + d(b, x).$$

Hence, $(e, f) \in \Theta$. Therefore, in any graph

$$\theta \subseteq \Theta. \tag{2.6}$$

On the other hand, $(e, f) \in \Theta$ and $(e, f) \notin \theta$ for any two distinct edges of the cycle C_3. Thus, $\theta \neq \Theta$ in this case. This is due to the fact that C_3 is not a bipartite graph as the next theorem demonstrates.

Theorem 2.22. *A connected graph is bipartite if and only if $\theta = \Theta$.*

Proof. (Necessity.) Let G be a connected bipartite graph. By (2.6), we need to show that $(e, f) \in \Theta$ implies $(e, f) \in \theta$ for any two edges $e = ab$ and $f = xy$ of G. The proof is by contradiction. Suppose that $(e, f) \notin \theta$; that is, the edge f does not join semicubes W_{ab} and W_{ba}. By Theorem 2.18, we may assume that $x, y \in W_{ab}$ (see Figure 2.10). Then, by Lemma 2.19, we have

$$d(a, x) + d(b, y) = d(b, x) - 1 + d(a, y) + 1 = d(b, x) + d(a, y).$$

This contradicts our assumption that $(e, f) \in \Theta$.

(Sufficiency.) The proof is again by contradiction. Suppose that for a connected nonbipartite graph $G = (V, E)$ we have $\theta = \Theta$. By Theorem 2.18, there is an edge $e = ab$ such that $W_{ab} \cup W_{ba} \neq V$. Because the graph G is connected, there is an edge $f = xy$ with ends $x \notin W_{ab} \cup W_{ba}$ and $y \in W_{ab} \cup W_{ba}$. Clearly, we have $(e, f) \notin \theta$. On the other hand,

$$d(a, x) + d(b, y) \neq d(b, x) + d(a, y),$$

inasmuch as $d(a, x) = d(b, x)$, and $d(b, y) = d(a, y) \pm 1$, by Lemma 2.19. Therefore, $(e, f) \in \Theta$, a contradiction, because we assumed that $\theta = \Theta$.   □

The relation $\theta = \Theta$ is called the *theta relation*, provided of course that the graph under consideration is bipartite.

The theta relation on the edge set of a bipartite graph need not be transitive as illustrated by the graph $K_{2,3}$ in Figure 2.6. We show (Theorem 2.25) that, in general, intransitivity of this relation is due to the fact that some semicubes of the graph are not convex. For instance, the semicube W_{ba} of $K_{2,3}$ shown in Figure 2.9 is not convex.

First, we establish two other useful results.

Theorem 2.23. *If* $\{W_{xy}, W_{yx}\} = \{W_{uv}, W_{vu}\}$ *in a graph* G, *then* $xy\,\theta\,uv$.

Theorem 2.23 follows immediately from the definition of the relation θ (cf. Exercise 2.19).

Theorem 2.24. *Let* $G = (V, E)$ *be a bipartite graph such that all its semicubes are convex. Then two edges* xy *and* uv *are in relation* θ *if and only if* $\{W_{xy}, W_{yx}\} = \{W_{uv}, W_{vu}\}$.

Proof. (Necessity.) Suppose that $xy\,\theta\,uv$. We assume that the notation is chosen such that $u \in W_{xy}$ and $v \in W_{yx}$. Let $z \in W_{xy} \cap W_{vu}$. By Lemma 2.19, $d(z, u) = d(z, v) + d(v, u)$. Because $z, u \in W_{xy}$ and W_{xy} is a convex set, we have $v \in W_{xy}$, a contradiction to the assumption that $v \in W_{yx}$. Thus $W_{xy} \cap W_{vu} = \varnothing$. Two opposite semicubes in a bipartite graph form a partition of V, thus we have $W_{uv} = W_{xy}$ and $W_{vu} = W_{yx}$.

A similar argument shows that $W_{uv} = W_{yx}$ and $W_{vu} = W_{xy}$, if $u \in W_{yx}$ and $v \in W_{xy}$.

(Sufficiency.) Follows from Theorem 2.23. $\qquad\qquad\square$

Theorem 2.25. *Let* $G = (V, E)$ *be a connected bipartite graph. The following statements are equivalent.*

(i)   *All semicubes of* G *are convex.*
(ii) *The relation* θ *is an equivalence relation on* E.

Proof. (i) $\Rightarrow$ (ii). Follows from Theorem 2.24.

(ii) $\Rightarrow$ (i). Suppose that θ is transitive and there is a nonconvex semicube W_{ab}. Then there are two vertices $u, v \in W_{ab}$ and a shortest path P from u to v that intersects W_{ba}. This path contains two distinct edges e and f joining vertices of semicubes W_{ab} and W_{ba}. The edges e and f stand in the relation θ to the edge ab. By transitivity of θ, we have $e\,\theta\,f$. Therefore, by Theorem 2.22, $e\,\Theta\,f$. This contradicts the result of Theorem 2.12. Thus all semicubes of G are convex. $\qquad\qquad\square$

Now we introduce two other families of sets that are useful in studying structural properties of bipartite graphs. For any two adjacent vertices a, b of a graph G, we define the following sets.

$$F_{ab} = \{uw \in E(G) : u \in W_{ab},\ w \in W_{ba}\}.$$
$$U_{ab} = \{u \in W_{ab} : u \text{ is adjacent to a vertex in } W_{ba}\}.$$

It is clear that $F_{ab} = \{uw \in E(G) : (uw, ab) \in \theta\}$. The five sets W_{ab}, W_{ba}, U_{ab}, U_{ba}, and F_{ab} are called the *fundamental sets* of the graph G; an example is shown in Figure 2.11.

Theorem 2.26. *If semicubes* W_{ab} *and* W_{ba} *of a bipartite graph* G *are convex, then the set* F_{ab} *is a matching in* G *that defines an isomorphism between the induced subgraphs* $G[U_{ab}]$ *and* $G[U_{ba}]$.

Proof. Suppose that F_{ab} is not a matching. Then there are distinct edges ux and vx with say $u, v \in U_{ab}$ and $x \in U_{ba}$ (cf. Figure 2.11). By the triangle inequality, $d(u, v) \leq 2$. Because G does not have odd cycles, $d(u, v) \neq 1$. Hence, $d(u, v) = 2$, which implies that x lies between u and v. This contradicts our assumption that W_{ab} is a convex subset of G. Therefore, F_{ab} is a matching.

To show that F_{ab} defines an isomorphism between U_{ab} and U_{ba}, suppose that $xy \in F_{ab}$, $uv \in F_{ab}$, and $xu \in E(G)$, with $x, u \in U_{ab}$ and $y, v \in U_{ba}$. Because G does not have odd cycles, $d(v, y) \neq 2$. By the triangle inequality,

$$d(v, y) \leq d(v, u) + d(u, x) + d(x, y) = 3.$$

W_{ba} is convex, therefore $d(v, y) \neq 3$. It follows that $d(v, y) = 1$; that is, vy is an edge of G. The result follows by symmetry. $\qquad\square$

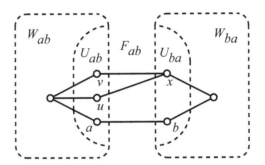

Figure 2.11. Fundamental sets in a graph.

The converse of Theorem 2.26 does not hold. Indeed, let G be the graph depicted in Figure 2.12. The set $F_{ab} = \{ab, xu, yv\}$ is a matching defining an isomorphism between the graphs induced by subsets $U_{ab} = \{a, x, y\}$ and $U_{ba} = \{b, u, v\}$. However, the set W_{ba} is not convex.

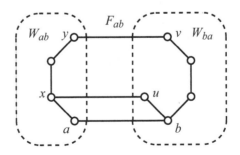

Figure 2.12. Counterexample.

2.4 Trees

Let us recall the definition of a tree from Chapter 1.

Definition 2.27. A connected graph G is called a *tree* if it does not contain cycles.

Graphs without cycles are known as *acyclic* graphs. Thus trees are connected acyclic graphs. Acyclic graphs are also called *forests*. The trees on six vertices are shown in Figure 2.13.

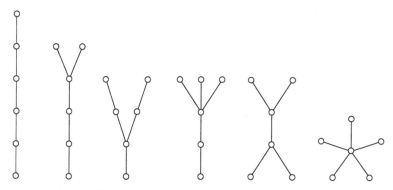

Figure 2.13. The trees on six vertices.

Theorem 2.28. *A nontrivial tree is a bipartite graph.*

Proof. We use the result of Theorem 2.3. Suppose there are a vertex v and an edge uw of a nontrivial tree T such that $d(v, u) = d(v, w)$. By concatenating a shortest vu-path, the edge uw, and a shortest wv-path, we obtain a closed odd walk W. By Theorem 1.5, W contains a cycle. This contradicts our assumption that T is a tree. □

The next theorem establishes a useful characterization of trees.

Theorem 2.29. *A graph is a tree if and only if there is a unique path between any two of its vertices.*

Proof. (Necessity.) Let T be a tree. Suppose that there are two vertices u and v in T such that there are two distinct uv-paths P' and P'' in T (see Figure 2.14). Let xy be an edge in one of these paths, say P', which is not an edge in the other path. Let us form an open xy-walk W in T by concatenating the ux-segment of P', the path P'', and the yv-segment of P'. Note that W does not contain the edge xy. By Theorem 1.4, the walk W contains an xy-path P in T. By adding the edge xy to the path P, we obtain a cycle. This contradicts our assumption that T is a tree.

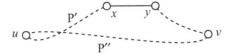

Figure 2.14. Proof of Theorem 2.29.

(Sufficiency.) Let G be a graph on more than one vertex such that there is a unique path connecting any two given vertices of G. Suppose that G is not a tree. Then it must contain a cycle. For any two distinct vertices of this cycle there are two distinct paths connecting these vertices. This contradiction shows that G is a tree.   □

Clearly, paths are finite trees. The ray and the double ray are examples of infinite trees. Note that we do not assume that the graph in Theorem 2.29 is finite.

Definition 2.30. A *leaf* of a graph G is a vertex of degree one.

By the result of Exercise 1.20, a finite tree must have at least one leaf. In fact, a stronger assertion holds for a nontrivial tree.

Lemma 2.31. *A finite tree $T = (V, E)$ on more than one vertex has at least two leaves. Deleting a leaf from the tree T produces a tree with $|V|-1$ vertices and $|E| - 1$ edges.*

Proof. Let $P = v_0 v_1 \cdots v_k$ be a path of maximum length in T. We may assume that $k > 1$. Let us show that the ends of P are leaves. Suppose that $\deg(v_k) > 1$ and let v be a neighbor of v_k different from v_{k-1}. We cannot have $v = v_i$, because otherwise we would have a cycle $v_i \cdots v_k v_i$. Therefore, $v_0 \ldots v_k v$ is a path of length $k + 1$. This is a contradiction, because P is assumed to be a path of maximum length. Hence, $\deg(v_k) = 1$; that is, v_k is a leaf. A similar argument shows that v_0 is a leaf.

Let $T - v$ be the graph obtained from T by deleting a leaf v. This graph is connected because v cannot belong to a path in T with ends in $T - v$. The graph $T-v$ does not contain cycles because there are no cycles in T. It follows that $T - v$ is a tree. Clearly, there are $|V| - 1$ vertices and $|E| - 1$ edges in the tree $T - v$.   □

Finite trees can be characterized as follows.

Theorem 2.32. *A finite connected graph $G = (V, E)$ is a tree if and only if*

$$|E| = |V| - 1; \tag{2.7}$$

that is, the number of edges of G is one less than the number of its vertices.

Proof. We use induction on $m = |V|$, the number of vertices of G.

(Necessity.) The case $m = 1$ is trivial. The inductive step follows immediately from the second statement of Lemma 2.31.

(Sufficiency.) We need to show that a connected graph $G = (V, E)$ satisfying condition (2.7) is acyclic. This is clearly true when it has one or two vertices. For the inductive step, suppose that it is true for any connected graph with $m - 1$ vertices and let G be a connected graph with m vertices. By (2.7), the graph G must have a vertex v with $\deg(v) = 1$. Indeed, otherwise we would have, by (1.1),

$$2|E| = \sum_{u \in V} \deg(u) \geq 2|V|,$$

which contradicts (2.7). By deleting the vertex v, we obtain a connected graph with $|V| - 1$ vertices and $|E| - 1$ edges, therefore satisfying condition (2.7). By induction, G is a tree (cf. Exercise 2.20). □

We now consider the relation Θ on the edge set of a tree T. Let $e = xy$ and $f = uv$ be two distinct edges of T, and notation is chosen such that $d(y, u)$ is the least number among the four distances between the ends of e and f (cf. Figure 2.5). Let $P = y \cdots u$ be the unique path connecting y and u (cf. Theorem 2.29). Inasmuch as $d(y, u) \leq d(x, u)$, we cannot have $x \in P$. Likewise, $v \notin P$. Therefore, $xPv = xy \cdots uv$ is the unique xv-path in T. By Theorem 2.12, we have

$$d(x, u) + d(y, v) = d(x, v) + d(y, u);$$

that is, $(e, f) \notin \Theta$. Hence, Θ is the identity relation on the set $V(T)$. On the other hand, if G is a connected graph containing a cycle, then, by Corollary 2.16, there are two distinct edges e and f with $e \Theta f$; that is, Θ is not the identity relation on connected graphs that are not trees. We obtained the following result.

Theorem 2.33. *A connected graph $G = (V, E)$ is a tree if and only if Θ is the identity relation on E.*

Let ab be an edge of a tree T. Because $\Theta = \theta$ is the identity relation, the edge ab is the unique edge joining vertices in semicubes W_{ab} and W_{ba}. The graphs W_{ab} and W_{ba} are obtained from T by deleting the edge ab. Clearly, these graphs are trees.

An edge e of a graph G is said to be a *cut-edge* if $G - e$ has more connected components than G (cf. Exercise 2.21). The argument in the previous paragraph shows that any edge of a tree is a cut-edge. The next theorem characterizes cut-edges in arbitrary graphs.

Theorem 2.34. *An edge e of a graph G is a cut-edge if and only if it belongs to no cycle in G.*

Proof. (Necessity.) The ends of e belong to different components of $G-e$, thus they cannot be connected by a path different from the edge e itself. Therefore, e does not belong to a cycle in G.

(Sufficiency.) If $e = xy$ is not a cut-edge, then there is an xy-path in $G - e$. By adding edge e to this path, we obtain a cycle of G containing e. Hence, an edge that does not belong to a cycle of G must be a cut-edge. □

Corollary 2.35. *A connected graph G is a tree if and only if every edge of G is a cut-edge.*

Similarly, a vertex v of a graph G is called a *cut-vertex* if $G - v$ has more components than G.

Theorem 2.36. *A vertex v of a tree T is a cut-vertex if and only if v is not a leaf.*

Proof. (Necessity.) Suppose that v is a cut-vertex of T. Then there are vertices x and y that belong to different components of T. By Theorem 2.29, there is a unique xy-path P in T. Clearly, v is different from x and y, and belongs to P. Hence, v is not a leaf.

(Sufficiency.) If v is not a leaf, then $\deg(v) \geq 2$. Therefore there are distinct vertices x and y adjacent to v. Because xvy is a unique path in T, the vertex v is a cut-vertex. □

A *spanning tree* of a graph G is a spanning subgraph (see Section 1.4) of G which is a tree.

Theorem 2.37. *A graph G is connected if and only if it has a spanning tree.*

Proof. (Necessity.) If G is a tree, it is a spanning tree of itself. Suppose that G is not a tree, and let e be an edge of a cycle of G. By Theorem 2.34, e is not a cut-edge. Hence, $G - e$ is a connected graph. By repeating this process of deleting edges in cycles until every edge that remains is a cut-edge, we obtain a spanning tree (cf. Corollary 2.35).

(Sufficiency.) Clearly, a graph with a spanning tree is connected. □

Example 2.38. A *wheel* W_n is a graph obtained from the cycle C_n by adding a new vertex and edges joining it to all vertices of C_n. Two spanning trees of the wheel W_6 are shown in Figure 2.15.

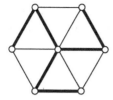 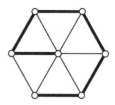

Figure 2.15. Two spanning trees of the wheel W_6.

2.5 Art Gallery Problem

We begin by recalling some concepts from plane geometry. Let $a_0, a_1, \ldots, a_n$ be points in the plane. A *polygonal curve* C in the plane is the union of closed line segments

$$C = [a_0, a_1] \cup [a_1, a_2] \cup \cdots \cup [a_{n-1}, a_n]$$

The curve C is said to be *simple* if no pair of its nonconsecutive segments shares a point. It is *closed* if $a_0 = a_n$. The complement $\mathbf{R}^2 \setminus C$ of a simple closed polygonal curve C is a union of two open connected sets (Jordan Curve Theorem; see Notes to Chapter 1). One of these sets is bounded and the other one is unbounded. The union P of the bounded component of $\mathbf{R}^2 \setminus C$ with C is called a *polygon* or *n-gon*. Points $a_1, \ldots, a_n$ are *vertices* of the polygon P and the respective line segments are *edges* of P. We denote the curve C by ∂P and call this curve the *boundary* of P. If a_i and a_j are two nonconsecutive vertices of a polygon P and $(a_i, a_j) \subseteq P \setminus \partial P$, then $[a_i, a_j]$ is called a *diagonal* of P.

Let x and y be two points in a polygon P. We say that y is *visible* from x in P if the line segment $[x, y]$ belongs to P. Note that $[x, y]$ may contain one or more points of ∂P. For instance, points of an edge are visible from its ends. The geometric concept of "visibility" is motivated by the Art Gallery Problem:

> How many guards are always sufficient and sometimes necessary to see (cover) every point in an art gallery?

We view a floor plan of the gallery as a simple polygon (see Figure 2.16) and call a "guard" an arbitrary vertex of the polygon. The problem is to select guards in such a way that they survey the gallery with n walls and no interior obstructions.

For a given n-gon P, let us define $G(P)$ to be the minimum number of guards who could see together all points in P, and define $g(n)$ to be the maximum value of $G(P)$ taken over all n-gons. The function $g(n)$ represents the maximum number of guards that are ever needed for an n-gon: $g(n)$ guards always suffice, and $g(n)$ guards are necessary for at least one n-gon.

By experimenting with "small" polygons (on 3–6 vertices), one can conjecture that $\lfloor n/3 \rfloor$ guards are needed to cover some n-gons. Let us consider a "comb" of k prongs and $n = 3k$ vertices shown in Figure 2.18. It is clear that no two upper vertices of prongs can be visible to a single guard. Thus, at least $k = n/3$ guards are necessary to cover the comb. It follows that $g(n) \geq \lfloor n/3 \rfloor$ for any $n \geq 3$ (cf. Exercise 2.38).

Figure 2.16. The floor plan of an art gallery.

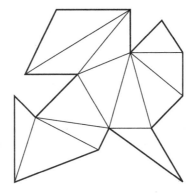

Figure 2.17. A triangulation of the polygon in Figure 2.16.

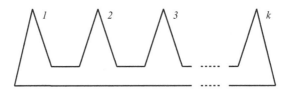

Figure 2.18. A comb with k prongs and $3k$ vertices.

In fact, $\lfloor n/3 \rfloor$ guards suffices to cover any n-gon (the "Art Gallery Theorem"). We begin by outlining the proof.

First, we "triangulate" the n-gon by adding diagonals until no more can be added (see Figure 2.17).

Second, we "colour" vertices of the resulting planar graph by three colours in such a way that no two adjacent vertices are assigned the same colour.

Finally, we place guards at the vertices of a least-used colour. At most $\lfloor n/3 \rfloor$ vertices are coloured by this colour, so no more than $\lfloor n/3 \rfloor$ guards are used. Because the vertices of any triangle have different colours, every triangle has a guard at one of its vertices and this guard sees every point in the triangle. Every point of the polygon belongs to some triangle. Thus every point of the polygon is visible by the guards.

There are two statements in this outline requiring proofs: it is not evident that a polygon can be triangulated and even less evident that its triangulation can be 3-coloured. Moreover, it is not obvious that a polygon must have a diagonal.

Theorem 2.39. *A polygon has at least one diagonal.*

Proof. Let a_1, a_2, a_3 be three consecutive vertices of a polygon P such that the angle $\angle a_1 a_2 a_3$ is less than $180°$ (cf. Exercise 2.39). If $[a_1, a_3]$ is a diagonal of P, we are done (cf. Figure 2.19, left). Otherwise, the closed triangle $a_1 a_2 a_3$ must contain at least one vertex of P. Let x be a vertex of P closest to a_2, where the distance is measured perpendicular to $[a_1, a_3]$ (see Figure 2.19, right). Then $[x, a_2]$ is a diagonal of P. $\square$

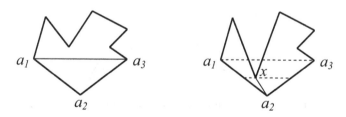

Figure 2.19. Proof of Theorem 2.39.

Now we prove the "Triangulation Theorem".

Theorem 2.40. *An n-gon may be partitioned into $n - 2$ triangles by adding $n - 3$ diagonals.*

Proof. The proof is by induction on n. The statement is trivial for a triangle. Let P be a polygon of $n \geq 4$ vertices and suppose that the statement of the theorem holds for any polygon with less than n vertices. Let d be a diagonal of P. It divides P into two smaller polygons P_1 and P_2. If P_i has n_i vertices,

$i \in \{1,2\}$, then $n_1 + n_2 = n + 2$ because both ends of d are shared between P_1 and P_2. Applying the induction hypothesis to each of the polygons P_1 and P_2 results in a triangulation for P of $(n_1 - 2) + (n_2 - 2) = n - 2$ triangles, and $(n_1 - 3) + (n_2 - 3) + 1 = n - 3$ diagonals, including d. $\qquad\square$

A partition of a polygon into triangles is called a *triangulation* of the polygon. Theorem 2.40 claims that any polygon can be triangulated by its diagonals.

We use graph theory to complete the proof of the "Art Gallery Theorem". For any graph G and a set X of cardinality k, a mapping $V(G) \to X$ is called a *vertex-colouring* (or *k-colouring*) of G. The elements of set X are called the available *colours*. A k-colouring of a graph is said to be *proper* if adjacent vertices have different colours. A graph is *k-colourable* if it has a proper k-colouring. The *chromatic number* $\chi(G)$ is the least k such that G is k-colourable.

Let G be the graph of a triangulation by diagonals of a polygon P. We define the *dual graph* of G as a planar graph with a vertex for each triangle of the triangulation and an edge connecting two vertices whose triangles share a diagonal (see Figure 2.20).

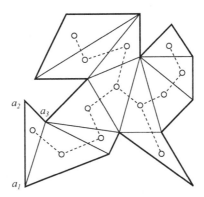

Figure 2.20. The dual graph of the polygon triangulation in Figure 2.17.

The dual graph of a triangulation is a tree. Indeed, otherwise it would have a cycle of triangles. This cycle encloses some vertices of the polygon, and therefore encloses points exterior to the polygon. This contradicts the definition of a simple polygon.

Theorem 2.41. *Let G be the graph of a triangulation of an n-gon P. Then G is 3-colourable. Moreover, $\chi(G) = 3$.*

Proof. By Theorem 2.31, the dual graph has a leaf. Let $a_1 a_2 a_3$ be a triangle defining this leaf (see Figure 2.20), where $[a_1, a_3]$ is a diagonal of P.

We prove the theorem by induction on n. The claim is trivial for $n = 3$. For $n \geq 4$, let us delete vertex a_2 and 3-colour the remaining triangulation. Now we put back the removed triangle and colour vertex a_2 by the colour not used on the diagonal $[a_1 a_3]$.

Inasmuch as three colours are needed to properly colour a triangle, we have $\chi(G) = 3$. $\qquad\qquad\qquad\qquad\qquad\qquad\qquad\qquad\qquad\qquad\qquad\qquad\qquad\qquad$ $\square$

In summary, we proved the following theorem.

Theorem 2.42. (Art Gallery Theorem) $\lfloor n/3 \rfloor$ *guards are sometimes necessary and always sufficient to cover a polygon of n vertices.*

Notes

In applications, bipartite graphs are useful in constructing models of interaction between two different types of objects. They are also reasonably simple graphs that inherit many difficult theoretical problems of general graph theory. A comprehensive study of bipartite graphs and examples of applications are found in Asratian et al. (1998).

Theorem 2.2 is arguably the best-known characterization of bipartite graphs. It was established by Dénis Kőnig in 1916 (see translation 10D in Biggs et al., 1986). In 1936 Dénis Kőnig published the first graph theory textbook *Theorie der endlichen und unendlichen Graphen* (English translation in Kőnig, 1989) marking the two-hundredth anniversary of Euler's paper on graphs (see Notes in Chapter 1). At least eight fundamental results in graph theory bear Kőnig's name (see subject index in the *Handbook of Combinatorics*, Graham et al., 1995).

Besides having obvious "geometric flavor", the concepts of betweenness, interval, and convexity also have their counterparts in algebra, specifically, in lattice theory (see, for instance, Birkhoff (1967) and Grätzer (2003)). Perhaps the first systematic attempt to use these concepts in graph theory was made by Henry Martyn Mulder in his PhD thesis (Mulder, 1980). The reader is also referred to the book by Imrich and Klavžar (2000). These notions are very useful in Chapters 3–7, where properties of various types of cubical graphs and their geometric representations are investigated.

The theta relation θ was originally introduced by Dragomir Ž. Djoković in the proof of his result concerning isometric embeddings of graphs into hypercubes (Djoković, 1973). Peter M. Winkler (1984) generalized Djoković's result using another binary relation on the edge set of a graph. The notation Θ for this relation was introduced later by Imrich and Klavžar (2000). Because $\Theta = \theta$ for bipartite graphs, the term "Djoković–Winkler relation" is used in Imrich and Klavžar (2000). Semicubes W_{ab} of a graph G appear as vertex sets $G(a, b)$ in Djoković's original paper (see also Deza and Laurent, 1997). The term "semicube" was coined by David Eppstein (2005). Semicubes and other fundamental sets are important tools in the theory of partial cubes (see Chapter 5). Theorems 2.25 and 2.26 reveal intimate relations between geometric and algebraic properties of fundamental sets in a graph. Specifically, the convexity property of semicubes plays a central role in characterizing partial cubes in Chapter 5.

Trees are arguably the most widely used graphs in pure mathematics and applications. Both textbooks, Bondy and Murty (2008) and West (2001), have chapters on trees. For a thorough treatment of applications, the reader is referred to Foulds (1992) and Roberts (1984).

For the history of the Art Gallery Problem, its generalizations, and related computer science problems, see the book by Joseph O'Rourke (1987). Vertex colouring and the chromatic number of a graph are important concepts of graph theory. The books by Bondy and Murty (2008) and West (2001) both have chapters entirely devoted to these notions.

Exercises

2.1. Prove that a walk in a bipartite graph $G[X, Y]$ is even (respectively, odd) if and only if its ends belong to the same part (respectively, different parts) of the bipartition (X, Y).

2.2. Prove that a graph is bipartite if and only if its connected components are bipartite.

2.3. Prove that the hexagonal lattice (see Figure 1.15) is bipartite.

2.4. Let G be a bipartite graph with n connected components. Show that there are exactly 2^{n-1} distinct bipartitions (X, Y) such that $G \cong G[X, Y]$.

2.5. Describe all intervals in the star $K_{1,X}$, that is, a complete bipartite graph in which one of the parts is a singleton.

2.6. Let $G = (V, E)$ be a connected graph. Show that, for any $u, v \in V$,

a) If $x \in I(u, v)$, then $I(u, x) \subseteq I(u, v)$.
b) If $x \in I(u, v)$, then $I(u, x) \cap I(x, v) = \{x\}$.
c) If $x \in I(u, v)$ and $y \in I(u, x)$, then $x \in I(y, v)$.

2.7. Let $G = (V, E)$ be a connected graph.

a) Show that for any three vertices $u, v, w \in V$ there is a vertex
$x \in I(u, v) \cap I(u, w)$ such that

$$I(x, v) \cap I(x, w) = \{x\}.$$

b) Show that this vertex z need not be unique.

2.8. a) Prove that the vertex set of a convex subgraph is convex.
b) Prove Theorem 2.8.

2.9. Show that the vertices of any two adjacent edges of a bipartite graph form a shortest path in the graph.

2.10. Prove Theorem 2.9 (Kay and Chartrand, 1965; Nebeský, 2008)

2.11. Describe the relation Θ for the path P_n, the grid $P_3 \,\square\, P_2$, the complete bipartite graphs $K_{2,2}$ and $K_{2,3}$, and the complete graph K_4. You need to identify explicitly those pairs of edges (e, f) that belong to Θ. Which of these relations are transitive?

2.12. Let G be a connected graph. For any two edges $e = xy$ and $f = uv$ of G we define

$$\mu(e, f) = |d(x, u) - d(x, v) - d(y, v) + d(y, u)|.$$

a) Show that the function μ is well defined and assumes values in $\{0, 1, 2\}$.
b) Prove that G is bipartite if and only if $\mu(e, f) \in \{0, 2\}$ for all $e, f \in E(G)$.

2.13. Prove Corollary 2.16.

2.14. Prove Theorem 2.18.

2.15. For an edge ab of a connected bipartite graph, prove that $I(u, a) \subseteq W_{ab}$ for any vertex u in W_{ab}. (Hint: Use Lemma 2.19.)

2.16. Let ab and uv be two edges of a connected bipartite graph such that $u \in W_{ab}$, and $v \in W_{ba}$. Prove that $d(a, u) = d(b, v)$.

2.17. Let u be a vertex of degree one in a connected graph G (a leaf), and let $e = uv$ be the unique edge with the end u. Show that for any other edge f of G we must have $(e, f) \notin \Theta$.

2.18. Describe the relation θ for the graphs listed in Exercise 2.11.

2.19. Prove Theorem 2.23.

2.20. Let $T = (V, E)$ be a tree. Show that adding a new vertex incident with a vertex of T produces a new tree with $|V| + 1$ vertices and $|E| + 1$ edges.

2.21. Let e be an edge of a graph G. Show that the number of connected components of the graph $G - e$ is either the same as the number of components of G or one more than that number.

2.22. A graph in which each vertex has even degree is called an *even graph*. Show that an even graph has no cut-edge.

2.23. Prove that a forest with n vertices and k components has exactly $n - k$ edges.

2.24. Show that there are exactly six nonisomorphic trees on six vertices (see Figure 2.13).

2.25. Let G be a connected graph with n vertices and m edges. Show that if G is not a tree then $m > n - 1$.

2.26. Describe all complete bipartite graphs that are trees.

2.27. Let $G = (V, E)$ be a connected graph. Prove that the following statements are equivalent:

a) G has a unique cycle.
b) $|V| = |E|$.
c) there is an edge $e \in E$ such that $G - e$ is a tree.

2.28. Prove that a graph is a tree if and only if it has exactly one spanning tree.

2.29. The *center* of a connected graph $G = (V, E)$ is the set of vertices u such that $\max_{v \in V} d(u, v)$ is as small as possible.

a) Let T be a tree on more than two vertices, and let T' be a tree obtained from T by deleting all its leaves. Show that T and T' have the same centers.
b) Show that the center of a tree consists of a single vertex or the ends of an edge.

2.30. Let G and H be connected nontrivial graphs. Show that $G \square H$ has no cut-vertex.

2.31. A graph $G = (V, E)$ is called *antipodal* if for every vertex v there is a vertex u such that $V = I(u, v)$. Show that the Cartesian product of two antipodal graphs is antipodal.

2.32. Let T be a tree with vertex set $\{v_1, \ldots, v_k\}$, $k \geq 2$. Show that the number of leaves of T is given by

$$1 + \frac{1}{2} \sum_{i=1}^{k} |\deg(v_i) - 2|$$

2.33. A *median* of a triple of vertices $\{u, v, w\}$ of a connected graph G is a vertex in $I(u, v) \cap I(u, w) \cap I(v, w)$. A graph G is a *median graph* if every triple of vertices of G has a unique median.

a) Show that $K_{2,3}$ is not a median graph.
b) Let v be a vertex of the cube Q_3. Show that $Q_3 - v$ is not a median graph.
c) Show that grids are median graphs.

d) Prove that trees are median graphs.

e) Show that median graphs are bipartite. (Hint: Show first that a shortest odd cycle of a graph is an isometric subgraph.)

2.34. Let T be a tree with a fixed vertex v_0. Let us denote by P_v the unique v_0v-path in T. Prove that, for any edge xy of T, either $x \in P_y$ or $y \in P_x$.

2.35. Prove that every nontrivial connected graph has at least two vertices that are not cut-vertices.

2.36. Let G be a connected graph such that $\deg_G(v) \leq 2$ for all $v \in V(G)$. Show that G is either a path or a cycle.

2.37. Let G be a bipartite graph and C be a cycle in G. Prove that C is an isometric subgraph of G if and only if $d_C(u, v) = d_G(u, v)$ for any pair $\{u, v\}$ of opposite vertices of C.

2.38. Modify Figure 2.18 to show that $g(n) \geq \lfloor n/3 \rfloor$ for any n-gon.

2.39. Prove that a polygon must have an inner angle which is less than $180°$.

2.40. Prove that the sum of inner angles of an n-gon is $(n - 2)\pi$.

3

Cubes

In the first four sections of this chapter we introduce objects that are commonly known as "cubes" in geometry, algebra, and graph theory. A definition of a cube as a graph is given in Section 3.5. We continue by discussing properties of infinite Cartesian products, metric geometry of cubes, and their automorphisms. Finite cubes are characterized in the last section.

3.1 Cubes in Geometry

The term "cube" appears in various branches of mathematics referring often to similar but still different objects. In Sections 3.1–3.3 we discuss several instances of the term.

The cube in geometry is one of the Platonic solids (convex regular polyhedra). By definition, it is a rectangular parallelepiped whose three dimensions are congruent to each other. A parallelepiped itself is an instance of a prism. For our purposes, we adopt the definition of a prism to the special case of the cube as follows.

Let S be a square with sides of length $a > 0$ lying in a "horizontal" plane E (see Figure 3.1). At each point P of S, we set up a vertical line segment I of length a, perpendicular to E. The union Q of all these segments is called a cube.

In the modern mathematical language, this definition tells us that the cube Q is the Cartesian product $S \times I$ of the square S and the interval I. The square S itself is the Cartesian product of two copies of the interval I:

$$S = I \times I = I^2.$$

Thus we can define the cube Q as the Cartesian product

$$Q = I \times I \times I = I^3.$$

This observation paves the road to generalizing the notion of a cube.

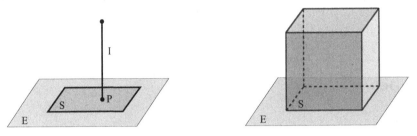

Figure 3.1. Definition of the cube.

Let I be a nondegenerate line segment. The *n-dimensional cube* I^n is the Cartesian product of $n \geq 2$ copies of the segment I:

$$I^n = \underbrace{I \times \cdots \times I}_{n \text{ factors}}.$$

As usual, we augment this definition by setting $I^1 = I$.

In the n-dimensional Euclidean space $\mathbf{R}^n$, the set

$$[0,1]^n = \{(x_1, \ldots, x_n) \in \mathbf{R}^n : 0 \leq x_i \leq 1, 1 \leq i \leq n\}$$

is called the *unit cube in* $\mathbf{R}^n$. In what follows, we assume that $I = [0,1]$, so I^n stands for the unit cube in $\mathbf{R}^n$.

A point $(x_1, \ldots, x_n) \in I^n$ is called a *vertex of* I^n if $x_i \in \{0,1\}$ for every $1 \leq i \leq n$. Thus, $\{0,1\}^n$ is the set of vertices of the unit cube I^n. Let K be a subset of $\{1, \ldots, n\}$ of cardinality k. Let us assign the value 0 or 1 to every coordinate x_i with $i \in K$. This assignment defines a subset of I^n which is called an $(n-k)$-dimensional *face* of the cube I^n. For $K = \varnothing$ we have I^n as the unique n-dimensional face of itself. All other faces of I^n are called *proper faces*. The zero-dimensional faces of the cube I^n are its vertices and the 1-dimensional faces are its *edges*. The $(n-1)$-dimensional faces are called *facets*.

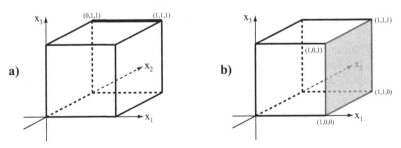

Figure 3.2. An edge and a face of the unit cube I^3.

Example 3.1. Let $J = \{1, 2, 3\}$ and $K = \{2, 3\}$. Let us fix the values $x_2 = 1$ and $x_3 = 1$. This assignment defines an edge of I^3:

$$\{(x_1, x_2, x_3) \in I^3 : x_2 = 1, \ x_3 = 1\},$$

which is a line segment in $\mathbf{R}^3$ with ends $(0, 1, 1)$ and $(1, 1, 1)$ (see Figure 3.2a). Let now $K = \{1\}$ and $x_1 = 1$. With this setting we have a facet

$$\{(x_1, x_2, x_3) \in I^3 : x_1 = 1\}.$$

This facet is a square in $\mathbf{R}^3$ with vertices $(1, 0, 0)$, $(1, 0, 1)$, $(1, 1, 1)$, and $(1, 1, 0)$ (see Figure 3.2b).

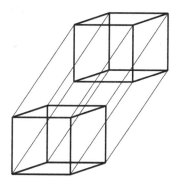

Figure 3.3. Constructing the unit cube I^4.

Example 3.2. We construct the unit cube I^4 as follows (see Figure 3.3). Take the unit cube I^3 in $\mathbf{R}^4$ in the 3-dimensional subspace defined by the first three coordinates. Move this cube "vertically" along the fourth coordinate by one unit. Connect every point in the first cube to a point in the second cube by a "vertical" unit line segment parallel to the fourth coordinate. The resulting 4-dimensional solid is the cube I^4. One can say that I^4 is a right prism with the base I^3 and height one.

3.2 Boolean Cubes

An n-fold Cartesian product of the set $\{0, 1\}$ with itself, that is, $\{0, 1\}^n$ is called a *Boolean cube*. It should be stressed that the numbers 0 and 1 in this definition are just labels of the elements of a two-element set. In many applications, the role of the set $\{0, 1\}$ is played by other two-element sets, such as {NO, YES}, {FALSE, TRUE}, {OFF, ON}, and so on, and the distinction is irrelevant. However, it is often convenient to view the elements of $\{0, 1\}$ as numerical quantities and perform arithmetic operations on these elements to model operations arising in applications.

The Boolean cube can also be defined as the power set of a set. The *power set* $\mathcal{P}(X)$ (notation 2^X is also popular) of a set X is the set of all subsets of the set X. The *characteristic function* $\chi_A : X \to \{0,1\}$ of a subset A of the set X assigns 1 to the elements of A and 0 to the elements of $X \setminus A$:

$$\chi_A(x) = \begin{cases} 1, & \text{if } x \in A, \\ 0, & \text{if } x \notin A, \end{cases} \quad \text{for } x \in X.$$

On the other hand, any function $f : X \to \{0,1\}$ is a characteristic function of the subset $\{x \in X : f(x) = 1\}$. Thus we have a one-to-one correspondence between the power set $\mathcal{P}(X)$ and the set of characteristic functions on X. Because elements of the Boolean cube are $0/1$-vectors,[1] they are functions from the set $J = \{1, \ldots, n\}$ to the set $\{0,1\}$; that is, they are characteristic functions of subsets of J. Therefore we have a bijection

$$h : \{0,1\}^n \to \mathcal{P}(J)$$

from the Cartesian power $\{0,1\}^n$ onto the power set $\mathcal{P}(J)$ defined by

$$h(x) = \{i \in J : x_i = 1\}, \tag{3.1}$$

where $x = (x_1, \ldots, x_n) \in \{0,1\}^n$. Note that

$$h^{-1}(A) = (x_1, \ldots, x_n)$$

for $A \subseteq J$, where $x_i = 1$ if $i \in A$ and $x_i = 0$, otherwise.

Definition 3.3. A binary relation P on a nonempty set A is called a *partial order* if it is irreflexive and transitive; that is,

(i) $(a, a) \notin P$, and
(ii) $(a, b) \in P$ and $(b, c) \in P$ implies $(a, c) \in P$,

for all $a, b, c \in A$. A set A with a partial order P on it is said to be a *partially ordered set* or *poset* for short and denoted by (A, P). Two posets (A, P) and (B, Q) are *isomorphic* if there is a bijection $f : A \to B$ such that

$$(x, y) \in P \quad \text{if and only if} \quad (f(x), f(y)) \in Q.$$

We often use term "poset" for the set A itself if (A, P) is a partially ordered set.

Both sets $\{0,1\}^n$ and $\mathcal{P}(J)$ are posets. The set $\{0,1\}^n$ is partially ordered by the binary relation $<$ defined by

---

[1] That is, vectors in $\mathbf{R}^n$ whose coordinates take values in $\{0, 1\}$.

$$x < y \quad \text{if and only if} \quad \begin{cases} x_i \le y_i, & \text{for all } 1 \le i \le n, \text{ and} \\ x_j < y_j, & \text{for some } 1 \le j \le n. \end{cases}$$

The set $\mathcal{P}(J)$ is partially ordered by the set inclusion relation $\subset$. As posets, the sets $\{0,1\}^n$ and $\mathcal{P}(J)$ are isomorphic since

$$x < y \quad \text{if and only if} \quad h(x) \subset h(y),$$

for all $x, y \in \{0,1\}^n$ (cf. Exercise 3.6). We denote by $\mathcal{B}_n$ the two isomorphic posets $(\{0,1\}^n, <)$ and $(\mathcal{P}(J), \subset)$.

Finite posets are conveniently represented by their diagrams as follows. Let $(A, <)$ be a finite poset. We represent each element of the set A by a point in the plane and draw a line segment (edge) upward from x to y if $x < y$ and there is no z such that $x < z < y$. The resulting drawing is a graph called the *Hasse diagram* of the poset.

The Hasse diagram of the Boolean cube $\mathcal{B}_3$ is depicted in Figure 3.4 with labelings corresponding to the two definitions of $\mathcal{B}_3$ given in foregoing paragraphs. Clearly, the bijection h is an isomorphism of these two graphs (cf. Exercise 3.10).

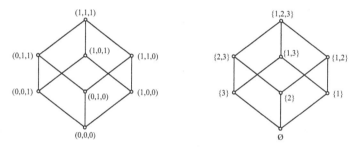

Figure 3.4. Two labelings of the Hasse diagram of $\mathcal{B}_3$.

3.3 Boolean Lattices

A partially ordered set L is called a *lattice* if every pair of elements of L has a unique lowest upper bound, called the *meet*, and a unique greatest lower bound, called the *join*. The meet and join of elements x and y of L are denoted by $x \wedge y$ and $x \vee y$, respectively. They are binary operations on the poset L. A lattice L is said to be *distributive* if

$$x \wedge (y \vee z) = (x \wedge y) \vee (x \wedge z) \text{ for all } x, y, z \in L.$$

A lattice is *bounded* if it has a least element $\mathbf{0}$ and a greatest element $\mathbf{1}$. An unary operation $x \to \overline{x}$ on a lattice L with $\mathbf{0}$ and $\mathbf{1}$ is called a *complementation* if

$$x \wedge \overline{x} = \mathbf{0} \text{ and } x \vee \overline{x} = \mathbf{1} \text{ for all } x \in L.$$

The element $\overline{x}$ is said to be a *complement* of x. A lattice endowed with a complementation is called a *complemented lattice*. By definition, a *Boolean lattice* is a complemented distributive lattice.

The set $\{0, 1\}$ considered as a poset with respect to the usual ordering $0 < 1$ is a complemented lattice. The operations $\wedge$, $\vee$, and $^-$ are given by the following rules.

$$x \wedge y = \min\{x, y\}, \quad x \vee y = \max\{x, y\} \text{ and } \overline{x} = 1 - x.$$

In fact, the poset $(\{0, 1\}, <)$ is a Boolean lattice (cf. Exercise 3.9). In logic, these operations are Boolean AND, OR, and NOT, respectively. They are extended to the Boolean cube $\{0, 1\}^n$ by writing

$$x \wedge y = (\min\{x_1, y_1\}, \dots, \min\{x_n, y_n\}),$$
$$x \vee y = (\max\{x_1, y_1\}, \dots, \max\{x_n, y_n\}),$$
$$\overline{x} = (1 - x_1, \dots, 1 - x_n),$$

for all $x = (x_1, \dots, x_n)$ and $y = (y_1, \dots, y_n)$ in $\{0, 1\}^n$.

Another important example of a Boolean lattice is the poset $(\mathcal{P}(J), \subset)$, where $J = \{1, \dots, n\}$. Operations meet and join are intersection $\cap$ and union $\cup$ of sets, respectively. The complement $\overline{A}$ of a set $A \subseteq J$ is the set difference $J \setminus A$. We also have $\mathbf{0} = \varnothing$ and $\mathbf{1} = J$.

We leave it to the reader to prove that the Boolean cubes $\{0, 1\}^n$ and $\mathcal{P}(J)$ with operations defined in the foregoing paragraphs are indeed Boolean lattices (cf. Exercise 3.10).

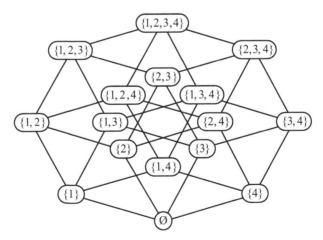

Figure 3.5. The Boolean lattice $\mathcal{B}_4 = (\mathcal{P}(\{1, 2, 3, 4\}), \subset)$.

Two Boolean lattices are said to be *isomorphic* if their defining posets are isomorphic. Inasmuch as $\{0,1\}^n$ and $\mathcal{P}(J)$ are isomorphic posets, the Boolean lattices $(\{0,1\}^n, \leq)$ and $(\mathcal{P}(J), \subseteq)$ are isomorphic. The bijection h (see (3.1)) defining an isomorphism of posets $\{0,1\}^n$ and $\mathcal{P}(J)$ preserves operations meet, join, and complementation; that is,

$$h(x \wedge y) = h(x) \cap h(y), \quad h(x \vee y) = h(x) \cup h(y), \quad h(\overline{x}) = \overline{h(x)}$$

for all $x, y \in \{0,1\}^n$.

The Hasse diagrams of the Boolean lattices $\mathcal{B}_3$ and $\mathcal{B}_4$ are depicted in Figures 3.4 and 3.5, respectively.

3.4 Finite Cubes

It is clear that the Hasse diagram of the Boolean lattice $\mathcal{B}_3$ in Figure 3.4 is the graph of the cube Q in Figure 3.1. This is also the same graph as the graph Q_3 in Figure 1.6. By moving in this direction, one can introduce the concept of the n-dimensional cube in graph theory either "geometrically" as the graph of the unit cube I^n, or "algebraically" as the Hasse diagram of the Boolean lattice $\mathcal{B}_n$. First, we introduce, for a given positive integer n, five graphs and show that they are isomorphic.

G_1 The graph of the unit cube I^n. The vertices and edges of G_1 are vertices and edges of the cube I^n.

G_2 The vertex set is the Boolean cube $\{0,1\}^n$. Two vertices $(x_1, \ldots, x_n)$ and $(y_1, \ldots, y_n)$ are adjacent in G_2 if and only if there is j such that $x_j \neq y_j$ and $x_i = y_i$ for all $i \neq j$.

G_3 The vertex set is the power set $\mathcal{P}(X)$ with $|X| = n$. Two vertices A and B are adjacent if and only if the sets A and B differ by precisely one element.

G_4 The Hasse diagram of the Boolean lattice $\mathcal{B}_n$.

G_5 The n-fold Cartesian product $K_2^n = K_2 \square \cdots \square K_2$, where K_2 is the complete graph on two vertices (by definition, $K_2^1 = K_2$).

Theorem 3.4. *The five graphs G_1–G_5 are isomorphic.*

Proof. $G_1 \cong G_2$. These two graphs have the same vertex set $\{0,1\}^n$. An edge of I^n is defined by choosing a number $j \in J$ and assigning, in arbitrary way, numbers 0 and 1 to coordinates x_i with $i \neq j$. It follows that an edge of I^n is completely determined by the choice of $j \in J$ and two vertices $(x_1, \ldots, x_n)$ and $(y_1, \ldots, y_n)$ with $y_i = x_i$ for $i \neq j$, $x_j = 0$, and $y_j = 1$. Thus we have a one-to-one correspondence between the edge sets of graphs G_1 and G_2.

$G_2 \cong G_3$. Let h be the bijection defined by (3.1). Two vertices x and y of G_2 are adjacent if and only if there is $j \in J$ such that $x_i = y_i$ for $i \neq j$ and either $x_j = 1$ and $y_j = 0$, or $x_j = 0$ and $y_j = 1$. Therefore, either

$h(y) = h(x) \cup \{j\}$ or $h(x) = h(y) \cup \{j\}$; that is, the sets $h(x)$ and $h(y)$ differ by exactly one element j. It follows that h is a graph isomorphism of G_2 onto G_3.

$G_3 \cong G_4$. We may assume that $X = J = \{1, \ldots, n\}$. Then G_3 and G_4 have the same vertex set $\mathcal{P}(J)$. If two vertices A and B are adjacent in G_3, then either $B = A \cup \{i\}$ or $A = B \cup \{i\}$ for some $i \in J$. Clearly, these vertices form an edge in the Hasse diagram of $\mathcal{B}_n$. Conversely, if A and B are adjacent in G_4, then they differ by a singleton, because (by the definition of Hasse diagram) there is no set C such that either $A \subset C \subset B$ or $B \subset C \subset A$.

$G_5 \cong G_2$. We may assume that $K_2 = (\{0, 1\}, \{01\})$. The graphs G_2 and G_5 have the same vertex set $\{0, 1\}^n$. Two vertices $(x_1, \ldots, x_n)$ and $(y_1, \ldots, y_n)$ are adjacent in $G_5 = K_2^n$ if and only if there is $1 \leq j \leq n$ such that $x_i = y_i$ for $i \neq j$, and x_j and y_j are adjacent in the graph K_2 (cf. Exercise 1.35). The vertices x_j and y_j are adjacent in K_2 if and only if $x_j \neq y_j$. Hence the graphs G_2 and G_5 are isomorphic.   □

A graph G is called the n-cube if it is isomorphic to one of the graphs G_1–G_5. As before (cf. Section 1.2), we do not distinguish between isomorphic graphs. Up to isomorphism, there is a unique n-cube which we denote by Q_n. Of course, in any particular problem, one is free to use any of the five graphs G_1–G_5 as a faithful representation of Q_n.

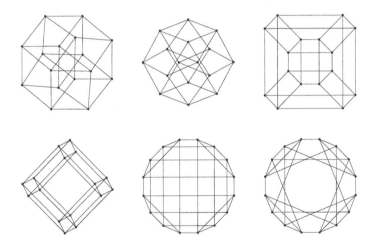

Figure 3.6. Drawings of the 4-cube Q_4.

Six drawings of the 4-cube graph created with the *Mathematica*® computer algebra system are shown in Figure 3.6. Each of these graphs is isomorphic to the graph of the unit cube I^4 depicted in Figure 3.3. A 3-dimensional model of the 4-cube graph is displayed in Figure 3.7.

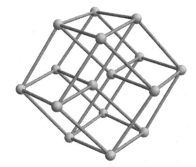

Figure 3.7. 3D-drawing of the 4-cube Q_4.

3.5 Arbitrary Cubes

The vertex set of every graph in the family $\{G_i\}_{1 \leq i \leq 5}$ introduced in the previous section is either the Cartesian power $\{0,1\}^n$ or the power set $\mathcal{P}(J)$ where $J = \{1, \ldots, n\}$. Accordingly, we consider two ways of defining cubes in graph theory.

Let X be an arbitrary set. For a set Y, the *Cartesian power* Y^X is the set of all functions $f : X \to Y$. If $Y = \{0,1\}$, then the power $\{0,1\}^X$ is the set of all characteristic functions on X. A characteristic function f on X defines a unique subset $\{x \in X : f(x) = 1\}$ of X, therefore we have a mapping

$$h : \{0,1\}^X \to \mathcal{P}(X) \qquad (3.2)$$

defined by

$$h(f) = \{x \in X : f(x) = 1\} \text{ for } f \in \{0,1\}^X.$$

Clearly, h is a bijection. The inverse mapping h^{-1} is given by

$$h^{-1}(A) = \chi_A \text{ for } A \subseteq X.$$

For a finite set X, the bijection h is the same as in (3.1).

Now we define two graphs, K_2^X and $\mathfrak{P}(X)$, with vertex sets $\{0,1\}^X$ and $\mathcal{P}(X)$, respectively.

We say that two distinct elements f and g of $\{0,1\}^X$ are adjacent in K_2^X if there is $y \in X$ such that

$$f(x) = g(x) \text{ for all } x \neq y.$$

Because $f \neq g$, we have $f(y) \neq g(y)$. The graph K_2^X is a natural generalization of the finite power K_2^n (the graph G_5 from Section 3.4) and called the *Cartesian power* of K_2.

Two sets A and B in $\mathcal{P}(X)$ are said to be adjacent in $\mathfrak{P}(X)$ if there is $a \in X$ such that

$$A \triangle B = \{a\},$$

where $A \triangle B = (A \setminus B) \cup (B \setminus A)$ is the *symmetric difference* of the sets A and B. If X is a finite set, then $\mathfrak{P}(X)$ is the graph G_3 from the previous section. As in the proof of Theorem 3.4, it is easy to verify that the bijection h establishes an isomorphism of graphs K_2^X and $\mathfrak{P}(X)$.

It seems that the graph $\mathfrak{P}(X)$ is a proper candidate for being called a cube because, for a finite set X of cardinality n, the graph $\mathfrak{P}(X)$ is the n-cube. However, there is a drawback associated with this approach. If X is an infinite set, then $\mathfrak{P}(X)$ is disconnected. Indeed, let A be a subset of an infinite set X, and $B = X \setminus A$ be its complement. At least one of the subsets A and B is infinite, therefore it is impossible to connect vertices A and B in $\mathfrak{P}(X)$ by a path. (Recall that a path is a *finite* sequence of pairwise adjacent vertices.) Another downside of calling $\mathfrak{P}(X)$ a cube is the size of its vertex set; it is uncountable if X is countably infinite. Nevertheless, we use the graph $\mathfrak{P}(X)$ in our definition of the cube on X.

Let us take a closer look at connected components of $\mathfrak{P}(X)$. For a given subset $A \subseteq X$, we denote by $H(A)$ the connected component of $\mathfrak{P}(X)$ containing vertex A. Recall that the vertex set of $H(A)$ consists of all vertices of $\mathfrak{P}(X)$ that are connected to A by a path.

The proof of the following lemma is schematically illustrated by the drawing in Figure 3.8.

Lemma 3.5. *A set S is a vertex in $H(A)$ if and only if $S \triangle A$ is a finite set.*

Proof. (Necessity.) Because S is a vertex in $H(A)$, there is a path in $H(A)$ connecting S to A. Consecutive vertices in this path differ by a singleton and the number of vertices in the path is finite, therefore we must have a finite number of elements in each of the sets $S \setminus A$ and $A \setminus S$. Hence, $S \triangle A$ is a finite set.

(Sufficiency.) Suppose that $S \triangle A$ is a finite set. By removing (if necessary) one by one all elements of $S \setminus A$ and then adding (again if necessary) one by one all elements of $A \setminus S$, we create a path connecting S to A. $\square$

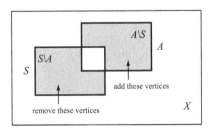

Figure 3.8. Constructing a path connecting S to A in Lemma 3.5.

Corollary 3.6. *The vertex set of the graph $H(\varnothing)$ is the set of all finite subsets of the set X.*

Let $p : V(H(A)) \to V(H(\varnothing))$ be a mapping defined by

$$p(S) = S \triangle A \text{ for } S \in V(H(A)). \tag{3.3}$$

By Lemma 3.5, $S \triangle A$ is a finite set for $S \in V(H(A))$. Therefore p is a well-defined mapping.

Lemma 3.7. *The mapping p is a bijection.*

Proof. It is clear that, for any two sets P and Q,

$$P = Q \text{ if and only if } P \triangle Q = \varnothing. \tag{3.4}$$

Let S and T be two vertices of $H(A)$. Because

$$(S \triangle A) \triangle (T \triangle A) = S \triangle T$$

(cf. Exercise 3.13), $S \triangle A = T \triangle A$ if and only if $S = T$, by (3.4). Hence, p is one-to-one.

If S is a vertex of $H(\varnothing)$, then, by Lemma 3.5, $S \triangle A$ is a vertex of $H(A)$ and

$$p(S \triangle A) = (S \triangle A) \triangle A = S.$$

Therefore, p is onto. $\qquad\square$

Lemma 3.8. *The mapping q defined by*

$$q(S) = S \triangle A, \text{ for } S \in V(H(\varnothing))$$

is the inverse of the bijection p; that is, $q = p^{-1}$.

Proof. We have

$$(p \circ q)(S) = (S \triangle A) \triangle A = S, \text{ for } S \in V(H(\varnothing)),$$

and

$$(q \circ p)(S) = (S \triangle A) \triangle A = S, \text{ for } S \in V(H(A)).$$

Therefore, $q = p^{-1}$. $\qquad\square$

Theorem 3.9. *The graphs $H(A)$ and $H(\varnothing)$ are isomorphic. Accordingly, the connected components of the graph $\mathfrak{P}(X)$ are pairwise isomorphic.*

Proof. We show that the bijection p defined by (3.3) establishes one-to-one correspondence between the edge sets of $H(A)$ and $H(\varnothing)$. Let ST be an edge of $H(A)$, so the set $S \triangle T$ is a singleton. Because

$$p(S) \triangle p(T) = (S \triangle A) \triangle (T \triangle A) = S \triangle T,$$

the vertices $p(S)$ and $p(T)$ are adjacent in $H(\varnothing)$. The converse is obtained by the same argument applied to the inverse p^{-1}. $\qquad\square$

Let X be a set. We denote $\mathcal{P}_f(X)$ the set of all finite subsets of X.

Definition 3.10. The *cube on* X (also known as a *hypercube*) is the graph $\mathcal{H}(X)$ whose vertex set is $\mathcal{P}_f(X)$; two sets P, Q are adjacent in $\mathcal{H}(X)$ if their symmetric difference $P \bigtriangleup Q$ is a singleton. If X is a finite set of cardinality n, then the graph $\mathcal{H}(X)$ is the *n-cube* Q_n. The *dimension* of the cube $\mathcal{H}(X)$ is the cardinality $|X|$ of the set X.

By Theorem 3.9, any connected component of the graph $\mathfrak{P}(X)$ is isomorphic to the cube on X. We also note that two cubes $\mathcal{H}(X)$ and $\mathcal{H}(Y)$, where $|X| = |Y|$, are isomorphic (cf. Exercise 3.14). Therefore, the graph $\mathcal{H}(X)$ is completely defined by the cardinality of the set X.

Inasmuch as graphs K_2^X and $\mathfrak{P}(X)$ are isomorphic, any connected component of the Cartesian power K_2^X is the cube on X. Specifically, we choose the set of all characteristic functions of finite subsets of X as the vertex set of the connected component containing the zero function on X. We call this component the *weak Cartesian power* of K_2. The mapping h in (3.2) establishes an isomorphism between the weak Cartesian power of the graph K_2 and the cube $\mathcal{H}(X)$.

Theorem 3.11. *The cube $\mathcal{H}(X)$ is a bipartite graph.*

Proof. If PQ is an edge of $\mathcal{H}(X)$, then $P \bigtriangleup Q$ is a singleton. Therefore, the numbers $|P|$ and $|Q|$ have opposite parity. The desired bipartition of $\mathcal{P}_f(X)$ is given by the families of sets

$$\{Y \in \mathcal{P}_f(X) : |Y| \text{ is even}\} \text{ and } \{Y \in \mathcal{P}_f(X) : |Y| \text{ is odd}\}.$$

The result follows. $\qquad\qquad\qquad\qquad\qquad\qquad\qquad\qquad\qquad\qquad\qquad$ $\square$

3.6 Cartesian Products

Let Y be a set and I be a set that we consider as a set of indices. By definition, a family $\{y_i\}_{i \in I}$ of elements of Y is a function from I to Y. Thus elements of the power Y^I (cf. Section 3.5) are families $\{y_i\}_{i \in I}$.

Let $\{Y_i\}_{i \in I}$ be a family of sets indexed by the set I and let $Y = \bigcup_{i \in I} Y_i$. The *Cartesian product* $\prod_{i \in I} Y_i$ is the subset of the power Y^I consisting of the families $\{y_i\}_{i \in I}$ such that $y_i \in Y_i$ for every $i \in I$. The sets Y_i are called *factors* of $\prod_{i \in I} Y_i$ and elements y_i are called *coordinates* of $\{y_i\}_{i \in I}$. If $Y_i = Y$ for all $i \in I$, then $\prod_{i \in I} Y_i = Y^I$. Note that $\prod_{i \in \varnothing} Y_i$ is the singleton $\{\varnothing\}$.

The mappings

$$p_k : \prod_{i \in I} Y_i \to Y_k$$

defined by $\{y_i\}_{i \in I} \mapsto y_k$ are called *projections*.

Let $\{G_i = (V_i, E_i)\}_{i \in I}$ be a family of graphs indexed by the set I. The Cartesian product

$$G = \prod_{i \in I} G_i$$

is defined as follows.

(i) $V(G) = \prod_{i \in I} V(G_i)$ is the Cartesian product of the vertex sets of the graphs G_i.

(ii) $E(G)$ is the set of all unordered pairs uv of distinct vertices of G to which there exists $k \in I$ such that $u_k v_k \in E_k$ and $u_i = v_i$ for $i \in I \setminus \{k\}$.

The graphs G_i are called *factors* of G. If all factors of the Cartesian product $\prod_{i \in I} H_i$ are isomorphic to some graph H, then this product is the *Cartesian power* H^I.

As we observed before (Theorem 1.7(i) and Exercise 1.33), a finite Cartesian product is connected if and only if its factors are connected. An infinite product of nontrivial graphs contains vertices that differ in infinitely many coordinates and ends of every edge of this graph differ in exactly one coordinate, therefore two such vertices cannot be connected by a path. It follows that an infinite Cartesian product of nontrivial graphs must be disconnected. Specifically, as we noted in Section 3.5, the Cartesian power K_2^X is disconnected if X is an infinite set.

One way of introducing a connected version of the Cartesian product of the family $\{G_i\}_{i \in I}$ of connected graphs is to define it as a connected component of its Cartesian product as was done in Section 3.5. To specify a component, it suffices to choose a single vertex of it. We denote the component of the Cartesian product $\prod_{i \in I} G_i$ containing a vertex a by

$$(G, a) = \prod_{i \in I} (G_i, a_i).$$

The notations (G, a) and (G_i, a_i) signify graphs with specified vertices a and a_i that are called *roots* of the respective graphs. If a vertex of a graph is specified as a root, the graph itself is called *rooted*. If I is a finite set, all graphs (G, a) are the same and there is no need to specify roots in this case.

We summarize the foregoing discussion as a definition.

Definition 3.12. The *weak Cartesian product* $(G, a) = \prod_{i \in I}(G_i, a_i)$ of a family of rooted graphs $\{(G_i, a_i)\}_{i \in I}$ is a graph with the vertex set

$$V = \left\{ u \in \prod_{i \in I} V(G_i) : u_i \neq a_i \text{ for at most finitely many } i \in I \right\}$$

and the edge set

$$E = \{uv : u, v \in V, u_k v_k \in E(G_k), \text{ for some } k \in I, \text{ and } u_i = v_i \text{ for } i \neq k\}.$$

The reader should verify that (G, a) is indeed a connected component of the Cartesian product $\prod_{i \in I} G_i$ containing a (cf. Exercise 3.16).

In the special case when $G_i = K_2$ for all $i \in I$, all connected components of the Cartesian product $\prod_{i \in I} G_i = K_2^I$ are isomorphic to the cube $\mathcal{H}(I)$. It is indeed a rather special case as the following example demonstrates.

We denote by ω the sequence $(0, 1, 2, \ldots)$ of all natural numbers (the first infinite ordinal).

Example 3.13. Let P be the Cartesian power P_3^ω of the path P_3 with vertex set $\{-1, 0, 1\}$, and edges $\{-1, 0\}$ and $\{0, 1\}$ (see Figure 3.9a). Then weak Cartesian products

$$(P, b) = \prod_{i \in \omega} (P_3, b_i) \text{ and } (P, c) = \prod_{i \in \omega} (P_3, c_i),$$

where $b_i = 0$ for all $i \in \omega$, and $c_i = 0$ if i is even and $c_i = 1$ otherwise, are not isomorphic graphs.

To justify the statement of Example 3.13, we need some properties of automorphism groups of the graphs (P, b) and (P, c).

Some general definitions and notations are in order. For an arbitrary weak Cartesian product $(G, a) = \prod_{i \in I} (G_i, a_i)$, we define sets of vertices

$$L_k(x) = \{u \in (G, a) : u_i = x_i, \ i \neq k\}, \text{ for } k \in I \text{ and } x \in (G, a).$$

The sets $L_k(x)$ are called *layers* of the graph (G, a). It is clear that two layers $L_k(y)$ and $L_k(x)$ are equal if and only if $y \in L_k(x)$ (or equivalently, if $x \in L_k(y)$). Otherwise, $L_k(x) \cap L_k(y) = \varnothing$ and such layers are said to be *parallel*. Evidently, any layer $L_k(x)$ is either equal or parallel to the layer $L_k(a)$. Two layers $L_i(x)$ and $L_k(y)$ with $i \neq k$ are said to be *orthogonal*.

Three layers of the Cartesian power P_3^3 are shown in Figure 3.9b. Two layers $L_1(x)$ and $L_1(y)$ with $x = (0, 0, 0)$ and $y = (1, 1, 1)$ are parallel. The layer $L_3(z)$ with $z = (-1, -1, -1)$ is orthogonal to the layers $L_1(x)$ and $L_1(y)$. There are two "vertical" and six "horizontal" layers in the graph shown in Figure 1.10. Note also that the layers of the cube $Q_n = K_2^n$ are its edges (cf. Exercise 3.18).

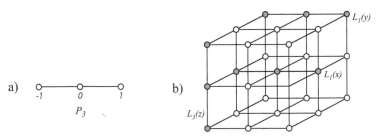

a) P_3 b) $L_3(z)$ $L_1(x)$ $L_1(y)$

Figure 3.9. The path P_3 and three layers in P_3^3.

We denote by the same symbol $L_k(x)$ the graph induced by the layer $L_k(x)$. It is clear that this graph is isomorphic to the graph G_k under projection p_k (cf. Exercise 3.19; cf. Figures 3.9 and 1.10).

Let α be an automorphism of the weak Cartesian product

$$(G, a) = \prod_{i \in \omega} (P, a_i),$$

where P is a path.

Lemma 3.14. *The image of a layer $L_i(x)$ under the automorphism α is the layer $L_j(\alpha(x))$ for some $j \in \omega$. Accordingly, the restriction $\alpha|_{L_i(x)}$ of the automorphism α to the layer $L_i(x)$ is an isomorphism from $L_i(x)$ onto the layer $L_j(\alpha(x)) = \alpha(L_i(x))$.*

Proof. Let u and v be two distinct vertices of $L_i(x)$. Suppose that vertices $\alpha(u)$ and $\alpha(v)$ differ in more than one coordinate. We may assume that

$$\alpha(u) = (u_0, u_1, u_2, \ldots) \text{ and } \alpha(v) = (v_0, v_1, v_2, \ldots)$$

with $u_0 \neq v_0$ and $u_1 \neq v_1$. Let u' and v' be two vertices given by

$$u' = (v_0, u_1, u_2, \ldots) \text{ and } v' = (u_0, v_1, v_2, \ldots).$$

Note that the four vertices $\alpha(u)$, $\alpha(v)$, u', and v' are distinct.
We have

$$d(\alpha(u), \alpha(v)) = \sum_{i=0}^{\infty} |u_i - v_i| = |u_0 - v_0| + \sum_{i=1}^{\infty} |u_i - v_i|$$
$$= d(\alpha(u), u') + d(u', \alpha(v)),$$

where d is the graph distance on (G, a) (cf. Exercise 3.20). It follows that $u' \in I(\alpha(u), \alpha(v))$. Similar calculation shows that

$$v' \in I(\alpha(u), \alpha(v)), \quad \alpha(u) \in I(u', v'), \quad \alpha(v) \in I(u', v').$$

Thus we have (cf. Exercise 3.22b)

$$u'' = \alpha^{-1}(u') \in I(u, v), \quad v'' = \alpha^{-1}(v') \in I(u, v)$$

and

$$u \in I(u'', v''), \quad v \in I(u'', v'')$$

The four vertices u, v, u'', and v'' of $L_i(x)$ are distinct and $L_i(x) \cong P$, thus we have a contradiction.

It follows that all vertices in the image of $L_i(x)$ differ in precisely one coordinate. Hence, $\alpha(L_i(x)) \subseteq L_j(\alpha(x))$ for some $j \in \omega$. Because α is an automorphism, we have $\alpha(L_i(x)) = L_j(\alpha(x))$. By the same argument, the restriction $\alpha|_{L_i(x)}$ of the automorphism α to the layer $L_i(x)$ is an isomorphism from $L_i(x)$ onto the layer $L_j(\alpha(x))$. $\qquad\square$

Let (P, b) be the first weak Cartesian product in Example 3.13. Recall that $b = (0, 0, \ldots, 0, \ldots)$ is the "zero" vertex of P_3^ω.

Theorem 3.15. $\alpha(b) = b$ *for any automorphism α of (P, b). In words: vertex b is a fixed vertex under actions of the automorphism group $\mathrm{Aut}(P, b)$.*

Proof. Let a be a vertex of (P, b) different from b; that is, there is $i \in \omega$ such that $a_i \neq 0$. By Lemma 3.14, the image of the layer $L_i(a)$ under an automorphism α is a layer $L_j(\alpha(a))$ for some $j \in \omega$. Because $\alpha|_{L_i(a)}$ is an isomorphism and all layers of (P, b) are isomorphic to the path P_3 in Figure 3.9, we must have $(\alpha(a))_j \neq 0$. (In fact, $(\alpha(a))_j = \pm a_i$.) Therefore, for any vertex $a \neq b$ and any automorphism α, we have $\alpha(a) \neq b$. It follows that b is a fixed vertex under actions from $\mathrm{Aut}(P, b)$. $\qquad\square$

For the second graph (P, c) in Example 3.13, we have the following result. Recall that $c = (0, 1, 0, 1, \ldots, 0, 1, \ldots)$.

Theorem 3.16. *For any vertex a of the graph (P, c), there is an automorphism $\alpha \in \mathrm{Aut}(P, c)$ such that $\alpha(a) = c$.*

Proof. Suppose that $a_i = -1$ for some $i \in \omega$ and let β be the automorphism of P_3 that maps -1 into 1. By the result of Exercise 3.21b, β induces an automorphism γ_i of (P, c) such that $(\gamma_i(a))_i = 1$ and $(\gamma_i(a))_j = a_j$ for all $j \neq i$. Because a differs from c in at most a finite number of coordinates, there is an automorphism γ such that $(\gamma(a))_i \neq -1$ for all $i \in \omega$. For $d = \gamma(a)$, we define sets

$$A = \{i \in \omega : d_i = 0\} \quad \text{and} \quad B = \{i \in \omega : d_i = 1\}.$$

Both sets A and B are infinite, $A \cap B = \varnothing$, and $A \cup B = \omega$. Therefore there is a bijection $\sigma : \omega \to \omega$ that maps A onto the subset of ω consisting of even integers and maps B onto the subset of ω consisting of odd integers. By Exercise 3.21a, σ defines an automorphism δ such that $\delta(d) = c$. Let $\alpha = \delta \circ \gamma$ be the composition of automorphisms γ and δ. Then

$$\alpha(a) = \delta(\gamma(a)) = \delta(d) = c.$$

The result follows. $\qquad\square$

The automorphism group $\mathrm{Aut}(G)$ of a graph G is said to be *transitive* (or *vertex-transitive*) if there is an automorphism $\alpha \in \mathrm{Aut}(G)$ to any pair of vertices u, v of G such that $\alpha(u) = v$.

Theorem 3.17. *Let G and H be two isomorphic graphs. If $\mathrm{Aut}(G)$ is transitive, then the group $\mathrm{Aut}(H)$ is also transitive.*

Proof. Let u, v be a pair of vertices of H and φ be an isomorphism from G onto H. Because $\mathrm{Aut}(G)$ is transitive, there is an automorphism α of G such that $\alpha(\varphi^{-1}(u)) = \varphi^{-1}(v)$. Therefore, $(\varphi \circ \alpha \circ \varphi^{-1})(u) = v$. Clearly, $\varphi \circ \alpha \circ \varphi^{-1}$ is an automorphism of H. Thus, the group $\mathrm{Aut}(H)$ is transitive. $\qquad\square$

Theorem 3.16 asserts that the group $\mathrm{Aut}(P, c)$ is transitive. On the other hand, by Theorem 3.15, the group $\mathrm{Aut}(P, b)$ is not transitive. It follows from Theorem 3.17 that graphs (P, b) and (P, c) are not isomorphic.

We conclude this section by showing that the weak Cartesian product of a family of cubes is a cube.

Theorem 3.18. *Let $\{X_i\}_{i\in I}$ be a family of pairwise disjoint sets. The weak Cartesian product $G = \prod_{i\in I}(\mathcal{H}(X_i), \varnothing)$ is the cube $\mathcal{H}(\cup_{i\in I}X_i)$.*

Proof. A vertex A of G is a family of finite sets $\{A_i\}_{i\in I}$ such that $A_i \subseteq X_i$ and $A_i \neq \varnothing$ for at most finitely many $i \in I$. Thus, $\cup_{i\in I}A_i$ is a finite subset of $\cup_{i\in I}X_i$ and the mapping $\varphi : V(G) \to \mathcal{P}_f(\cup_{i\in I}X_i)$ given by

$$\varphi : A \mapsto \cup_{i\in I}A_i$$

is well defined. Moreover, this mapping is a bijection because any set in $\mathcal{P}_f(\cup_{i\in I}X_i)$ is uniquely represented as a finite union of finite subsets of sets in the family $\{X_i\}_{i\in I}$. We leave it to the reader to verify that φ is an isomorphism from G onto $\mathcal{H}(\cup_{i\in I}X_i)$ (cf. Exercise 3.17). $\qquad\square$

3.7 Metric Structures on Cubes

The first theorem of this section gives a simple formula for the graph distance on the cube $\mathcal{H}(X)$.

Theorem 3.19. *The graph distance d on the vertex set $\mathcal{P}_f(X)$ of the cube $\mathcal{H}(X)$ is given by*

$$d(P,Q) = |P \triangle Q|, \ \text{ for } P, Q \in \mathcal{P}_f(X). \tag{3.5}$$

Proof. Because any two vertices in a path connecting P to Q differ by one element, the length of this path cannot be less than $|P \triangle Q|$. On the other hand, the construction in the sufficiency part of the proof of Lemma 3.5 yields a path of length $|P \triangle Q|$ (see Figure 3.8). The result follows. $\qquad\square$

Lemma 3.20. *If P and Q are finite subsets of a set X, then*

$$|P \triangle Q| = \sum_{x \in P \cup Q} |\chi_P(x) - \chi_Q(x)|, \tag{3.6}$$

where χ_P and χ_Q are characteristic functions of subsets P and Q, respectively.

Proof. It is clear that $\chi_P(x) \neq \chi_Q(x)$ if and only if either $x \in P \setminus Q$ or $x \in Q \setminus P$. Therefore, $|\chi_P(x) - \chi_Q(x)|$ equals 1 for $x \in P \triangle Q$ and is zero otherwise. $\qquad\square$

It is a common practice to write equation (3.6) as

$$|P \triangle Q| = \sum_{x \in X} |\chi_P(x) - \chi_Q(x)| \tag{3.7}$$

assuming that the summation on the right side is taken over those $x \in X$ for which the summands are not zero.

For $X = \{1, \ldots, n\}$, the characteristic functions are $0/1$-vectors in $\mathbf{R}^n$. These vectors form the vertex set $\{0, 1\}^n$ of the n-cube Q_n. From (3.5) and (3.7) we obtain the following formula for the graph distance on Q_n,

$$d(x, y) = \sum_{i=1}^{n} |x_i - y_i|, \tag{3.8}$$

where $x = (x_1, \ldots, x_n)$ and $y = (y_1, \ldots, y_n)$ are vertices of the n-cube. The sum on the right side of (3.8) is known as the *Hamming distance* on the Boolean cube $\{0, 1\}^n$.

Theorem 3.21. *Vertex C lies between vertices A and B in the graph $\mathcal{H}(X)$ if and only if*

$$A \cap B \subseteq C \subseteq A \cup B. \tag{3.9}$$

First, we "prove" Theorem 3.21 informally using the Venn diagram in Figure 3.10a. The union $A \cup B \cup C$ is partitioned into seven pairwise disjoint sets that are labeled a, b, $\ldots$, g in Figure 3.10a. For instance, $a = A \cap B \cap C$ and $b = (A \cap B) \setminus C$. We have

$$|A \triangle B| = |c| + |f| + |d| + |g|,$$
$$|A \triangle C| = |f| + |b| + |e| + |d|,$$
$$|C \triangle B| = |c| + |e| + |b| + |g|.$$

Therefore,

$$|A \triangle C| + |C \triangle B| = |A \triangle B| + 2(|b| + |e|).$$

By (3.5),

$$d(A, C) + d(C, B) = d(A, B) + 2(|b| + |e|).$$

It follows that C lies between A and B if and only if $b = e = \varnothing$, which is equivalent to (3.9) (see Figure 3.10b).

a)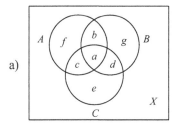   b)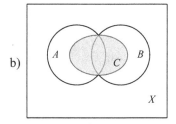

Figure 3.10. Proof of Theorem 3.21.

Now we give a rigorous proof of Theorem 3.21. The advantage of this formal proof is that it can be generalized and used in other applications (see Notes at the end of this chapter).

Proof. We use the following identities that hold for any $P, Q \in \mathcal{P}_f(X)$ (cf. Exercise 3.23),

$$|P \cap Q| + |P \cup Q| = |P| + |Q|, \tag{3.10}$$

$$|P \triangle Q| = |P| + |Q| - 2|P \cap Q|. \tag{3.11}$$

(Necessity.) Suppose that

$$d(A, C) + d(C, B) = d(A, B);$$

that is, by (3.5),

$$|A \triangle C| + |C \triangle B| = |A \triangle B|.$$

By 3.11, we have

$$|A \triangle C| + |C \triangle B| = |A| + |C| - 2|A \cap C| + |C| + |B| - 2|C \cap B|$$

and

$$|A \triangle B| = |A| + |B| - |A \cap B|.$$

The quantities on the left sides are equal, thus we obtain

$$|C| + |A \cap B| = |A \cap C| + |C \cap B|. \tag{3.12}$$

By (3.10) and (3.12), we have

$$|C \cup (A \cap B)| = |C| + |A \cap B| - |A \cap B \cap C|$$
$$= |A \cap C| + |C \cap B| - |A \cap B \cap C|$$

and

$$|C \cap (A \cup B)| = |(C \cap A) \cup (C \cap B)| = |A \cap C| + |C \cap B| - |A \cap B \cap C|.$$

Therefore,

$$|C| \leq |C \cup (A \cap B)| = |C \cap (A \cup B)| \leq |C|.$$

Because C is a subset of $C \cup (A \cap B)$ and $C \cap (A \cup B)$ is a subset of C, we have (cf. Exercise 3.23)

$$C \cup (A \cap B) = C = C \cap (A \cup B),$$

which is equivalent to (3.9).

(Sufficiency.) Suppose that

$$A \cap B \subseteq C \subseteq A \cup B.$$

Then

$$|(A \cup C) \cap (B \cup C)| = |(A \cap B) \cup C| = |C|,$$
$$|(A \cap C) \cup (B \cap C)| = |(A \cup B) \cap C| = |C|.$$

The quantities on the left sides of the above identities are equal, therefore we have

$$|A \bigtriangleup B| = |A \cup B| - |A \cap B| = |(A \cup C) \cup (B \cup C)| - |(A \cap C) \cap (B \cap C)|$$
$$= |(A \cup C) \cup (B \cup C)| + |(A \cup C) \cap (B \cup C)|$$
$$\quad - |(A \cap C) \cup (B \cap C)| - |(A \cap C) \cap (B \cap C)|$$
$$= |A \cup C| + |B \cup C| - |A \cap C| - |B \cap C|$$
$$= |A \bigtriangleup C| + |B \bigtriangleup C|.$$

Therefore, $d(A, C) + d(C, B) = d(A, B)$. □

Let A be an infinite subset of a set X. As we showed in Theorem 3.9, the graphs $H(A)$ and $\mathcal{H}(X)$ are isomorphic. Therefore they are isometric metric spaces. On the other hand, they are quite different as families of sets with respect to set operations. For instance, the vertex set $\mathcal{P}_f(X)$ of the cube $\mathcal{H}(X)$ has the least element $\varnothing$, whereas the vertex set of $H(A)$ does not have one. However, the assertion of Theorem 3.21 holds for the graph $H(A)$.

Theorem 3.22. *Vertex C lies between vertices A and B in the graph $H(A)$ if and only if*

$$A \cap B \subseteq C \subseteq A \cup B.$$

Proof. First we note that our "informal" proof based on the Venn diagram in Figure 3.10 works in this case, because all sets in the proof are finite (see Lemma 3.5).

A rigorous proof is obtained by showing that

$$A \cap B \subseteq C \subseteq A \cup B$$

in $V(H(A))$ is equivalent to

$$p(A) \cap p(B) \subseteq p(C) \subseteq p(A) \cup p(B)$$

in $\mathcal{P}_f(X)$, where p is an isomorphism of $H(A)$ onto $\mathcal{H}(X)$ defined by (3.3). We leave the details of the proof to the reader (cf. Exercise 3.24). □

Let us recall that the interval $I(A, B)$ between two vertices A and B of the cube $\mathcal{H}(X)$ is defined as (see Section 2.2):

$$I(A, B) = \{C \in \mathcal{P}_f : d(A, C) + d(C, B) = d(A, C)\}.$$

By (3.21), it is equivalent to

$$I(A, B) = \{C \in \mathcal{P}_f : A \cap B \subseteq C \subseteq A \cup B\}.$$

Clearly, $I(A, B) = I(A \cap B, A \cup B)$. We denote by the same symbol $I(A, B)$ the subgraph of $\mathcal{H}(X)$ induced by the interval $I(A, B)$.

Theorem 3.23. *The graph $I(A, B)$ is isomorphic to the n-cube Q_n, where $n = d(A, B)$.*

Proof. Because $I(A, B) = I(A \cap B, A \cup B)$, we may assume that $A \subseteq B$. The function $\psi : Y \mapsto Y \setminus A$ is a bijection from the interval $I(A, B)$ onto the interval $I(\varnothing, B \setminus A)$, which is isomorphic to the cube $\mathcal{H}(B \setminus A)$. Because

$$d(Y, Y') = |Y \triangle Y'| = |(Y \setminus A) \triangle (Y' \setminus A)| = d(\psi(Y), \psi(Y')),$$

the function ψ is an isomorphism. Since $n = d(A, B) = |B \setminus A|$, we have $\mathcal{H}(B \setminus A) \cong Q_n$.   □

Intervals are convex subsets of vertices of $\mathcal{H}(X)$. Indeed, let P and Q be two vertices in an interval $I(A, B)$ and R be a vertex lying between P and Q. Inasmuch as $A \cap B \subseteq P$, $A \cap B \subseteq Q$ and $P \subseteq A \cup B$, $Q \subseteq A \cup B$, we have

$$A \cap B \subseteq P \cap Q \subseteq R \subseteq P \cup Q \subseteq A \cup B.$$

Therefore, $R \in I(A, B)$.

It turns out that all convex subsets of a cube are interval-like subsets. In order to make sense of this statement, we first establish two properties of convex subsets of a cube.

Lemma 3.24. *Let $\mathcal{C}$ be a convex subset of $\mathcal{H}(X)$ and $\{A_i\}_{1 \leq i \leq n}$ be a finite family of vertices in $\mathcal{C}$. Then $\cup_{i=1}^{n} A_i$ is a vertex in $\mathcal{C}$.*

Proof. The proof is by induction on n. The case $n = 1$ is trivial. Suppose that the statement of the lemma holds for any set of vertices of $\mathcal{C}$ with $n - 1$ elements. Then $B = \cup_{i=1}^{n-1} A_i \in \mathcal{C}$, by the induction hypothesis. We have

$$B \cap A_n \subseteq \cup_{i=1}^{n} A_i = B \cup A_n.$$

Thus the union $\cup_{i=1}^{n} A_i$ lies between $B \in \mathcal{C}$ and $A_n \in \mathcal{C}$. The result follows, because $\mathcal{C}$ is convex.   □

Lemma 3.25. *Let $\mathcal{C}$ be a convex subset of $\mathcal{H}(X)$ and $\mathcal{A}$ be a nonempty family of vertices in $\mathcal{C}$. Then the intersection $\cap \mathcal{A}$ is a vertex in $\mathcal{C}$.*

Proof. Clearly, $\cap \mathcal{A}$ is a finite set. Let A_1 be an element of $\mathcal{A}$ and

$$n_1 = |A_1 \setminus \cap \mathcal{A}|.$$

If $n_1 = 0$, then $\cap \mathcal{A} = A_1$ and we are done. Otherwise, there is $A_2 \in \mathcal{A}$ such that

$$n_2 = |(A_1 \cap A_2) \setminus \cap \mathcal{A}| < n_1.$$

The intersection $A_1 \cap A_2$ lies between A_1 and A_2. Therefore it is a vertex in $\mathcal{C}$. If $n_2 = 0$, then $\cap \mathcal{A} = A_1 \cap A_2$ and we are done. By continuing this process, we obtain a decreasing sequence $n_1 > n_2 > \cdots$ of nonnegative integers. Therefore, $n_k = 0$ for some k. Then $\cap \mathcal{A} = A_1 \cap \cdots \cap A_k$ which is a vertex in $\mathcal{C}$. □

Let $\mathcal{C}$ be a convex subset of a cube $\mathcal{H}(X)$. By Lemma 3.25, $A = \cap \mathcal{C}$ is a vertex in $\mathcal{C}$. Let $Y = \cup \mathcal{C}$ be the union of all sets in $\mathcal{C}$.

Theorem 3.26. $\mathcal{C} = \{P \in \mathcal{P}_f(X) : A \subseteq P \subseteq Y\}.$

Proof. Let

$$\mathcal{I}(A, Y) = \{P \in \mathcal{P}_f(X) : A \subseteq P \subseteq Y\}.$$

Evidently, $\mathcal{C} \subseteq \mathcal{I}(A, Y)$. Let P be a vertex in $\mathcal{I}(A, Y)$. Because $P \subseteq Y = \cup \mathcal{C}$, for every element $x \in P$ there is a set $C_x \in \mathcal{C}$ such that $x \in C_x$. Then

$$A \subseteq P \subseteq \bigcup_{x \in P} C_x.$$

By Lemma 3.24, $\bigcup_{x \in P} C_x$ is a vertex in $\mathcal{C}$, because P is a finite set. Therefore, $P \in \mathcal{C}$ which implies $\mathcal{I}(A, Y) \subseteq \mathcal{C}$. It follows that $\mathcal{C} = \mathcal{I}(A, Y)$. □

Note that $\mathcal{I}(A, Y)$ is the interval $I(A, Y)$, if Y is a finite set. If $Y = \cup \mathcal{C}$ is an infinite set, the set $\mathcal{I}(A, Y)$ can be regarded as "unbounded" interval. Then one can say that any convex set in a cube is an interval, "bounded", if Y is a finite set, and "unbounded" otherwise. In conclusion of this section, we describe the semicubes of a cube and the theta relations on its vertex set.

Theorem 3.27. *Let A and B be two adjacent vertices of a cube $\mathcal{H}(X)$ with $A \triangle B = \{x\}$, $x \in X$. By symmetry, we may assume that $A = B \cup \{x\}$. Then*

$$W_{AB} = \{R \in \mathcal{P}_f(X) : x \in R\} \quad \text{and} \quad W_{BA} = \{R \in \mathcal{P}_f(X) : x \notin R\}.$$

Proof. Semicubes form a partition of $\mathcal{P}_f(X)$, thus it suffices to prove that

$$W_{AB} = \{R \in \mathcal{P}_f(X) : x \in R\}.$$

If $C \in W_{AB}$, then, by Lemma 2.19, A lies between C and B; that is,

$$C \cap B \subseteq A \subseteq C \cup B. \tag{3.13}$$

Because $x \in A$, $x \notin B$, it follows from the second inclusion in (3.13) that $x \in C$. Hence, $W_{AB} \subseteq \{R \in \mathcal{P}_f(X) : x \in R\}$.

Conversely, let $C \in \{R \in \mathcal{P}_f(X) : x \in R\}$. Inasmuch as $x \in C$, we have

$$C \cap B = C \cap (A \setminus \{x\}) \subseteq A \subseteq C \cup (A \setminus \{x\}) = C \cap B;$$

that is, A lies between R and C. Therefore, by Lemma 2.19,

$$\{R \in \mathcal{P}_f(X) : x \in R\} \subseteq W_{AB}.$$

We proved that $W_{AB} = \{R \in \mathcal{P}_f(X) : x \in R\}$. □

It follows from Theorem 3.27 that pairs of mutually opposite semicubes are completely defined by the elements of the set X. This result is illustrated by the drawing in Figure 3.11 (cf. Figure 2.8).

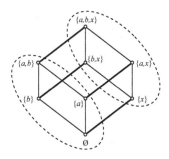

Figure 3.11. A pair of mutually opposite semicubes in Q_3.

Theorem 3.28. *Let E be the edge set of a cube $\mathcal{H}(X)$ and AB and PQ be two edges in E. Then $(AB, PQ) \in \theta$ if and only if $A\triangle B = P\triangle Q$. Accordingly, the relation θ is an equivalence relation on E and the set of equivalence classes E/θ (the quotient set) of θ is in one-to-one correspondence with the set X.*

Proof. (Necessity.) Suppose that $(PQ, AB) \in \theta$ and let $A\triangle B = \{x\}$. Because ends of PQ belong to opposite semicubes W_{AB} and W_{BA}, and $P\triangle Q$ is a singleton, we must have

$$P\triangle Q = \{x\} = A\triangle B,$$

by Theorem 3.27.

(Sufficiency.) Suppose that $A\triangle B = P\triangle Q = \{x\}$ for some $x \in X$. By Theorem 3.27, vertices P, Q must belong to opposite semicubes W_{AB} and W_{BA}. Hence, $(PQ, AB) \in \theta$.

Clearly, θ is an equivalence relation on E. It remains to note that the map $PQ \mapsto P\triangle Q$ defines a one-to-one correspondence between the quotient set E/θ and X. $\qquad\square$

3.8 The Automorphism Group of a Cube

In this section we describe the automorphism group of the cube on a set X. First, we introduce two subgroups of the group $\mathrm{Aut}(\mathcal{H}(X))$.

For $A \in \mathcal{P}_f(X)$ we define a mapping $\alpha_A : \mathcal{P}_f(X) \to \mathcal{P}_f(X)$ by

$$\alpha_A(S) = S\triangle A, \text{ for } S \in \mathcal{P}_f(X). \tag{3.14}$$

This mapping is an automorphism of $\mathcal{H}(X)$ (cf. Lemmas 3.7 and 3.8 and Theorem 3.9). The automorphisms α_A form a subgroup K of the group $\mathrm{Aut}(\mathcal{H}(X))$ under the group operation $\alpha_A \circ \alpha_B = \alpha_{A\triangle B}$, inverse $\alpha_A^{-1} = \alpha_A$, and the identity $e = \alpha_\varnothing$.

The set $\mathcal{P}_f(X)$ itself is a commutative group under the operation of symmetric difference of sets. The empty set is the identity element of this group.

It is clear that the subgroup K is isomorphic to the group $\mathcal{P}_f(X)$. Inasmuch as $A \triangle A = \varnothing$, all elements of $\mathcal{P}_f(X)$ have order 2. Groups with this property are known as *Boolean groups* or *elementary Abelian 2-groups* in group theory.

Let $A = \{x_1, \ldots, x_n\}$. Then

$$A = \{x_1\} \triangle \cdots \triangle \{x_n\}.$$

Therefore, $\alpha_A = \alpha_{x_1} \circ \cdots \circ \alpha_{x_n}$, where α_{x_i} stands for $\alpha_{\{x_i\}}$. It follows that the family $\{\alpha_x\}_{x \in X}$ is a set of generators of the subgroup K. These generators satisfy relations:

(i) $\alpha_x \circ \alpha_y = \alpha_y \circ \alpha_x$,
(ii) $\alpha_x^2 = e$,

for all $x, y \in X$. It turns out that these two relations completely characterize the group $\mathcal{P}_f(X) \cong K$.

Theorem 3.29. *Let G be a group generated by a family of elements $\{g_i\}_{i \in J}$ satisfying the relations (and only these relations):*

(i) $g_i g_y = g_j g_x$,
(ii) $g_i^2 = e$,

for all $i, j \in J$. Then G is isomorphic to $\mathcal{P}_f(J)$.

The proof is straightforward and left to the reader (cf. Exercise 3.28).

Another subgroup of $\mathrm{Aut}(\mathcal{H}(X))$ is induced by permutations of X. Let σ be a permutation of X, that is, a bijection from X onto itself. The permutation σ defines a mapping $\hat{\sigma} : \mathcal{P}_f(X) \to \mathcal{P}_f(X)$ by

$$\hat{\sigma}(S) = \sigma(S) = \{\sigma(x) : x \in S\}, \quad \text{for } S \in \mathcal{P}_f(X).$$

It is clear that $\hat{\sigma}$ is an automorphism of $\mathcal{H}(X)$ and that the set of all these automorphisms form a subgroup of $\mathrm{Aut}(\mathcal{H}(X))$ isomorphic to the *symmetric group* $S(X)$ of all permutations of the set X. We denote this subgroup by H.

Let α be an arbitrary automorphism of the cube $\mathcal{H}(X)$ and $A = \alpha(\varnothing)$. Consider the automorphism $\beta = \alpha_A^{-1} \circ \alpha$. The vertex $\varnothing$ is a fixed vertex of this automorphism:

$$\beta(\varnothing) = (\alpha_A^{-1} \circ \alpha)(\varnothing) = \alpha_A^{-1}(A) = A \triangle A = \varnothing.$$

The vertices $\varnothing$ and $\{x\}$ of $\mathcal{H}(X)$ are adjacent for every $x \in X$, thus the vertex $\beta(\{x\})$ is adjacent to the vertex $\beta(\varnothing) = \varnothing$. Therefore, $\beta(\{x\})$ is a singleton $\{y\}$. Because β is an automorphism, the correspondence $x \mapsto y$ is a permutation σ_β of X (cf. Exercise 3.29). Let us show that $\hat{\sigma}_\beta = \beta$.

An automorphism preserves the betweenness relation, thus we have

$$\varnothing \subseteq \{x\} \subseteq S \text{ if and only if } \varnothing \subseteq \beta(\{x\}) \subseteq \beta(S),$$

for any $S \in \mathcal{P}_f(X)$. Therefore, $x \in S$ if and only if $\sigma_\beta(x) \in \beta(S)$. It follows that $\widehat{\sigma}_\beta(S) = \beta(S)$ for all $S \in \mathcal{P}_f(X)$.

We proved that $\alpha = \alpha_A \circ \beta = \alpha_A \circ \widehat{\sigma}_\beta$; that is, any automorphism α of $\mathcal{H}(X)$ is a composition of an automorphism from the subgroup K with an automorphism from the subgroup H. Thus, $\mathrm{Aut}(\mathcal{H}(X)) = KH$.

Lemma 3.30. K *is a normal subgroup of* $\mathrm{Aut}\,\mathcal{H}(X)$.

Proof. We have

$$(\widehat{\sigma} \circ \alpha_A)(S) = \widehat{\sigma}(S \triangle A) = \widehat{\sigma}(S) \triangle \widehat{\sigma}(A) = (\alpha_{\widehat{\sigma}(A)} \circ \widehat{\sigma})(S),$$

for all $S \in \mathcal{P}_f(X)$. Therefore,

$$\widehat{\sigma} \circ \alpha_A \circ \widehat{\sigma}^{-1} = \alpha_{\widehat{\sigma}(A)},$$

for any $\alpha_A \in K$ and $\widehat{\sigma} \in H$.

Let $\gamma = \alpha_B \circ \widehat{\sigma}$ be an arbitrary element of $\mathrm{Aut}\,H(X)$. We need to show that $\gamma \circ \alpha_A \circ \gamma^{-1}$ belongs to the subgroup K. Indeed,

$$(\alpha_B \circ \widehat{\sigma}) \circ \alpha_A \circ (\alpha_B \circ \widehat{\sigma})^{-1} = \alpha_B \circ \widehat{\sigma} \circ \alpha_A \circ \widehat{\sigma}^{-1} \circ \alpha_B = \alpha_{\widehat{\sigma}(A)} \in K,$$

inasmuch as $\widehat{\sigma} \circ \alpha_A \circ \widehat{\sigma}^{-1} = \alpha_{\widehat{\sigma}(A)}$ and the group K is commutative. □

Because K is a normal subgroup of $\mathrm{Aut}(\mathcal{H}(X))$, $K \cap H = \{e\}$ (cf. Exercise 3.31), and $\mathrm{Aut}(\mathcal{H}(X)) = KH$, the group $\mathrm{Aut}(\mathcal{H}(X))$ is the *semidirect product* of the subgroups K and H:

Theorem 3.31. *The automorphism group of the cube* $\mathcal{H}(X)$ *is the semidirect product of the subgroup* K *and the subgroup* H.

For any two vertices P, Q of the cube $\mathcal{H}(X)$, we have

$$(\alpha_P \circ \alpha_Q)(P) = P \triangle Q \triangle P = Q.$$

Therefore the group $\mathrm{Aut}(\mathcal{H}(X))$ is vertex-transitive. In fact, the automorphism group of a cube has a stronger transitivity property.

Theorem 3.32. *Let* P *and* Q *be two vertices of a cube* $\mathcal{H}(X)$, *and let* $\{P_1, \ldots, P_n\}$ *and* $\{Q_1, \ldots, Q_n\}$ *be sets of neighbors of* P *and* Q, *respectively. There is an automorphism* γ *of* $\mathcal{H}(X)$ *such that* $\gamma(P) = Q$ *and* $\gamma(P_i) = Q_i$ *for all* $1 \le i \le n$.

Proof. P_i is adjacent to P, thus we have $P_i \triangle P = \{x_i\}$ for some $x_i \in X$. Therefore, $\alpha_P(P_i) = \{x_i\}$. Likewise, $\alpha_Q(Q_i) = \{y_i\}$ for some element y_i of X. Clearly, there is a permutation σ of X that sends each x_i into y_i for $1 \le i \le n$. Then, for $\gamma = \alpha_Q \circ \widehat{\sigma} \circ \alpha_P$, we have $\gamma(P) = Q$ and

$$\gamma(P_i) = (\alpha_Q \circ \widehat{\sigma} \circ \alpha_P)(P_i) = Q_i, \quad \text{for all } 1 \le i \le n.$$

Hence, γ is the required automorphism. □

3.9 Embeddings

Let $\varphi : \mathcal{H}(X) \to \mathcal{H}(Y)$ be an embedding of one cube into another. In this section, we show that φ must be isometric. We start by establishing a special case of the claim.

Lemma 3.33. *If $\varphi(\varnothing) = \varnothing$, then*

$$|\varphi(A)| = |A|, \quad \text{for any } A \in \mathcal{P}_f(X);$$

that is, $d(\varnothing, \varphi(A)) = d(\varnothing, A)$.

Proof. The proof is by induction on $n = |A|$. If $n = 1$, then A is a singleton in X. Because φ is an embedding and $\varphi(\varnothing) = \varnothing$, it maps the edge with ends $\varnothing$ and A into the edge with ends $\varnothing$ and $\varphi(A)$. It follows that $\varphi(A)$ is a singleton in Y, so $|\varphi(A)| = 1$.

Suppose that the claim holds for all $k < n$, and let A be a subset of X of cardinality $n \geq 2$. Let A_1 and A_2 be two distinct subsets of A of cardinality $n - 1$ each. Then $A_{12} = A_1 \cap A_2$ is an $(n - 2)$-element subset of A which is adjacent to both A_1 and A_2. By the induction hypothesis,

$$|\varphi(A_1)| = |\varphi(A_2)| = n - 1 \text{ and } |\varphi(A_{12})| = n - 2.$$

$\varphi(A_{12})$ is adjacent to both $\varphi(A_1)$ and $\varphi(A_2)$, thus we have, for $i \in \{1, 2\}$,

$$\varphi(A_{12}) \triangle \varphi(A_i) = \{x_i\},$$

where $x_1 \neq x_2$ and $x_i \notin \varphi(A_{12})$. By the same argument,

$$\varphi(A) \triangle \varphi(A_i) = \{y_i\}, \quad \text{for } i \in \{1, 2\},$$

where $y_1 \neq y_2$ and $y_i \notin \varphi(A)$. Therefore,

$$\{x_1, x_2\} = \varphi(A_1) \triangle \varphi(A_2) = \varphi(A) \triangle \varphi(A_1) \triangle \varphi(A) \triangle \varphi(A_2) = \{y_1, y_2\}.$$

Suppose that $y_1 = x_1$. Then $\varphi(A) \triangle \varphi(A_1) = \{x_1\}$, which implies

$$\varphi(A) = \{x_1\} \triangle \varphi(A_1) = \varphi(A_{12}).$$

This contradicts our assumption that φ is an embedding. Therefore, $y_1 = x_2$, and we have $\varphi(A) \triangle \varphi(A_1) = \{x_2\}$, or, equivalently, $\varphi(A) = \{x_2\} \triangle \varphi(A_1)$. Inasmuch as $x_2 \notin \varphi(A_1)$, it implies

$$\varphi(A) = \{x_2\} \cup \varphi(A_1) = \{x_2\} \cup \varphi(A_{12}) \cup \{x_1\}.$$

Because $x_i \notin \varphi(A_{12})$ and $|\varphi(A_{12})| = n - 2$, we have $|\varphi(A)| = n = |A|$. $\square$

Now let ψ be an arbitrary embedding of $\mathcal{H}(X)$ into $\mathcal{H}(Y)$. For vertices $P, Q \in \mathcal{H}(X)$, we denote by $R = \psi(P)$, $S = \psi(Q)$ their images in $\mathcal{H}(Y)$ under mapping ψ. Because α_P and α_R are automorphisms of cubes $\mathcal{H}(X)$ and $\mathcal{H}(Y)$, respectively, $\varphi = \alpha_R \circ \psi \circ \alpha_P$ is an embedding of $\mathcal{H}(X)$ into $\mathcal{H}(Y)$. We have

$$\varphi(\varnothing) = (\alpha_R \circ \psi \circ \alpha_P)(\varnothing) = (\alpha_R \circ \psi)(P) = \alpha_R(R) = \varnothing$$

and

$$\varphi(P \triangle Q) = (\alpha_R \circ \psi \circ \alpha_P)(P \triangle Q) = (\alpha_R \circ \psi)(Q) = R \triangle S.$$

Therefore, $|R \triangle S| = |P \triangle Q|$, by Lemma 3.33. It follows that φ is an isometric embedding. This completes our proof of the claim:

Theorem 3.34. *Any embedding of a cube into a cube is an isometric embedding.*

If a cube $\mathcal{H}(X)$ is embeddable into a cube $\mathcal{H}(Y)$, then $|\mathcal{P}_f(X)| \le |\mathcal{P}_f(Y)|$. It follows (cf. Exercise 3.32) that

$$\dim(\mathcal{H}(X)) = |X| \le |Y| = \dim(\mathcal{H}(Y)).$$

On the other hand, if $|X| \le |Y|$, then there is a one-to-one function $X \to Y$ that induces an embedding $\mathcal{H}(X) \to \mathcal{H}(Y)$. Therefore any family of pairwise nonisomorphic cubes $\{\mathcal{H}(X)\}$ is well-ordered by the cardinal numbers $|X|$ of the sets X.

3.10 Characterizations of Cubes

There are several properties of cubes that can be used to characterize these graphs. Two of these properties are established in the following theorem.

Theorem 3.35. *Let A and B be two vertices of a cube and let $n = d(A, B)$. Then:*

(i) *There are precisely $n!$ shortest AB-paths in the cube.*
(ii) *There are precisely n neighbors of A that lie on a shortest path connecting A to B:*
$$|I(A, B) \cap N(A)| = n.$$

Proof. By Theorems 3.23 and 3.32, we may assume that $A = \varnothing$, $B = X$ in the cube $\mathcal{H}(X)$, where $X = \{1, \dots, n\}$.

(i) For a permutation σ of the set X, we define a path $P(\sigma)$ in $\mathcal{H}(X)$ connecting $\varnothing$ and X as

$$P(\sigma) = \{\varnothing, \{\sigma(1)\}, \{\sigma(1), \sigma(2)\}, \dots, \{\sigma(1), \dots, \sigma(n-1)\}, X\}.$$

The length of $P(\sigma)$ is $n = d(\varnothing, X)$, therefore this path is a shortest path from $\varnothing$ to X. Clearly, the paths $P(\sigma)$ defined by different permutations are different. On the other hand, any shortest path P from $\varnothing$ to X in $\mathcal{H}(X)$

is a nested sequence of $n+1$ subsets of X. It is clear that $P = P(\sigma)$ for some permutation σ of the set X. Thus, there is a one-to-one correspondence between the set of all shortest paths connecting the empty set with X and the set of all permutations of the n-element set X. This proves claim (i) of the theorem.

(ii) This is an obvious property of the finite cube $\mathcal{H}(X)$.   □

In fact, the two properties in Theorem 3.35 are equivalent as the next theorem asserts.

Theorem 3.36. *Let G be a connected graph. The following conditions are equivalent.*

(i) *For each pair of vertices x and y, the number of shortest paths between x and y is $d(x,y)!$.*
(ii) *For each pair of vertices x and y, the number of those neighbors of x that lie on at least one shortest xy-path is $d(x,y)$:*

$$|I(x,y) \cap N(x)| = d(x,y).$$

Proof. (i) $\Rightarrow$ (ii). By (i), $I(x,y)$ is a finite set. Let $n = |I(x,y) \cap N(x)|$. For any vertex $u \in I(x,y) \cap N(x)$, there are

$$d(u,y)! = (d(x,y) - 1)!$$

shortest uy-paths. Clearly, any shortest xy-path contains exactly one vertex of $I(x,y) \cap N(x)$. It follows that $n(d(x,y) - 1)! = d(x,y)!$. Hence, $n = d(x,y)$.

(ii) $\Rightarrow$ (i). Suppose that $|I(x,y) \cap N(x)| = d(x,y)$ for $x,y \in V(G)$. We prove by induction on $n = d(x,y)$ that the number of shortest paths between x and y is $n!$. The claim is trivial if $n = 1$. Suppose it holds for all pairs u,v with $d(u,v) = n - 1$ and let x,y be two vertices such that $d(x,y) = n$. By the induction hypothesis, there are $(n-1)!$ distinct shortest uy-paths for any vertex u of $I(x,y) \cap N(x)$. Because any shortest xy-path contains exactly one vertex of $I(x,y) \cap N(x)$, there are exactly $(n-1)!n = n!$ shortest xy-paths in the graph G.   □

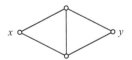

Figure 3.12. $|I(x,y) \cap N(x)| = d(x,y)$; the graph is not a cube.

A connected graph satisfying either of two conditions of Theorem 3.36 need not be a cube. A simple counterexample is the graph in Figure 3.12.

However, connected bipartite graphs satisfying condition (ii) (equivalently, condition (i)) of Theorem 3.36 are cubes. To prove this assertion, we first introduce some useful notions.

Let $G = (V, E)$ be a connected graph, u_0 be a fixed vertex of G, and k be a nonnegative integer. We define

$$N_k = \{v \in V : d(v, u_0) = k\},$$
$$B_k = \{v \in V : d(v, u_0) \leq k\}.$$

It is clear that $N_0 = B_0 = \{u_0\}$, $N_1 = N(u_0)$, and $B_k = \cup_{i=0}^{k} N_i$. We denote by G_k the subgraph of G induced by B_k.

Let us assume now that $G = (V, E)$ is a bipartite graph satisfying condition (ii) of Theorem 3.36. Our goal is to show that G is isomorphic to the cube $\mathcal{H}(N_1)$. For this, we show that a mapping $\varphi : V \to \mathcal{P}_f(N_1)$ defined by

$$\varphi(x) = N_1 \cap I(u_0, x)$$

is the required isomorphism. Let $\mathcal{P}_k$ be the family of subsets of N_1 defined as

$$\mathcal{P}_k = \{A \subseteq N_1 : |A| \leq k\}$$

and H_k be the subgraph of $\mathcal{H}(N_1)$ induced by $\mathcal{P}_k$. To prove that φ is an isomorphism, it suffices to show that the restriction φ_k of φ to the set B_k is an isomorphism of G_k onto H_k for all $k \geq 0$. The claim clearly holds for $k = 0, 1$, because $\varphi(u_0) = \varnothing$ and $\varphi(u) = \{u\}$ for $u \in N(u_0)$ (see Figure 3.13).

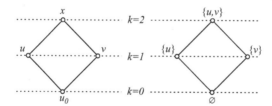

Figure 3.13. Isomorphism $\varphi_2 : G_2 \to H_2$.

Let us consider the case when $k = 2$. Condition (ii) of Theorem 3.36 implies that for a given $x \in N_2$ there is a unique pair of vertices $u, v \in N_1$ adjacent to x. The distance between any two distinct vertices of N_1 is 2 (the graph G is bipartite), therefore the same condition implies that for any two vertices $u, v \in N_1$ there is a unique vertex $x \in N_2$ adjacent to both u and v. It follows that there is one-to-one correspondence between vertices in N_2 and pairs of distinct vertices in N_1. Hence, φ_2 is an isomorphism between G_2 and H_2 (see Figure 3.13).

Let $k \geq 3$. Suppose that φ_{k-1} is an isomorphism and proceed by induction. For a k-element subset A of N_1, let $A_1, \ldots, A_k$ be the $(k-1)$-element subsets of

A and let $a_i = \varphi_{k-1}^{-1}(A_i)$ for $1 \leq i \leq k$. It is clear that $d(A_i, A_j) = 2$ and that $A_i \cap A_j$ is the unique vertex in $\mathcal{P}_{k-1}$ adjacent to both A_i and A_j. Inasmuch as φ_{k-1} is an isomorphism, we have $d(a_i, a_j) = 2$. Moreover, $\varphi_{k-1}^{-1}(A_i \cap A_j)$ is the unique vertex in N_{k-2} adjacent to both a_i and a_j. By condition (ii), there is a unique vertex $s(a_i, a_j) \in N_k$ adjacent to both a_i and a_j. We claim that all the vertices $s(a_i, a_j)$ are the same. Evidently, it suffices to show that $x = s(a_1, a_2)$ is adjacent to a_3.

Let $B_1 = A_1 \cap A_2$, $B_2 = A_1 \cap A_3$, $B_3 = A_2 \cap A_3$, and $C = A_1 \cap A_2 \cap A_3$, and let b_1, b_2, b_3, and c be the respective vertices of G (see Figure 3.14).

Suppose that x is not adjacent to a_3. Because $d(x, b_2) = 2$, there is a unique vertex $y_1 \neq a_1$ in N_{k-1} adjacent to both x and b_2. Likewise, there is a unique vertex $y_2 \neq a_1$ in N_{k-1} adjacent to both x and b_3. Because x is not adjacent to a_3 and, by the induction hypothesis, G_{k-1} is isomorphic to H_{k-1}, the vertices y_1, a_1, a_2, a_3, y_2 are pairwise distinct (see Figure 3.14). The four neighbors y_1, a_1, a_2, y_2 of x each lie on the shortest path connecting x to c. Because $d(x, c) = 3$, this contradicts condition (ii) of Theorem 3.36. We proved that x is a unique vertex in N_k adjacent to each of $a_1, \ldots, a_k$.

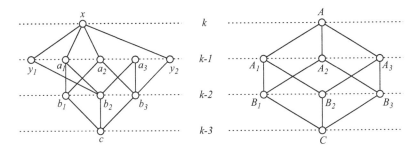

Figure 3.14. Isomorphism $\varphi_k : G_k \rightarrow H_k$.

We now prove that φ_k is an isomorphism from G_k to H_k. Inasmuch as $A_i = \varphi_{k-1}(a_i) = I(a_i, u_0) \cap N_1$, we have

$$\varphi_k(x) = I(x, u_0) \cap N_1 = \left(\{x\} \cup \bigcup_{i=1}^{k} I(a_i, u_0) \right) \cap N_1 = \bigcup_{i=1}^{k} A_i = A.$$

Thus, for any k-element subset A of N_1, there is a vertex $x \in N_k$ such that $\varphi_k(x) = A$. On the other hand, let us suppose that $\varphi(y) = A$ for some vertex $y \in N_k$. By condition (ii), there are precisely k distinct vertices $y_1, \ldots, y_k$ in N_{k-1} adjacent to y. We have

$$\varphi_{k-1}(y_i) = I(y_i, u_0) \cap N_1 \subset I(y, u_0) \cap N_1 = A.$$

$|\varphi_{k-1}(y_i)| = k - 1$, therefore it follows that $\varphi_{k-1}(y_i) = A_j$ for some $1 \leq j \leq k$; that is, $y_i = a_j$ for some j. Because φ_{k-1} is an isomorphism, we have

$$\{y_1, \ldots, y_k\} = \{a_1, \ldots, a_k\}.$$

Because x is a unique vertex in N_k adjacent to each of $a_1, \ldots, a_k$, we conclude that $y = x$. Therefore, $\varphi_k|_{N_k}$ is a one-to-one correspondence between the sets N_k and $\mathcal{P}_k \setminus \mathcal{P}_{k-1}$.

Let $x \in N_k$ and $y \in N_{k-1}$ be two adjacent vertices of G_k. We have

$$\varphi_k(y) = \varphi_{k-1}(y) = I(y, u_0) \cap N_1 \subset I(x, u_0) \cap N_1 = \varphi_k(x).$$

Therefore the vertices $\varphi_k(x)$ and $\varphi_k(y)$ are adjacent in G_k. As we have already shown, for any two adjacent sets $A, B \in H_k$ with $|A| = k$ (so $B = A_i$ for some index i), their inverse images under φ are adjacent in G_k. This completes the proof of our assertion that φ is an isomorphism from G onto $\mathcal{H}(N_1)$.

We established the following result characterizing cubes in the class of bipartite graphs.

Theorem 3.37. *Let G be a connected bipartite graph. The following conditions are equivalent.*

(i) *G is a cube.*
(ii) *For each pair of vertices x and y of G, the number of shortest paths between x and y is $d(x, y)!$.*
(iii) *For each pair of vertices x and y of G, the number of those neighbors of x that lie on at least one shortest xy-path is $d(x, y)$.*

Notes

The five Platonic solids have been known since antiquity, perhaps more than a thousand years before Plato wrote about them in the dialogue *Timaeus*, circa 360 BC. The first mathematical description of these solids was given by Euclid in Book XIII of his *Elements*. The name of each figure is derived from the number of its faces: respectively 4, 6, 8, 12, and 20. Although regular solids of various kinds were thoroughly studied and classified in the twentieth century, the ancient symmetry patterns of the five Platonic solids still play a fundamental role in modern mathematics, physics, and chemistry. A good account of the theory of regular polytopes is found in Coxeter (1973). For a general theory of convex polytopes, the reader is referred to Grünbaum (2003).

The five Platonic solids and their planar graphs are shown in Figures 3.15, 3.16, and 3.17, respectively.

Boolean cubes are named after George Boole (1815–1864), an English mathematician and philosopher, who formalized propositional logic. As the inventor of Boolean algebra, which is the basis of all modern computer arithmetic, Boole is regarded as one of the founders of the field of computer science. There are numerous applications of Boolean cubes and Boolean functions (mappings $f : \mathcal{B}_n \to \{0, 1\}$) in logic, artificial intelligence, electrical engineering, game theory, reliability theory, and combinatorics. Studies in the axiomatics of Boolean algebras at the end of the nineteenth century have led Charles Peirce and Ernst Schröder to introduce the lattice concept. In the mid-thirties of the last century, Garrett Birkhoff (1911–1966) started the general development of lattice theory. His monograph *Lattice Theory* (Birkhoff, 1967), the first edition of which was published in 1940, is the classic in the field.

Figure 3.15. 3D models of the cube, tetrahedron, and octahedron.

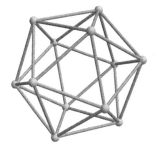 

Figure 3.16. 3D models of the icosahedron and dodecahedron.

Figure 3.17. Graphs of Platonic solids.

Theorem 3.4 illustrates the notion of *cryptomorphism* in mathematics: the capacity of concepts to be faithfully translated from one mathematical language to another. The five graphs in Section 3.4 are defined geometrically (G_1), combinatorially (G_2 and G_3), and algebraically (G_4 and G_5). However, the resulting object—the n-cube—is the same graph. The term "cryptomor-phism" was coined by Garrett Birkhoff (1967) and is used to describe the relation between objects that are equivalent (perhaps in some informal sense) but not clearly equivalent (*crypto* is "hidden" in Greek).

The concept of weak Cartesian product was introduced by Gert Sabidussi (1960) in connection with his studies of prime graph factorization. Our defi-nition of the cube $\mathcal{H}(X)$ is a special case of the weak Cartesian power K_2^X. In this, we follow the original paper by Dragomir Djoković (1973), where the cubes are defined as in our Definition 3.10. In his paper, Djoković attributes this definition to Richard Rado (1906–1989). We recommend the book by Im-rich and Klavžar (2000) as an ultimate source of information on graph prod-ucts.

The Hamming distance is named after Richard Hamming (1915–1998), who introduced it in his fundamental paper (Hamming, 1950) about error-detecting and error-correcting codes. This distance is widely used in several disciplines including telecommunication, information theory, coding theory, cryptography, and biological systematics. The Hamming distance is a pri-mary geometric structure, along with the betweenness relation, used in our treatment of cubical graphs and partial cubes later in the book.

The Boolean cube $\mathcal{B}_n$ is an n-dimensional vector space over the field $\mathbf{F}_2$ (the finite field of characteristic two). The transformations α_A (see Section 3.8) form the translation group K of $\mathcal{B}_n$. Elements of the group H can be regarded as "orthogonal" transformations of this vector space. In this sense, Theo-rem 3.31 is an analogue of the result from Euclidean geometry: The group of motions of n-dimensional Euclidean space is a semidirect product of the translation group by the orthogonal group $O(n)$.

The diverse applications of cubes have resulted in many ways of charac-terizing them. Arguably, the original characterizations were given by Stephan Foldes (1977) (Theorem 3.37). Some other characterizations are found in Mul-der (1980), Bandelt and Mulder (1983), Scapellato (1990), and Brešar et al. (2005). See also a survey in Harary et al. (1988).

Exercises

3.1. Prove that the unit cube I^n is a convex set; that is, for any two points $(x_1, \ldots, x_n)$ and $(y_1, \ldots, y_n)$ in I^n, the line segment

$$\{(z_1, \ldots, z_n) \in \mathbf{R}^n : z_i = tx_i + (1-t)y_i,\ 0 \le t \le 1,\ 1 \le i \le n\}$$

is a subset of I^n.

3.2. Show that the unit cube I^n is the intersection of the halfspaces

$$H_i^+ = \{(x_1, \ldots, x_n) \in \mathbf{R}^n : x_i \ge 0\},\ \ H_i^- = \{(x_1, \ldots, x_n) \in \mathbf{R}^n : x_i \le 1\}$$

$(1 \le i \le n)$; that is, $I^n = \bigcap_{i=1}^n (H_i^+ \cap H_i^-)$.

3.3. Show that I^n has 2^n vertices and $(n-1)2^n$ edges.

3.4. Find the number of k-dimensional faces of the cube I^n.

3.5. Show that a k-dimensional face of the cube I^n is congruent to I^k.

3.6. Let $J = \{1,\dots,n\}$. Prove that posets $(\{0,1\}^n, <)$ and $(\mathcal{P}(J), \subset)$ are isomorphic.

3.7. Show that, in any poset P, the operations of meet and join satisfy the following laws, whenever the expressions referred to exist.

a) $x \wedge x = x$, $x \vee x = x$.
b) $x \wedge y = y \wedge x$, $x \vee y = y \vee x$.
c) $x \wedge (y \wedge z) = (x \wedge y) \wedge z$, $x \vee (y \vee z) = (x \vee y) \vee z$.
d) $x \wedge (x \vee y) = x \vee (x \wedge y) = x$.

3.8. Show that, in a distributive lattice L,

$$x \vee (y \wedge z) = (x \vee y) \wedge (x \vee z), \quad \text{for all } x, y, z \in L.$$

3.9. Show that $(\{0,1\}, <)$ is a Boolean lattice.

3.10. Show that the sets $\{0,1\}^n$ and $\mathcal{P}(J)$ with operations meet, join, and complementation defined in Section 3.3 are Boolean lattices.

3.11. The family of sets

$$L = \{\varnothing, \{a\}, \{b\}, \{c\}, \{a,b,c\}\}$$

is a poset with respect to the set inclusion relation. Describe the meet and join operations in L. Show that the Hasse diagram of L is isomorphic to the graph $K_{2,3}$. Is the lattice L distributive? Boolean?

3.12. Let P_3 be the path from Example 3.13. Show that the grid $P = P_3 \,\square\, P_3$ is a bounded distributive lattice with respect to the order on the vertex set of P defined by

$$(x, y) \leq (u, v) \text{ if and only if } x \leq y, \ u \leq v.$$

Is this lattice Boolean?

3.13. Show that the operation of symmetric difference $\triangle$ on $\mathcal{P}(X)$ is commutative and associative; that is,

a) $A \triangle B = B \triangle A$,
b) $A \triangle (B \triangle C) = (A \triangle B) \triangle C$,

for all $A, B, C \in \mathcal{P}(X)$.

3.14. Prove that cubes $\mathcal{H}(X)$ and $\mathcal{H}(Y)$ are isomorphic if and only if the sets X and Y are of the same cardinality, $|X| = |Y|$.

3.15. Let X and Y be two arbitrary sets. Show that

a) $\mathcal{H}(X) \cap \mathcal{H}(Y) = \mathcal{H}(X \cap Y)$,
b) $\mathcal{H}(X) \cup \mathcal{H}(Y) \subseteq \mathcal{H}(X \cup Y)$, and
c) $\mathcal{H}(X) \cup \mathcal{H}(Y) = \mathcal{H}(X \cup Y)$ if and only if $X \subseteq Y$ or $Y \subseteq X$.

3.16. Show that a connected component of a Cartesian product $\prod_{i \in I} G_i$ containing vertex a is the weak Cartesian product $\prod_{i \in I}(G_i, a_i)$. Also show that $\prod_{i \in I}(G_i, a_i) = \prod_{i \in I}(G_i, b_i)$ if and only if the vertices a and b belong to the same component of $\prod_{i \in I} G_i$.

3.17. Complete the proof of Theorem 3.18.

3.18. The definition of a layer assumes a particular representation of the graph as a weak Cartesian product (graph factorization).

a) Show that $Q_{m+n} \cong Q_m \square Q_n$.
b) What are the layers of the product $Q_{m+n} \cong Q_m \square Q_n$?

3.19. Show that the restriction $p_k|_{L_k(x)}$ of the projection p_k to the layer $L_k(x)$ defines an isomorphism $L_k(x) \cong G_k$.

3.20. Show that the graph distance d on the weak Cartesian product of a family of connected rooted graphs $\{(G_i, a_i)\}$ is given by

$$d(u, v) = \sum_{i \in I} d_{G_i}(u_i, v_i),$$

where the sum on the right side is taken over a finite number of nonzero summands. (Use the results of Theorem 1.7 and Exercise 1.34.)

3.21. Let $(G, a) = \prod_{i \in \omega}(P_n, a_i)$ be a weak Cartesian product.

a) Show that a permutation $\sigma : \omega \to \omega$ such that the vertex $\sigma(a) = (a_{\sigma(0)}, a_{\sigma(1)}, \ldots)$ differs from $a = (a_0, a_1, \ldots)$ in at most finitely many coordinates defines an automorphism

$$(x_0, x_1, \ldots, x_k, \ldots) \mapsto (x_{\sigma(0)}, x_{\sigma(1)}, \ldots, x_{\sigma(k)}, \ldots)$$

of (G, a).
b) Show that an automorphism $\alpha : P_n \to P_n$ of the kth factor of (G, a) defines an automorphism

$$(x_0, x_1, \ldots, x_k, \ldots) \mapsto (x_0, x_1, \ldots, \alpha(x_k), \ldots)$$

of (G, a).

3.22. Let α be an automorphism of a connected graph G. Show that

a) α is an isometry of $V(G)$ onto itself.
b) The image of an interval $I(u, v)$ under α is the interval $I(\alpha(u), \alpha(v))$.

3.23. Let P and Q be finite subsets of a set X. Prove that

a) If $P \subseteq Q$ and $|P| = |Q|$, then $P = Q$.
b) $|P \cap Q| + |P \cup Q| = |P| + |Q|$.
c) $|P \triangle Q| = |P| + |Q| - 2|P \cap Q|$.

3.24. Complete the proof of Theorem 3.22.

3.25. Let $G = (V, E)$ be a graph. The *convex hull* $C(Y)$ of a set $Y \subseteq V$ is the smallest convex set of vertices containing Y.

a) Give an example of a graph G and a finite set $Y \subseteq V$ such that the set $C(Y)$ is infinite.
b) Show that the convex hull of a finite subset of vertices of the cube $\mathcal{H}(X)$ is a finite set.

3.26. The *diameter* $d(G)$ of a finite graph $G = (V, E)$ is the maximum distance between any pair of vertices. The *total distance* $td(G)$ is

$$td(G) = \frac{1}{2} \sum_{u,v \in V} d(u, v).$$

The *average length* of a graph G is

$$\ell_{av}(G) = \frac{\sum_{u,v \in V} d(u, v)}{|V|(|V| - 1)} = \frac{2td(G)}{|V|(|V| - 1)}.$$

Show that

a) $d(Q_n) = n$.
b) $td(Q_n) = n2^{2n-2}$.
c) $\ell_{av}(Q_n) = \dfrac{n2^{n-1}}{2^n - 1}$.
d) $\ell_{av}(C_{2n}) = \dfrac{n^2}{2n - 1}$.

3.27. Find the average lengths (cf. Exercise 3.26) of the graphs P_n, C_n, $K_{2,3}$, and K_n.

3.28. Prove Theorem 3.29.

3.29. Let β be an automorphism of $\mathcal{H}(X)$ such that $\beta(\varnothing) = \varnothing$. Prove that β defines a permutation on the set X.

3.30. Show that automorphisms of $\mathcal{H}(X)$ preserve the symmetric difference of sets, that is

$$\alpha(A \bigtriangleup B) = \alpha(A) \bigtriangleup \alpha(B) \text{ for } \alpha \in \text{Aut}(\mathcal{H}(X)) \text{ and } A, B \in \mathcal{P}_f(X).$$

3.31. Show that $K \cap H = \{e\}$, where K and H are subgroups of $\text{Aut}(\mathcal{H}(X))$ introduced in Section 3.8.

3.32. Prove that $|\mathcal{P}_f(X)| \leq |\mathcal{P}_f(Y)|$ if and only if $|X| \leq |Y|$.

3.33. Two uv-paths P and Q in a graph are said to be *internally disjoint* if $P \cap Q = \{u, v\}$. Show that the maximum number of internally disjoint shortest AB-paths in the cube $\mathcal{H}(X)$ is $d(A, B)$.

4

Cubical Graphs

Cubical graphs are subgraphs of cubes. We use c-valuations on graphs to characterize cubical graphs and investigate their properties.

4.1 Definitions and Examples

Let us recall (Section 1.4) that an embedding of a graph G into graph H is an edge-preserving, one-to-one mapping $\varphi : V(G) \rightarrow V(H)$. In this case we also say that G is *embeddable* into H. The image $\varphi(G)$ of the graph G is a subgraph of H that is isomorphic to G.

Definition 4.1. A graph G is said to be *cubical* if it is embeddable in some cube $\mathcal{H}(X)$.

Cubes are bipartite graphs (Theorem 3.11) and a cubical graph is isomorphic to a subgraph of a cube, therefore a cubical graph must be bipartite.

Cubical graphs should be distinguished from *cubic* graphs, those graphs in which the degree of every vertex is three. For example, the Petersen graph (see Figure 1.19) is cubic but not cubical (it is not even bipartite).

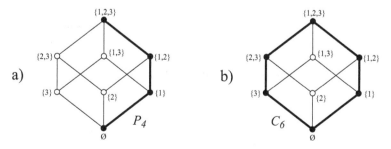

Figure 4.1. Embeddings of the path P_4 and the cycle C_6 into the cube Q_3.

Let $X = \{1, \ldots, n\}$, $X_0 = \varnothing$, and $X_k = \{1, \ldots, k\}$ for $0 < k \le n$. It is clear that $P = X_0, X_1, \ldots, X_n$ is a path in $\mathcal{H}(X) \cong Q_n$ connecting vertices $X_0 = \varnothing$ and $X_n = X$. This path is isomorphic to P_{n+1} (cf. Figure 4.1a). Therefore, any path is a cubical graph. In fact, the ray and the double ray are also cubical graphs (cf. Exercise 4.1).

Let $P' = \overline{X}_0, \overline{X}_1, \ldots, \overline{X}_n$, where $\overline{X}_k$ is the complement of the set X_k, be another path in the same cube $\mathcal{H}(X)$. By concatenating the paths P and P' at $X_n = \overline{X}_0$, we obtain a cycle of length $2n$ (cf. Figure 4.1b). Thus, any even cycle C_{2n} is a cubical graph. On the other hand, odd cycles are not bipartite and therefore are non-cubical graphs.

As we observed, the path P_{n+1} can be embedded into the n-dimensional cube Q_n. Actually, a path can be embedded in a cube of much lesser dimension. For instance, an embedding of the path P_8 into Q_3 is shown in Figure 1.7 and an embedding of P_4 into Q_2 is shown in Figure 2.3. Suppose that the path P_n is embedded into the cube Q_m. Then we must have $n \le 2^m$, the number of vertices of Q_m. Hence, $m \ge \log_2 n$. Because m is an integer, we have[1]

$$m \ge \lceil \log_2 n \rceil. \tag{4.1}$$

Definition 4.2. The *cubical dimension* $\dim_c(G)$ of a finite cubical graph G is the minimum dimension of a cube in which G is embeddable.

For the path P_n, we have, by (4.1), $\dim_c(P_n) \ge \lceil \log_2 n \rceil$. To prove that

$$\dim_c(P_n) = \lceil \log_2 n \rceil,$$

we first show that there is a spanning path in Q_m, that is, a path containing all vertices of Q_m. A spanning path in a graph is called a *Hamilton path*; a graph with a Hamilton path is called *traceable*.

Theorem 4.3. *The cube Q_m is a traceable graph.*

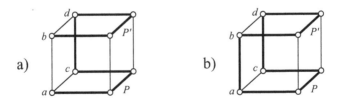

Figure 4.2. Proofs of Theorems 4.3 and 4.4.

Proof. We actually prove that there is a Hamilton path with initial vertex $a = (0, 0, \ldots, 0)$ and terminal vertex $b = (0, \ldots, 0, 1)$. The proof is by induction

---

[1] The *ceiling* $\lceil x \rceil$ of a real number x is the smallest integer not less than x; the *floor* $\lfloor x \rfloor$ of x is the largest integer not greater than x.

on m. The cases $m = 1, 2$ are trivial. Suppose that the result is true for Q_{m-1}. Consider the layer $L_1(a)$ in the Cartesian product $Q_m = Q_{m-1} \square K_2$. This layer is isomorphic to the graph Q_{m-1}. By the induction hypothesis, there is a Hamilton path P in $L_1(a)$ with initial vertex a and terminal vertex $c = (0, \ldots, 0, 1, 0)$. The last coordinate of all vertices in P is 0. By changing this coordinate to 1 for all vertices in P, we create a path P' in the layer $L_1(b)$ with initial vertex b and terminal vertex $d = (0, 1, \ldots, 1)$ (cf. Figure 4.2a). Let us add the edge cd to the union $P \cup P'$. The resulting path is a Hamilton path in Q_m with initial vertex a and terminal vertex b. □

It follows that the path P_{2^m} is embeddable into the cube Q_m. A Hamilton path in the cube Q_4 is depicted in Figure 4.3a (cf. Figure 4.2a). This path is isomorphic to P_{16}.

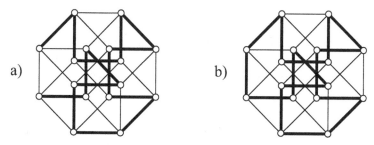

a)   b)

Figure 4.3. A Hamilton path and a Hamilton cycle in Q_4.

For a given n, there is a unique integer m such that $2^{m-1} < n \leq 2^m$. Clearly, the path P_n is not embeddable into the cube Q_{m-1}. Because the path P_n is isomorphic to a segment of the path P_{2^m}, and P_{2^m} is embeddable into Q_m, the path P_n itself is embeddable into Q_m. Thus, $\dim_c(P_n) = \lceil \log_2 n \rceil$.

We use essentially the same arguments to show that the cubical dimension of an even cycle C_{2n} is $\lceil \log_2 2n \rceil$. More definitions are in order.

A spanning cycle of a graph is called a *Hamilton cycle* of the graph. A graph is said to be *hamiltonian* if it contains a Hamilton cycle. A graph on n vertices is *pancyclic* if it contains a cycle of length k for every $3 \leq k \leq n$. A bipartite graph $G[X, Y]$ with $|X| = |Y| = n$ (each part contains n vertices) is said to be *bipancyclic* if it contains a cycle of length $2k$ for every $2 \leq k \leq n$.

Theorem 4.4. *For $m \geq 2$, the cube Q_m is a bipancyclic graph. Accordingly, it is hamiltonian.*

Proof. As in the proof of Theorem 4.3, we factor $Q_m = Q_{m-1} \square K_2$ and consider two parallel layers $L_1(a)$ and $L_1(b)$, where $a = (0, 0, \ldots, 0)$ and $b = (0, \ldots, 0, 1)$ (cf. Figure 4.2b). By Theorem 4.3, for any $2 \leq k \leq 2^{m-1}$, there is a path P of length $k - 1$ in $L_1(a)$ with initial vertex a. Let c be the terminal vertex of P. By "lifting" P along the edge ab we obtain a path P' in the layer

$L_1(b)$ with initial vertex b. Let d be the terminal vertex of P'. By adding edges ab and cd to the union $P \cup P'$, we obtain a cycle of length $2k$ in Q_m.   □

It is clear that the cycle C_{2n} is not embeddable into the cube Q_{m-1} if $2n > 2^{m-1}$. On the other hand, by Theorem 4.4, this cycle is embeddable into Q_m if $2n \leq 2^m$. It follows that the cubical dimension of C_{2n} is $\lceil \log_2 2n \rceil$:

$$\dim_c(C_{2n}) = \lceil \log_2 2n \rceil.$$

A nontrivial path P_n is a tree with two leaves. In fact, any tree is a cubical graph. Indeed, let T be a tree and v_0 be a fixed vertex of T, so we have a rooted tree (T, v_0). By Theorem 2.29, for any vertex v of T, there is a unique path P_v connecting v_0 to v. Let $X = E(T)$ and let $\varphi : V(T) \to \mathcal{P}_f(X)$ be a mapping defined by

$$v \mapsto E(P_v).$$

For any edge xy of T, either $x \in P_y$ or $y \in P_x$ (cf. Exercise 2.34). It follows that $E(P_x) \,\triangle\, E(P_y) = \{xy\}$. Hence, φ is an embedding of T into $\mathcal{H}(X)$.

If T is a finite tree on n vertices, the construction from the foregoing paragraph gives an upper bound for the cubical dimension of the tree:

$$\dim_c(T) \leq n - 1.$$

The cubical dimension of a finite tree T on n vertices attains its maximum value when T is a star (cf. Exercise 4.10) and its minimum for a path (cf. Exercise 4.11):

$$\dim_c(K_{1,n-1}) = n - 1 \qquad \text{and} \qquad \dim_c(P_n) = \lceil \log_2 n \rceil,$$

where $n = |V(T)|$.

Figure 4.4. Complete bipartite graph $K_{2,3}$ is not cubical.

A cubical graph is bipartite, but the converse does not hold. The smallest counterexample (cf. Exercise 4.4) is the complete bipartite graph $K_{2,3}$. Suppose there is an embedding $\varphi : K_{2,3} \to \mathcal{H}(X)$ for some set X, and let vertices of $K_{2,3}$ be labeled as in Figure 4.4. Without loss of generality (cf. Exercises 4.2 and 4.3), we may assume that $\varphi(x) = \varnothing$. Then $\varphi(u)$, $\varphi(v)$, and $\varphi(w)$ must be distinct singletons and $\varphi(y)$ must be a subset of X that differs from each of these singletons by exactly one element. Clearly, such a subset does not exist.

A connected bipartite graph is either a tree or contains a cycle. All trees on six vertices are shown in Figure 2.13. It is not difficult to obtain a list of the trees on less than six vertices from graphs in Figure 2.13.

The connected bipartite graphs on six or fewer vertices containing cycles are depicted in Figure 4.5. The graph B10 is the complete bipartite graph $K_{2,3}$. It is easy to see (cf. Exercise 4.9) that the graphs B22 and B24–B27 contain subgraphs isomorphic to $K_{2,3}$. Therefore they are not cubical. (The graphs in Figure 4.5 are labeled according to the classification in the book *An Atlas of Graphs* by Reed and Wilson (1998).)

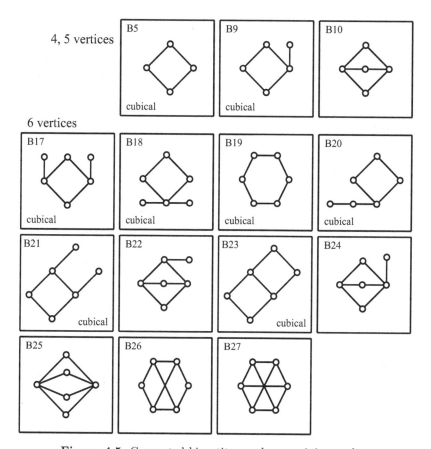

Figure 4.5. Connected bipartite graphs containing cycles.

4.2 A Criterion

Let $\mathcal{H}(X)$ be a cube. For every edge PQ of $\mathcal{H}(X)$, the set $P \triangle Q$ is a singleton. On the other hand, for every singleton $\{x\}$ in $\mathcal{P}_f(X)$ there is an edge PQ such

that $P \triangle Q = \{x\}$. (Consider, for instance, $P = \varnothing$, $Q = \{x\}$.) Thus we have
a mapping c from the edge set of $\mathcal{H}(X)$ onto the set X given by:

$$c : PQ \mapsto x, \quad PQ \in E(\mathcal{H}(X)), \quad P \triangle Q = \{x\}. \tag{4.2}$$

In general, for any graph G and a set X, a mapping $E(G) \to X$ is called
an *edge-colouring* of the graph G; the elements of the set X are called the
available *colours*. An edge-colouring of a graph is said to be *proper* if adjacent
edges have different colours. The set X in the foregoing paragraph is the set of
colours for the proper edge-colouring c of the cube $\mathcal{H}(X)$ (cf. Exercise 4.21).

The edge-colouring c of the cube $\mathcal{H}(X)$ induces an edge-colouring of any
subgraph of $\mathcal{H}(X)$. We denote this induced colouring by the same letter c.

Let us consider an example. Let $X = \{1, 2, 3\}$ and $\mathcal{H}(X) \cong Q_3$ be the
cube on X. A Hamilton path P_8 and a Hamilton cycle C_8 in Q_3 together with
their edge-colourings are shown in Figure 4.6 (cf. Figure 4.2). Note that every
colour appears an even number of times in C_8, whereas the colour 3 appears
only once in the path P_8. It turns out that these properties are crucial for
cubical graphs.

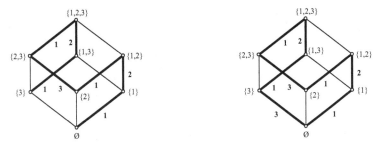

Figure 4.6. Edge-colourings of a Hamilton path and a Hamilton cycle in Q_3.

Definition 4.5. An edge-colouring $E(G) \to X$ of a connected graph G is
called a *c-valuation* on the graph G if the following conditions are satisfied.

(i) For each path of G there is a colour that appears an odd number of times
     in that path.
(ii) In each cycle of G every colour appears an even number of times.

It can be easily seen that edge-colourings of the Hamilton path and cycle
in Figure 4.6 are *c*-valuations.

Two adjacent edges of a graph form a path. If the graph admits a *c*-
valuation, then, by condition (i), the colours of two adjacent edges must be
different. Thus a *c*-valuation on a graph is a proper edge-colouring.

Theorem 4.6. *Let f be a c-valuation on a graph G. Then*

(i) *For each open walk in G there is a colour that appears an odd number of times in that walk.*

(ii) *In each closed walk in G every colour appears an even number of times.*

Proof. (i) Let $W = v_0 v_1 \cdots v_n$ be an open walk in G. The proof is by induction on n. The case $n = 1$ is trivial. Suppose that the claim holds for all open walks of length less than n. If W does not contain repeated vertices, then it is a path and the claim holds by the definition of c-valuation. Otherwise, let us consider the family $\mathcal{S}$ of all segments $v_i \cdots v_j$ with $v_i = v_j$. These segments are partially ordered by inclusion. Let $S = v_p \ldots v_q$ be a minimal segment in the family $\mathcal{S}$. Clearly, S is a closed walk in G. If $q = p + 2$, then we have two occurrences of a single colour in S. Otherwise, S is a cycle and, by property (ii), every colour appears an even number of times in S. By the induction hypothesis, there is a colour that appears an odd number of times in the walk $v_0 \cdots v_p v_{q+1} \cdots v_n$. The result follows, because any colour appears an even number of times in S.

(ii) Let W be a closed walk in G. If W is a cycle, we are done. Otherwise, as in part (i), we select a minimal closed segment $v_p \cdots v_q = v_p$ in W and delete vertices $v_{p+1} \cdots v_{q-1}$ from W, resulting in a closed walk of length less than n. An obvious inductive argument proves claim (ii). $\square$

Theorem 4.7. *The function c defined by (4.2) is a c-valuation on $\mathcal{H}(X)$.*

Proof. Let $W = X_1, \ldots, X_n$ be a walk in $\mathcal{H}(X)$ and a be a fixed element of the set X. Let us define a sequence (a_i) by

$$
a_i = \begin{cases} 1, & \text{if } a \in X_i \setminus X_{i-1}, \\ -1, & \text{if } a \in X_{i-1} \setminus X_i, \\ 0, & \text{otherwise,} \end{cases} \quad \text{for } 1 < i \le n.
$$

It is clear that we cannot have $a_i = a_{i+1} = \pm 1$ for some i. Suppose that, for given $i < j$, we have $a_i = 1$, $a_j = 1$, and $a_k = 0$ for all $i < k < j$. We have $a \in X_i$, because $a_i = 1$. Because $a_{i+1} = 0$ and $a \in X_i$, we must have $a \in X_i \cap X_{i+1} \subseteq X_{i+1}$. By repeating this argument we obtain $a \in X_{j-1}$, which contradicts our assumption that $a_j = 1$. Therefore, $a_k = -1$ for some $i < k < j$. Essentially the same argument shows that we cannot have $a_i = -1$, $a_j = -1$, and $a_k = 0$ for all $i < k < j$. It follows that the occurrences of numbers 1 and -1 alternate in the sequence (a_i).

Suppose that the walk W is open; that is, $X_n \ne X_1$. Therefore the set $X_1 \triangle X_n$ is not empty. Assume first that $X_n \setminus X_1 \ne \varnothing$, and let a be an element of $X_n \setminus X_1$. Inasmuch as $a \notin X_1$, $a \in X_n$, the first and last occurrences of a nonzero element in the sequence (a_i) are 1. Hence, there are an odd number of occurrences of nonzero elements in (a_i). It follows that there are an odd number of edges in W with the colour a. If $X_n \setminus X_1 = \varnothing$, then we choose $a \in X_1 \setminus X_n$ and apply the same argument to the walk $X_n, \ldots, X_1$, obtaining the same result. This establishes property (i) in Definition 4.5.

Suppose now that the walk $\mathcal{W}$ is closed; that is, $X_n = X_1$. Let a be the colour of the edge $X_1 X_2$. Because $X_n = X_1$, the first and last nonzero terms in the sequence (a_i) must have opposite signs. Hence, there is an even number of nonzero terms in the sequence (a_i). It follows that the colour a occurs an even number of times in the closed walk $\mathcal{W}$. By repeating this argument for each of the remaining edges of $\mathcal{W}$, we prove property (ii) in Definition 4.5. □

Theorem 4.8. *A connected graph G is a cubical graph if and only if there is a c-valuation on G.*

Proof. (Necessity.) We may assume that G is a subgraph of some cube $\mathcal{H}(X)$. It is clear that the restriction of the c-valuation on $\mathcal{H}(X)$ to the subgraph G is a c-valuation on G (cf. Figure 4.6).

(Sufficiency.) Let $f : E(G) \to X$ be a c-valuation on G with some set of colours X. Let u be a fixed vertex of G. We denote by $C(v, P)$ the set of colours occurring an odd number of times in a uv-path P. Let us show that $C(v, P) = C(v, P')$ for any two uv-paths P and P'. Because $P \cup P'$ is a closed walk in G, a colour that appears an odd number of times in P must also occur an odd number of times in P' (Theorem 4.6). Thus, $C(v, P) \subseteq C(v, P')$. By symmetry, $C(v, P') \subseteq C(v, P)$. Therefore, $C(v, P) = C(v, P')$. It follows that the function $\varphi(v) = C(v, P)$ is a well-defined mapping from the vertex set $V(G)$ into $\mathcal{P}_f(X)$. In particular, we have $\varphi(u) = \varnothing$.

Let us show that φ is a one-to-one mapping. Let v and w be two distinct vertices of G. Consider the vw-walk $\mathcal{W}$ obtained by concatenating a vu-path P with a uw-path P'. By Theorem 4.6, there is a colour that appears an odd number of times in $\mathcal{W}$. This colour must appear an odd number of times either in P or in P', but not in both. Therefore, $\varphi(v) \neq \varphi(w)$. It follows that φ is one-to-one.

Finally, we show that φ is an embedding. Let vw be an edge of G and let us add this edge to the union of paths P and P' introduced in the previous paragraph. We obtain a closed walk $\mathcal{W} = P \cup P' \cup \{vw\}$. By Theorem 4.6, the colour $f(vw)$ must appear an odd number of times in one and only one of paths P and P'. The numbers of appearances of any other colour in P and P', respectively, must have the same parity. It follows that $\varphi(v) \triangle \varphi(w) = \{f(vw)\}$. Thus, $\varphi(v)$ and $\varphi(w)$ are adjacent in $\mathcal{H}(X)$. □

Example 4.9. Let T be a tree. It is clear that the identity mapping of the edge set $E(T)$ onto itself is a c-valuation on T. Hence, T is a cubical graph (see Section 4.1).

Theorem 4.10. *Let G be a finite connected cubical graph. The cubical dimension $\dim_c(G)$ is the minimum number of colours needed for a c-valuation on the graph G.*

Proof. Let $f : E(G) \to X$ be a *c*-valuation on G. By the sufficiency part of the proof of Theorem 4.8, there is an embedding $\varphi : G \to \mathcal{H}(X)$. Therefore, $\dim_c(G) \le |X|$. On the other hand, by definition of the cubical dimension, there is an embedding of G into a cube $\mathcal{H}(X)$ with $|X| = \dim_c(G)$. This embedding defines a *c*-valuation on G with $|X|$ colours inherited from the *c*-valuation on $\mathcal{H}(X)$ defined by (4.2). Hence, the minimum number of colours $|X|$ needed for a *c*-valuation on G is the cubical dimension of G.   □

4.3 Dichotomic Trees

As we have shown, the cubical dimension of a tree T on n vertices varies exponentially between $\lceil \log_2 n \rceil$ and $n-1$. The exact answer depends of course on the structure of the tree T rather than on its simple numerical parameters such as the number of vertices. In this section we find the cubical dimension of a finite dichotomic tree.

A *dichotomic tree* DT_n (also known as a *perfect binary tree*) is defined recursively as follows. We set $DT_1 = P_3 = aub$ (cf. Figure 4.7). The middle vertex u of DT_1 is its root. The tree DT_n is obtained from DT_1 by identifying the roots of two copies of DT_{n-1} with leaves a and b of DT_1, respectively. The edges added at each stage are depicted by heavy lines in Figure 4.7; the root of each tree is labeled by u.

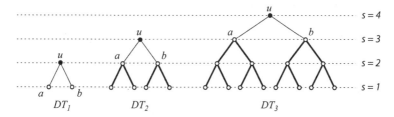

Figure 4.7. Dichotomic trees.

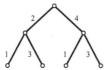

Figure 4.8. A *c*-valuation on DT_2.

It is easy to see that $\dim_c(DT_1) = 2$ and that DT_2 is not embeddable into the cube Q_3 (cf. Exercise 4.18). Let f be the function on the edge set of

DT_2 with values shown in Figure 4.8. Clearly, f is a c-valuation on DT_2. By Theorem 4.10, $\dim_c(DT_2) = 4$.

We need a particular labeling of vertices of the dichotomic tree DT_n for constructing a c-valuation that defines the cubical dimension of this tree.

The *height* $h(v)$ of a vertex v of a rooted tree is the distance between v and the root of the tree. If $s = n + 1 - h(v)$, we say that v is on the *s*th *level* of the tree (cf. Figure 4.7). We also define the level of an edge of DT_n as the level of its lower end. Note that each vertex on level s is the root of the tree DT_{s-1}.

It is easy to see that there are 2^{n-s+1} vertices and the same number of edges on the s-th level. Let us enumerate vertices on the sth level left-to-right by integers r ranging from 1 to 2^{n-s+1}. Then every vertex of DT_n is uniquely represented by a pair (r, s), where $1 \leq s \leq n + 1$ and $1 \leq r \leq 2^{n-s+1}$. This labeling is illustrated by the drawing of the dichotomic tree DT_3 in Figure 4.9.

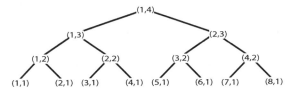

Figure 4.9. A labeling of the dichotomic tree DT_3.

The set of edges of DT_n is the disjoint union of two sets

$$\{(2r - 1, s)(r, s + 1) : 1 \leq s \leq n, \ 1 \leq r \leq 2^{n-s}\} \qquad (\text{"left" edges})$$

and

$$\{(2r, s)(r, s + 1) : 1 \leq s \leq n, \ 1 \leq r \leq 2^{n-s}\} \qquad (\text{"right" edges})$$

(see Figure 4.9).

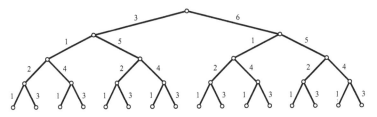

Figure 4.10. A c-valuation on DT_4.

We define a function f on the edge set of DT_n as follows. The values of f on the first two levels are (cf. Figure 4.10):

$$f((2r-1,1)(r,2)) = 1, \quad f((2r,1)(r,2)) = 3 \quad \text{(edges on level } s = 1\text{)}$$
$$f((2r-1,2)(r,3)) = 2, \quad f((2r,2)(r,3)) = 4 \quad \text{(edges on level } s = 2\text{)}.$$

Now, for $m \geq 2$, we define values on odd and even levels s by

$$f((2r-1,2m-1)(r,2m)) = 1 \qquad \text{(left edges, odd level)}$$
$$f((2r,2m-1)(r,2m)) = 2m+1 \quad \text{(right edges, odd level)}$$

and

$$f((2r-1,2m)(r,2m+1)) = 2m-1 \quad \text{(left edges, even level)}$$
$$f((2r,2m)(r,2m+1)) = 2m+2 \quad \text{(right edges, even level)},$$

respectively.

The edge-colouring of the graph DT_4 defined by function f is shown in Figure 4.10. Note that f maps the edge set of DT_n onto the set of colours $\{1,\ldots,n+2\}$.

We want to show that f is a c-valuation on DT_n for $n > 2$ (we have already shown that it is, if n equals 1 or 2.). First, we establish a property of the edge-colouring defined by this function.

Let (r,s) be a vertex of DT_n. The colour of the right edge $(2r,s-1)(r,s)$ is $s+2$. Indeed, if $s = 2m$, then

$$f((2r,2m-1)(r,2m)) = 2m+1 = s+2$$

and, for $s = 2m+1$, we have

$$f(2r,2m)(r,2m+1) = 2m+2 = s+2.$$

Thus the colour of all right edges on level $s-1$ is $s+2$. Clearly, this colour is greater than the colour of the left edges on the same level and the colours of all edges on the lower levels.

We prove that f is a c-valuation by contradiction. Because a tree has no cycles, the assumption that there is a path P in DT_n in which every colour appears an even number of times should lead to the desired contradiction.

Let (r,s) be a unique vertex of P with the maximum value of the level. If (r,s) is not an end of P, then the colour $s+2$ of the right edge $(2r,s-1)(r,s)$ appears only once in P. This is already a contradiction.

Otherwise, we may assume that (r,s) is the initial vertex of P. By the definition of f, any even colour appears on a unique level of DT_n. Therefore the colours of edges of P are odd numbers. It follows that the terminal vertex of P is on the level greater than two. Let $2k+1$ be the minimum odd colour different from 1 appearing in P (note that $2k+1 > 3$). By the definition of f and our assumption that every colour appears even number of times in P, the colour $2k+1$ appears on levels $2k-1$ and $2k+2$ in the path P. Inasmuch as $2k$ is an even level and P has an edge on level $2k$, the colour of this edge must be $2k - 1$ (there are no even colours of edges of P). This contradicts the assumption that $2k + 1$ is the minimum colour of edges in P.

We proved that f is a c-valuation on DT_n with $n + 2$ colours. By Theorem 4.10, $\dim_c(DT_n) \leq n + 2$.

To prove that the cubical dimension of DT_n is $n + 2$, we show that DT_n is not embeddable into the cube Q_{n+1}.

The proof is again by contradiction. We may assume that $n > 2$. Suppose that DT_n is isomorphic to a subgraph of Q_{n+1}. There are $(n + 1)2^n$ edges of Q_{n+1} (cf. Exercise 3.3) and DT_n has $2^{n+1} - 2$ edges (cf. Exercise 4.17). Therefore there are

$$(n + 1)2^n - 2^{n+1} + 2 = (n - 1)2^n + 2$$

edges of Q_{n+1} that are not edges of DT_n. Two leaves of DT_n are on the same distance n from the root. Cubes are bipartite graphs, thus there is no edge of Q_{n+1} joining two leaves; for otherwise, we would have a cycle of odd length $2n + 1$ in Q_{n+1}. Therefore there are n edges of Q_{n+1} adjacent to every leaf of DT_n that are not edges of DT_n. Hence,

$$n2^n \leq (n - 1)2^n + 2,$$

which is equivalent to $n \leq 1$. This contradiction completes the proof of the following result.

Theorem 4.11. For $n > 1$, the cubical dimension of the dichotomic tree DT_n is $n + 2$.

4.4 Operations on Cubical Graphs

In this section we consider properties of three basic operations on graphs: intersection, union, and Cartesian product.

The intersection $G \cap H$ (if it is defined) of a cubical graph G with a graph H is a subgraph of G and therefore is a cubical graph.

The union of two cubical graphs is not necessarily cubical as the example in Figure 4.11 illustrates.

Figure 4.11. Union of two paths is not cubical.

However, another operation of union, called "disjoint union", yields a cubical graph from cubical graphs. We call this operation the "sum" and define it as follows.

The *sum* (*disjoint union*) $A + B$ of two sets A and B is the union

$$(\{1\} \times A) \cup (\{2\} \times B).$$

Let $G_1 = (V_1, E_1)$ and $G_2 = (V_2, E_2)$ be two graphs. Their *sum* $G_1 + G_2$ has $V_1 + V_2$ as the vertex set; two vertices (i, u) and (j, v) are adjacent in $G_1 + G_2$ if and only if $i = j$ and $uv \in E_i$. This concept is illustrated by the example shown in Figure 4.12. We leave it to the reader to show that the sum of two cubical graphs is a cubical graph (cf. Exercise 4.13).

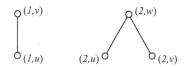

Figure 4.12. The sum of two cubes from Figure 4.11.

Let G_1 and G_2 be subgraphs of cubes $\mathcal{H}(X_1)$ and $\mathcal{H}(X_2)$, respectively, where X_1 and X_2 are disjoint sets. By Theorem 3.18, the Cartesian product $G_1 \square G_2$ is a subgraph of the cube $\mathcal{H}(X_1 \cup X_2)$. Therefore, the Cartesian product of two cubical graphs is a cubical graph (cf. Exercise 1.36). This result can be easily generalized as follows (cf. Exercise 4.16).

Theorem 4.12. *The Cartesian product of a finite family of cubical graphs is a cubical graph.*

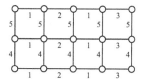

Figure 4.13. *c*-valuation on the grid $P_5 \square P_3$.

Example 4.13. A path is a cubical graph, therefore any grid $P_n \square P_m$ is a cubical graph. A *c*-valuation on the grid $P_5 \square P_3$ is shown in Figure 4.13 (cf. Exercise 4.14). By Theorem 4.10, $\dim_c(P_5 \square P_3) \leq 5$. On the other hand, it is not difficult to verify that one needs at least 5 colours for a *c*-valuation on the grid $P_5 \square P_3$. Hence, $\dim_c(P_5 \square P_3) = 5$ (cf. Exercise 4.15).

As we can see from Example 4.13, the cubical dimension of the grid $P_5 \square P_3$ is the sum of cubical dimensions of the factors P_5 and P_3:

$$\dim_c(P_5 \square P_3) = 5 = 3 + 2 = \dim_c(P_5) + \dim_c(P_3).$$

This is actually true for the Cartesian product $G_1 \square G_2$ of any two finite connected cubical graphs G_1 and G_2, as the following arguments show.

Let $n_i = \dim_c(G_i)$ for $i \in \{1, 2\}$. The graphs G_1 and G_2 are embeddable into cubes Q_{n_1} and Q_{n_2}, respectively. Therefore their product $G_1 \square G_2$ is embeddable into the cube $Q_{n_1+n_2} = Q_{n_1} \square Q_{n_2}$. Hence,

$$\dim_c(G_1 \square G_2) \leq n_1 + n_2 = \dim_c(G_1) + \dim_c(G_2).$$

To prove the opposite inequality, we need three properties of a c-valuation on $G_1 \square G_2$. In what follows, f is a given c-valuation on $G_1 \square G_2$ with the set of colours X. We assume that f maps $E(G_1 \square G_2)$ onto X and that graphs G_1 and G_2 are nontrivial.

Lemma 4.14. *Let u_1v_1 and u_2v_2 be edges of graphs G_1 and G_2, respectively. Then*

$$f((u_1, u_2)(v_1, u_2)) = f((u_1, v_2)(v_1, v_2)),$$
$$f((u_1, u_2)(u_1, v_2)) = f((v_1, u_2)(v_1, v_2)),$$

and

$$f((u_1, u_2)(v_1, u_2)) \neq f((u_1, u_2)(u_1, v_2)).$$

Proof. The four vertices (u_1, u_2), (v_1, u_2), (v_1, v_2), and (u_1, v_2) form a cycle in $G_1 \square G_2$ (cf. Figure 4.14). f is a c-valuation, thus each colour must appear an even number of times in the cycle and no adjacent edges can have the same colour. The only possible colouring of the four edges in the cycle is shown in Figure 4.14, where

$$a = f((u_1, u_2)(v_1, u_2)) = f((u_1, v_2)(v_1, v_2))$$

and

$$b = f((u_1, u_2)(u_1, v_2)) = f((v_1, u_2)(v_1, v_2)),$$

with $a \neq b$. □

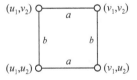

Figure 4.14. Proof of Lemma 4.14.

Lemma 4.15. *Let uv be an edge of G_1. Then, for any two vertices x and y of G_2,*

$$f((u, x)(v, x)) = f((u, y)(v, y)).$$

A similar statement holds for edges of G_2 and vertices of G_1.

Proof. Let

$$x_1 = x, x_2, \ldots, x_k = y$$

be an xy-path in the graph G_2. By Lemma 4.14,

$$f((u, x_i)(v, x_i)) = f((u, x_{i+1})(v, x_{i+1})), \quad \text{for } 1 \leq i < k.$$

The result follows. $\square$

The edges $(u, x)(v, x)$, where $uv \in E(G_1)$ and $x \in V(G_1)$, are the elements of $p_1^{-1}(uv)$, where p_1 is the projection of $G_1 \square G_2$ onto G_1. By Lemma 4.15, the function f is constant on $p_1^{-1}(uv)$. Therefore the equation

$$f_1(uv) = f((u, x)(v, x))$$

defines a function on $E(G_1)$.

Lemma 4.16. *The function f_1 is a c-valuation on G_1.*

Proof. Let $L_1((u, x))$ be a layer in $G_1 \square G_2$. The projection p_1 is an isomorphism from the layer $L_1((u, x))$ onto the graph G_1. Moreover,

$$f_1(uv) = f((u, x)(v, x)), \quad \text{for all } uv \in E(G_1).$$

The restriction of f to the layer $L_1((u, x))$ is a c-valuation, therefore the function f_1 is a c-valuation. (The drawing in Figure 4.13 is instructive.) $\square$

In a similar way, a c-valuation f_2 is defined on the graph G_2.

Let us consider the sets $X_1 = f_1(E(G_1))$ and $X_2 = f_2(E(G_2))$. It is clear that $X = X_1 \cup X_2$. Moreover, by Lemma 4.14, $X_1 \cap X_2 = \varnothing$. Hence, the pair $\{X_1, X_2\}$ is a partition of the set X.

By Theorem 4.10, we may assume that f takes values in a set X of cardinality $|X| = \dim_c(G_1 \square G_2)$. Then, by the same Theorem 4.10,

$$\dim_c(G_1 \square G_2) = |X| = |X_1| + |X_2| \geq \dim_c(G_1) + \dim_c(G_2).$$

We proved the following theorem.

Theorem 4.17. *Let G_1 and G_2 be finite connected cubical graphs. Then*

$$\dim_c(G_1 \square G_2) = \dim_c(G_1) + \dim_c(G_2).$$

4.5 Vertex- and Edge-Pastings of Graphs

We now consider another operation on graphs which is called "pasting" (or "gluing").

Let $G_1 = (V_1, E_1)$ and $G_2 = (V_2, E_2)$ be two graphs, $H_1 = (U_1, F_1)$ and $H_2 = (U_2, F_2)$ be two isomorphic subgraphs of G_1 and G_2, respectively, and $\psi : U_1 \rightarrow U_2$ be a bijection defining an isomorphism between H_1 and H_2. The bijection ψ defines an equivalence relation R on the sum $V_1 + V_2$ as follows: any element in $(V_1 \setminus U_1) \cup (V_2 \setminus U_2)$ is equivalent to itself only and elements $u_1 \in U_1$ and $u_2 \in U_2$ are equivalent if and only if $u_2 = \psi(u_1)$. We say that the quotient set $V = (V_1 + V_2)/R$ is obtained by *pasting together the sets V_1 and V_2 along the subsets U_1 and U_2* (cf. Figure 4.15).

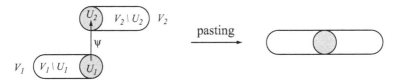

Figure 4.15. Pasting together sets V_1 and V_2.

The graphs H_1 and H_2 are isomorphic, therefore a pasting of the sets V_1 and V_2 can be naturally extended to a pasting of sets of edges E_1 and E_2 resulting in the set E of edges joining vertices in V. We say that the graph $G = (E, V)$ is obtained by *pasting together the graphs G_1 and G_2 along the isomorphic subgraphs H_1 and H_2*. The pasting construction allows for identifying in a natural way the graphs G_1 and G_2 with subgraphs of G, and the isomorphic graphs H_1 and H_2 with a common subgraph H of both graphs G_1 and G_2. We often follow this convention below.

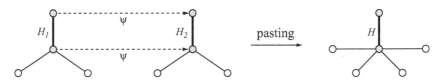

Figure 4.16. Pasting together two trees along two edges.

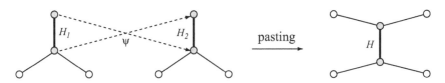

Figure 4.17. Another pasting of the same trees.

Note that in the above construction the resulting graph G depends not only on graphs G_1 and G_2 and their isomorphic subgraphs H_1 and H_2 but also on the bijection ψ defining an isomorphism from H_1 onto H_2 (see the drawings in Figures 4.16 and 4.17)

Pasting together two cubical graphs does not necessarily produce a cubical graph as the pasting in Figure 4.18 illustrates. However, there are two special instances of the pasting operation that yield cubical graphs from cubical graphs.

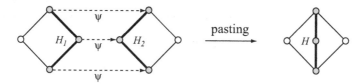

Figure 4.18. Pasting together two squares along isomorphic paths H_1 and H_2.

Let $G_1 = (V_1, E_1)$ and $G_2 = (V_2, E_2)$ be two connected graphs, $a_1 \in V_1$, $a_2 \in V_2$, and $H_1 = (\{a_1\}, \varnothing)$, $H_2 = (\{a_2\}, \varnothing)$. Let G be the graph obtained by pasting G_1 and G_2 along subgraphs H_1 and H_2. In this case we say that the graph G is obtained from graphs G_1 and G_2 by *vertex-pasting*. We also say that G is obtained from G_1 and G_2 by *identifying* vertices a_1 and a_2. Figure 4.19 illustrates this construction. Note that the vertex $a = \{a_1, a_2\}$ is a cut-vertex of G, because $G_1 \cup G_2 = G$ and $G_1 \cap G_2 = \{a\}$. (We follow our convention and identify graphs G_1 and G_2 with subgraphs of G.)

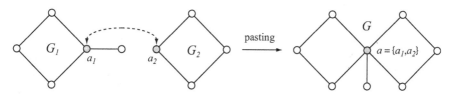

Figure 4.19. An example of vertex-pasting.

It is not difficult to see that a graph G obtained by vertex-pasting together bipartite graphs G_1 and G_2 is bipartite (cf. Exercise 4.25). Moreover, we have the following result.

Theorem 4.18. *A graph obtained by vertex-pasting together two cubical graphs is a cubical graph.*

Proof. Let f_1 and f_2 be c-valuations on G_1 and G_2, respectively. We may assume that the respective colour sets X_1 and X_2 are disjoint. Let us define a function f on $E(G)$ by

$$f(uv) = \begin{cases} f_1(uv), & \text{if } uv \in E(G_1), \\ f_2(uv), & \text{if } uv \in E(G_2). \end{cases}$$

Clearly, f is a c-valuation on G. By Theorem 4.8, G is a cubical graph.   □

The vertex-pasting construction can be generalized as follows. Let

$$\mathcal{G} = \{(G_i, a_i)\}_{i \in I}$$

be a family of connected rooted graphs, and let G be the graph obtained from the graphs G_i by identifying vertices in the set $\mathcal{A} = \{a_i\}_{i \in I}$. We say that G is obtained by *vertex-pasting together the graphs* G_i (along the set $\mathcal{A}$).

Example 4.19. Let $J = \{1, \dots, n\}$ with $n \geq 2$,

$$\mathcal{G} = \{G_i = (\{a_i, b_i\}, \{a_i b_i\})\}_{i \in J}, \text{ and } \mathcal{A} = \{a_i\}_{i \in J}.$$

Clearly, each G_i is K_2. By vertex-pasting these graphs along $\mathcal{A}$, we obtain the n-star graph $K_{1,n}$.

Inasmuch as the star $K_{1,n}$ is a tree it can also be obtained from K_1 by successive vertex-pasting as in Example 4.20.

Example 4.20. Let G_1 be a tree and $G_2 = K_2$. By vertex-pasting these graphs we obtain a new tree. Conversely, let G be a tree and v be its leaf. Let G_1 be a tree obtained from G by deleting the leaf v. Clearly, G can be obtained by vertex-pasting G_1 and K_2. It follows that any tree can be obtained from the graph K_1 by successive vertex-pasting of copies of K_2 (cf. Lemma 2.31).

We now consider another simple way of pasting two graphs together.

Let $G_1 = (V_1, E_1)$ and $G_2 = (V_2, E_2)$ be two connected graphs, $a_1 b_1 \in E_1$, $a_2 b_2 \in E_2$, and $H_1 = (\{a_1, b_1\}, \{a_1 b_1\})$, $H_2 = (\{a_2, b_2\}, \{a_2 b_2\})$. Let G be the graph obtained by pasting G_1 and G_2 along isomorphic subgraphs H_1 and H_2. In this case we say that the graph G is obtained from graphs G_1 and G_2 by *edge-pasting*. Figures 4.16, 4.17, and 4.20 illustrate this construction.

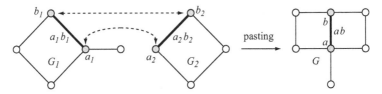

Figure 4.20. An example of edge-pasting.

Example 4.21. Let $\mathcal{H}(X_1)$ and $\mathcal{H}(X_2)$ be two cubes with $X_1 \cap X_2 = \varnothing$, and let x_1 and x_2 be fixed elements of X_1 and X_2, respectively. Consider the vertices $A_1 = A_2 = \varnothing$, $B_1 = \{x_1\}$, and $B_2 = \{x_2\}$ (cf. Figure 4.21 where the two cubes are depicted separately). Let us edge-paste the two cubes by identifying vertices A_1, B_1 with vertices A_2, B_2, respectively. The vertices of the resulting graph G are subsets of the set X obtained from the sets X_1 and X_2 by identifying elements x_1 and x_2. Let us replace elements $x_1 \in X_1$ and $x_2 \in X_2$ by the element $z = \{x_1, x_2\}$:

$$X_1' = (X_1 \setminus \{x_1\}) \cup \{z\}, \quad X_2' = (X_2 \setminus \{x_2\}) \cup \{z\}$$

Clearly, $\mathcal{H}(X_1') \cong \mathcal{H}(X_1)$, $\mathcal{H}(X_2') \cong \mathcal{H}(X_2)$, and $X = X_1' \cup X_2'$. The graph G is the union of cubes $\mathcal{H}(X_1')$ and $\mathcal{H}(X_2')$ and the intersection of these two cubes is the edge with ends $A = \varnothing$ and $B = \{z\}$ (cf. Figure 4.21).

As before, we identify the graphs G_1 and G_2 with subgraphs of the graph G and denote by $a = \{a_1, a_2\}$ and $b = \{b_1, b_2\}$ the two vertices obtained by pasting together vertices a_1, a_2 and b_1, b_2, respectively. The edge $ab \in E$ is obtained by pasting together edges $a_1 b_1 \in E_1$ and $a_2 b_2 \in E_2$ (see Figures 4.20 and 4.21). Then $G = G_1 \cup G_2$, $V_1 \cap V_2 = \{a, b\}$, and $E_1 \cap E_2 = \{ab\}$.

Theorem 4.22. *A graph G obtained by edge-pasting together bipartite graphs G_1 and G_2 is bipartite.*

Proof. Let C be a cycle in G. If $C \subseteq G_1$ or $C \subseteq G_2$, then the length of C is even, because the graphs G_1 and G_2 are bipartite. Otherwise, the vertices a and b separate C into two paths P_1 and P_2 each of odd length. Indeed, adding the edge ab to the path P_i yields a cycle in G_i ($i \in \{1, 2\}$). Each of these two cycles is of even length. Thus paths P_1 and P_2 are of odd length. It follows that C is a cycle of even length. Hence, G is bipartite. $\square$

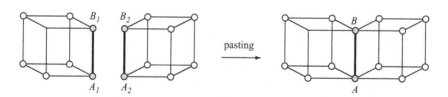

Figure 4.21. Edge-pasting two cubes.

In order to show that edge-pasting of cubical graphs produces a cubical graph, we first establish a property of the automorphism group of a cube (cf. Exercise 4.2). The automorphism group of a graph is said to be *edge-transitive* if, for any two edges of the graph, there is an automorphism that maps one edge onto another. By Theorem 3.32, we have the following results.

Theorem 4.23. *The automorphism group of a cube $\mathcal{H}(X)$ is edge-transitive. Moreover, for any two edges PQ and RS of $\mathcal{H}(X)$, there is an automorphism α of $\mathcal{H}(X)$ such that $\alpha(P) = R$ and $\alpha(Q) = S$.*

Corollary 4.24. *Let G be a graph embeddable into a cube $\mathcal{H}(X)$. For a given edge $uv \in E(G)$ and element $z \in X$, there is an embedding $\psi : G \to \mathcal{H}(X)$ such that $\psi(u) = \varnothing$ and $\psi(v) = \{z\}$.*

Proof. Let φ be an embedding of a graph G into a cube $\mathcal{H}(X)$. By Theorem 4.23, there is an automorphism α of $\mathcal{H}(X)$ such as $\alpha(\varphi(u)) = \varnothing$ and $\alpha(\varphi(v)) = \{z\}$. It follows that $\psi = \alpha \circ \varphi$ is an embedding of the graph G into the cube $\mathcal{H}(X)$ that maps vertices u, v into vertices $\varnothing, \{z\}$, respectively. $\square$

Theorem 4.25. *The graph G obtained by edge-pasting together two cubical graphs G_1 and G_2 along edges $a_1b_1 \in E(G_1)$ and $a_2b_2 \in E(G_2)$ is a cubical graph.*

Proof. By applying the construction from Example 4.21, we may assume that the graphs G_1 and G_2 are embeddable into cubes $\mathcal{H}(X_1)$ and $\mathcal{H}(X_2)$, respectively, and that $X_1 \cap X_2$ is a singleton, say $\{z\}$. By Corollary 4.24, we may also assume that there are embeddings $\psi_1 : G_1 \to \mathcal{H}(X_1)$ and $\psi_2 : G_2 \to \mathcal{H}(X_2)$ such that

$$\psi_1(a_1) = \psi_1(a_2) = \varnothing \text{ and } \psi_2(b_1) = \psi_2(b_2) = \{z\}.$$

Then the mapping ψ given by

$$\psi(u) = \begin{cases} \psi_1(u), & \text{if } u \in V(G_1), \\ \psi_2(u), & \text{if } u \in V(G_2) \end{cases}$$

is an embedding of G into the cube $\mathcal{H}(X)$. $\square$

4.6 Separation and Connectivity

In Chapter 2 we introduced the terms "cut-vertex" and "cut-edge". These are vertices and edges of a graph whose deletion increases the number of components of the graph. They are also known as *articulation points* and *bridges*, respectively.

Removing a cut-vertex from a connected graph G "separates" G into a number of components. For instance, deleting a vertex of degree $k > 1$ of a tree T results in k disjoint trees (cf. Exercise 4.26). A connected graph is said to be *nonseparable* if it does not have cut-vertices; otherwise it is called *separable*. A *block* of a graph G is a maximal nonseparable subgraph of G. For example, a nonseparable graph has just one block, namely the graph itself, and the blocks of a nontrivial tree are the copies of K_2 induced by its edges.

Example 4.26. Let G be the graph shown in Figure 4.22a. Vertices u, v, and w are the cut-vertices of G, and edges e, f, g, and h are its cut-edges. The blocks of G are depicted in Figure 4.22b.

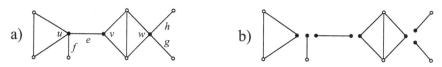

Figure 4.22. Cut-vertices, cut-edges, and blocks of a graph.

A *decomposition* of a graph G is a family $\{G_i\}_{i \in I}$ of subgraphs of G such that

$$E(G) = \cup_{i \in I} E(G_i), \quad E(G_i) \cap E(G_j) = \varnothing, \quad \text{for all } i \neq j.$$

Theorem 4.27. *Let G be a graph. Then:*

(i) *Any two blocks of G have at most one vertex in common.*
(ii) *The blocks of G form a decomposition of G.*
(iii) *Each cycle of G is contained in a block of G.*

Proof. (i) We prove the claim by contradiction. Suppose that there are blocks B_1 and B_2 with at least two common vertices. By the maximality property of blocks, the subgraph $B = B_1 \cup B_2$ properly contains both B_1 and B_2. For any vertex v of G, the graphs $B_1 - v$ and $B_2 - v$ are connected. Because the union of two connected graphs with a nonempty intersection is connected, the graph $B - v = (B_1 - v) \cup (B_2 - v)$ is connected. Hence, B has no cut vertices. This contradicts the maximality of B_1 and B_2.

(ii) Each edge induces a nonseparable subgraph K_2 of G and therefore belongs to a block. By (i), no edge lies in two blocks. Hence, blocks form a decomposition of G.

(iii) A cycle of G is a nonseparable subgraph and therefore is contained in a block of G. $\qquad \square$

Let $\mathcal{S}$ be the set of cut-vertices of a graph G, and $\mathcal{B}$ be the set of blocks of G. We form a bipartite graph $B(G)$ with bipartition $(\mathcal{B}, \mathcal{S})$ by joining a block B and a cut-vertex v if and only if $v \in B$. The graph $B(G)$ where G is the graph from Example 4.26 is shown in Figure 4.23.

Figure 4.23. The block tree of the graph G in Figure 4.22a.

Each path in G connecting vertices in distinct blocks gives rise to a unique path in $B(G)$ connecting the blocks. It follows that if G is connected, so is $B(G)$. A cycle in $B(G)$ would correspond to a cycle in G intersecting at least two blocks, contradicting Theorem 4.27(iii). Therefore, $B(G)$ is a tree, called the *block tree* of G. If G is a block, then $B(G)$ is the trivial graph K_1. Otherwise, the blocks of G corresponding to leaves of its block tree are referred to as its *end blocks*.

As any tree can be obtained from the trivial graph by successive vertex-pasting of copies of K_2 (Example 4.20), any connected graph G can be constructed by successive vertex-pasting of its blocks using the block tree structure. Indeed, let G_1 be an end block of G with a cut vertex v and G_2 be the union of the remaining blocks of G. Then G can be obtained from G_1 and G_2 by vertex-pasting along the vertex v. An obvious inductive argument proves the claim.

By using the block tree structure and the result of Theorem 4.18, one can easily prove the following result (cf. Exercise 4.28).

Theorem 4.28. *A graph is cubical if and only if its blocks are cubical.*

It is clear that blocks are "more strongly connected" than separable graphs, for disconnecting a block requires deleting at least two vertices from it. We say that a subset S of the set of vertices $V(G)$ of a graph G is a *vertex cut* if $G - S$ is disconnected. If $|S| = k$, then S is called a *k-vertex cut*. If G has at least one pair of nonadjacent vertices, the *connectivity* $\kappa(G)$ of G is the minimum integer k for which G has a k-vertex cut. Otherwise, by definition, $\kappa(G) = |V| - 1$. A graph G is said to be *k-connected* if $\kappa(G) \geq k$.

Some remarks are in order. A complete graph has no pair of nonadjacent vertices. Thus, $\kappa(K_n) = n - 1$. In particular, the connectivity of the trivial graph K_1 is zero. The empty set is a vertex cut of a disconnected graph. Therefore, the connectivity of a disconnected graph is zero. If a connected graph G has a cut-vertex, then $\kappa(G) = 1$. If G is a block with at least three vertices, then it is 2-connected; that is, $\kappa(G) \geq 2$. Thus a block is either K_1, or K_2, or a 2-connected graph on three or more vertices. The graph G from Example 4.26 has four blocks isomorphic to K_2 (bridges) and two blocks of connectivity two.

Example 4.29. Inasmuch as the star $K_{1,n}$ is a tree, we have $\kappa(K_{1,n}) = 1$. Let $K_{m,n}$ be the bipartite graph with $m, n > 1$ and bipartition (X, Y). Every induced subgraph of $K_{m,n}$ that has vertices in X and Y is connected. It follows that every vertex cut of $K_{m,n}$ must contain X or Y. Therefore, $\kappa(K_{m,n}) = \min\{m, n\}$.

The result of the following theorem is an important characterization of 2-connected graphs. Let us recall that two uv-paths P and Q in a graph are said to be *internally disjoint* if $P \cap Q = \{u, v\}$ (cf. Exercise 3.33).

Theorem 4.30. *A graph G is 2-connected if and only if every two distinct vertices u, v of G are connected by at least two internally disjoint uv-paths.*

Proof. (Necessity.) Let G be a 2-connected graph. The proof is by induction on $d(u, v)$.

Suppose that $d(u, v) = 1$; that is, uv is an edge. Because $|V(G)| \geq 3$ and G has no cut-vertices, the edge uv is not a cut-edge (cf. Exercise 4.29). By Theorem 2.34, the edge uv belongs to a cycle. It follows that u and v are connected by two internally disjoint paths.

For the induction step, let $d(u, v) = k > 1$ and assume that G has internally disjoint xy-paths whenever $d(x, y) = k - 1$. Let w be the vertex that precedes v on a uv-path of length k, so $d(u, w) = k - 1$. By the induction hypothesis, there are two internally-disjoint uw-paths P and Q. Also, because G is 2-connected, $G - w$ is connected and so contains a uv-path P'. Let x be the last vertex of P' that is also in $P \cup Q$ (cf. Figure 4.24). Because $u \in P \cup Q$, there is such a vertex. Without loss of generality, we may assume that $x \in P$. Then G has two internally disjoint paths, one composed of the ux-segment of P together with the xv-segment of P', and the other composed of Q together with the path wv.

(Sufficiency.) If any two vertices of G are connected by at least two internally disjoint paths then G is connected and has no cut-vertex. Hence, G is 2-connected. □

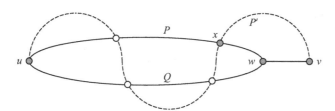

Figure 4.24. Proof of Theorem 4.30.

It is clear that the claim of Theorem 4.30 can be reformulated as follows.

Theorem 4.31. *A graph G is 2-connected if and only if every two distinct vertices u, v of G belong to a common cycle.*

We establish now two more characterizations of 2-connected graphs. Several other characterizations are the subject of Exercise 4.30.

It is convenient to introduce the operation of subdivision of an edge. We say that an edge $e = uv$ is *subdivided* when it is deleted and replaced by a path uwv, where w is a new vertex (cf. Figure 4.25).

Figure 4.25. Subdividing edge uv.

Lemma 4.32. *If G is a 2-connected graph, then the graph G' obtained by subdividing an edge of G is 2-connected.*

Proof. Let uwv be a subdivision of the edge uv of G. It is clear that no vertex of G is a cut-vertex of G'. Because G is 2-connected, there is a uv-path in G different from uv. Thus, w is not a cut-vertex of G'. It follows that G' is 2-connected. □

Theorem 4.33. *A graph G is 2-connected if and only if every two edges of G belong to a common cycle.*

Proof. (Necessity.) Let uv and xy be edges of a 2-connected graph G. Let us subdivide edges uv and xy by new vertices w and z. By Lemma 4.32, the resulting graph G' is 2-connected. By Theorem 4.31, vertices w and z lie on a common cycle C of G'. Inasmuch as w, z each have degree 2, this cycle must contain paths uwv and xzy. We replace these paths in C by edges uv and xy, respectively, to obtain the desired cycle in G.

(Sufficiency.) Any two distinct vertices u, v of G are ends of distinct edges. By assumption, they belong to a common cycle. By Theorem 4.31, the graph G is 2-connected. □

The next characterization of 2-connected graphs is based on a recursive procedure for generating any such graph from a cycle.

Let H be a subgraph of a graph G. An *ear* of H in G is a nontrivial path in G with ends in H and all other vertices in $V(G) \setminus V(H)$.

Lemma 4.34. *Let H be a nontrivial proper subgraph of a 2-connected graph G. Then H has an ear in G.*

Proof. Suppose first that $V(G) \setminus V(H) \neq \varnothing$ and let u and v be vertices lying in $V(H)$ and $V(G) \setminus V(H)$, respectively. G is 2-connected, thus there are internally disjoint vu-paths P and Q in G. Let x and y be the first vertices in P and Q, respectively, that belong to H. The union of vx- and vy-segments of P and Q is the desired ear of H (cf. Figure 4.26).

If $V(G) \setminus V(H) = \varnothing$, then H is a proper spanning subgraph of G. Then any edge in $E(G) \setminus E(H)$ is an ear of H in G. □

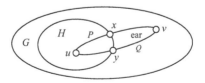

Figure 4.26. Proof of Lemma 4.34.

Lemma 4.35. *Let H be a 2-connected proper subgraph of a graph G, and let Q be an ear of H. Then $H \cup Q$ is 2-connected.*

Proof. It is clear that $H \cup Q$ is connected and does not have cut-vertices. The result follows. □

An *ear decomposition* of a finite 2-connected graph G is a sequence

$$(G_0, G_1, \ldots, G_k)$$

of 2-connected subgraphs of G such that:

(i) G_0 is a cycle.
(ii) $G_{i+1} = G_i \cup Q_i$, where Q_i is an ear of G_i in G, $0 \le i < k$.
(iii) $G_k = G$.

A ear decomposition of the cube Q_3 is shown in Figure 4.27. The initial cycle and the ear added at each step are indicated by heavy lines. Of course, there are other possible ear decompositions of the graph Q_3. For instance, one can start with a Hamilton cycle of this cube (cf. Figure 4.6).

Theorem 4.36. *A finite 2-connected graph has an ear decomposition. Furthermore, every cycle in a 2-connected graph is the initial cycle in some ear decomposition.*

Proof. Let G be a finite 2-connected graph. By Theorem 4.31, G contains a cycle C. Let $G_0 = C$ and suppose that a 2-connected subgraph G_i has been constructed by adding ears. If $G_i = G$, we are done. Otherwise, by Lemma 4.34, G_i has an ear Q in G. By Lemma 4.35, $G_{i+1} = G_i \cup Q$ is a 2-connected subgraph of G. The process ends when all of G has been absorbed. □

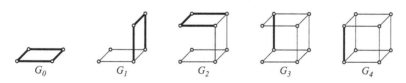

G_0 ⠀⠀⠀ G_1 ⠀⠀⠀ G_2 ⠀⠀⠀ G_3 ⠀⠀⠀ G_4

Figure 4.27. Building up the cube Q_3 by adding ears.

The result of Theorem 4.36 can be used to establish many properties of finite 2-connected graphs by induction or recursion. We apply it below to finding the maximum value of cubical dimension of finite 2-connected cubical graphs.

Let G be a finite 2-connected cubical graph and $(G_0, G_1, \ldots, G_k)$ be an ear decomposition of G with the initial cycle G_0 and ears $(Q_0, \ldots, Q_{k-1})$. The edge sets of ears and the initial cycle define a partition of the edge set of the graph G:

$$E(G) = E(G_0) \cup (\cup_{i=0}^{k-1} E(Q_i)).$$

By Theorem 4.8, there is a c-valuation f on G. For a given f, we use the above partition of $E(G)$ to obtain an upper bound for the cubical dimension of G by estimating the number of colours used by f in each element of the partition.

Let $m_0 = |V(G_0)|$ and $m_{i+1} = |V(Q_i)| - 2$ for $0 \leq i < k$. Clearly,

$$\sum_{i=0}^{k} m_i = |V(G)|.$$

G_0 is a cycle, therefore the number of colours in the colouring of G_0 defined by f is bounded by

$$\frac{1}{2}|E(G_0)| = \frac{1}{2}|V(G_0)| = \frac{m_0}{2}$$

(condition (ii) of Definition 4.5).

For every $0 \leq i < k - 1$, we estimate the number of new colours added to $f(E(G_i))$ when passing from G_i to $G_{i+1} = G_i \cup Q_i$. Let x and y be the ends of the path Q_i and let P be a path in G_i with ends x, y. The union $Q_i' = Q_i \cup P$ is a cycle in G. By condition (i) of Definition 4.5, there are colours appearing an odd number of times in Q_i. By condition (ii), these colours appear even number of times in Q_i', so they must appear in $P \subseteq G_i$. Therefore the number of new colours in Q_i is bounded by

$$\frac{1}{2}(|E(Q_i)| - 1) = \frac{1}{2}(|V(Q_i)| - 2) = \frac{m_{i+1}}{2}.$$

It follows that the number of colours in $f(E(G))$ is bounded by

$$\frac{1}{2}\sum_{i=0}^{k} m_i = \frac{1}{2}|V(G)|.$$

By Theorem 4.10,

$$\dim_c(G) \leq \frac{1}{2}|V(G)|.$$

The same inequality clearly holds for graphs K_1 and K_2. We established the following result (recall that the floor $\lfloor x \rfloor$ of x is the largest integer not greater than x).

Theorem 4.37. *Let G be a block and f be a c-valuation on G. Then*

$$|f(E(G))| \leq \left\lfloor \frac{1}{2}|V(G)| \right\rfloor.$$

Accordingly,

$$\dim_c(G) \leq \left\lfloor \frac{1}{2}|V(G)| \right\rfloor.$$

Corollary 4.38. *Let G be a connected cubical graph with k blocks, and let f be a c-valuation on G. Then*

$$\dim_c(G) \le |f(E(G))| \le \left\lfloor \frac{|V(G)| + k - 1}{2} \right\rfloor. \tag{4.3}$$

Proof. The proof is by induction on k. The case $k = 1$ is the result of Theorem 4.37.

Suppose that the claim holds for any graph with $k - 1$ blocks and let G be a graph with k blocks. Let B be an end block of G and H be the union of the remaining blocks of G. Then $H \cup B = G$ and $H \cap B$ is the trivial graph on a cut-vertex of G. By the induction hypothesis and Theorem 4.37, we have

$$|f(E(H))| \le \frac{|V(H)| + k - 2}{2} \qquad \text{and} \qquad |f(E(B))| \le \frac{|V(B)|}{2},$$

respectively. Therefore,

$$|f(E(G))| \le |f(E(H))| \le |f(E(B))|$$
$$\le \frac{|V(H)| + |V(B)| + k - 2}{2} = \frac{|V(G)| + k - 1}{2},$$

and the result follows by induction. $\qquad\qquad\qquad\qquad\qquad\qquad\qquad\square$

The cubical dimension of a connected cubical graph G on n vertices attains its maximum value $n - 1$ when G is a star $K_{1,n-1}$ (cf. Exercises 4.10 and 4.11). Indeed, by (4.3), for any cubical graph G on n vertices,

$$\dim_c(G) \le \frac{n + k - 1}{2} \le n - 1,$$

inasmuch as the number of blocks k is less than n (cf. Exercise 4.31).

4.7 Critical Graphs

If a graph G contains a noncubical subgraph, then of course G itself is not cubical. For instance, graphs containing odd cycles or the graph $K_{2,3}$ are not cubical. Odd cycles and $K_{2,3}$ are examples of critical graphs.

Definition 4.39. A graph G is said to be *critical* if

(i) G is not cubical.
(ii) Every proper subgraph of G is cubical.

In other words, critical subgraphs of a graph are minimal subgraphs that prevent the graph from being cubical. Clearly, no critical graph can be a proper subgraph of any other critical graph.

The importance of critical graphs is evidenced by the following result (cf. Exercise 4.39).

Theorem 4.40. *A finite graph is not cubical if and only if it contains a critical subgraph.*

Let G be a critical graph. By Theorem 4.28, at least one of the blocks of G is not cubical. It follows that G itself is a block. If the block G is not bipartite, then it must contain an odd cycle C_{2n-1}. Because G is critical it is itself the cycle C_{2n-1}. We summarize these observations as the following theorem.

Theorem 4.41. *A critical graph is a block. A critical nonbipartite graph is an odd cycle.*

The smallest bipartite block is the graph $K_{2,3}$ (cf. Exercise 4.4). The five bipartite blocks on six vertices are graphs B19, B23, and B25–B27 shown in Figure 4.5. Among these graphs only graphs B25–B27 are not cubical. These graphs are not critical because each of them contains $K_{2,3}$. Therefore there are no critical graphs on six vertices. By examining bipartite blocks listed in Reed and Wilson (1998), one can verify that there are no bipartite critical graphs on seven vertices either. In fact, there is only one critical graph on eight vertices.

The observations made in the foregoing paragraph could lead to a wrong conclusion that critical graphs are rare objects in the family of bipartite graphs. The main purpose of this section is to show that, in fact, there are exponentially many bipartite critical graphs on n vertices. Unfortunately, there is no known procedure for generating all critical graphs. In what follows, we present a particular construction yielding an exponential family of these graphs.

The graph $K_{2,3}$ and the unique critical graph G_8 on 8 vertices (cf. Exercise 4.40) are depicted in Figure 4.28.

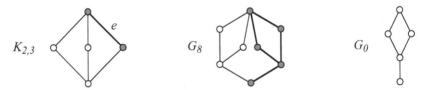

Figure 4.28. Two critical graphs, $K_{2,3}$ and G_8.

The graph G_8 can be obtained from $K_{2,3}$ by replacing the edge e with the graph G_0 shown in the same figure. This observation is instrumental as the following construction suggests.

Let G be a graph containing a 4-cycle $acbx$ with $\deg(c) = \deg(x) = 2$ as schematically shown in Figure 4.29. Let us replace the edge ax of G with the graph G_0 obtaining a new graph $T(G)$ in Figure 4.29.

Theorem 4.42. *Let T be the transformation shown in Figure 4.29. Then G is critical if and only if $T(G)$ is critical.*

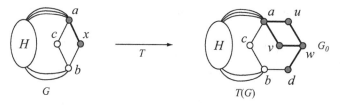

Figure 4.29. Transformation T.

Proof. We prove only necessity leaving the proof of sufficiency to the reader (cf. Exercise 4.41).

Suppose that G is critical and assume that $T(G)$ is cubical. Then, by Theorem 4.8, there is a c-valuation f on $T(G)$. Assuming that colours of av, au, and ac are 1, 2, and 3, respectively, we must have $f(uw) = 1$, $f(vw) = 2$, and $f(wd) = 3$. Then either $f(cb) = 1$, $f(bd) = 2$ or $f(cb) = 2$, $f(bd) = 1$. These two edge-colourings of the subgraph G_1 of $T(G)$ induced by the set of vertices $\{a, b, c, d, u, v, w\}$ are shown in Figure 4.30.

Note that G_1 is a subgraph of a 3-dimensional face Q_3 of the cube Q_n in which $T(G)$ is embeddable. The embedding of $T(G)$ into Q_n induces an embedding of the graph $G - x$ into the same cube. Let us map x into v if v is adjacent to b in Q_n. Otherwise, we map x into u (cf. Figure 4.30). Clearly, we obtained an embedding of G into Q_n. This contradicts our assumption that G is a critical graph. It follows that $T(G)$ is not a cubical graph.

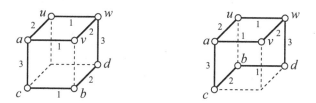

Figure 4.30. The two possible edge-colourings of G_1 defined by the c-valuation f.

We now show that $T(G)$ is a critical graph; that is, removing an edge e from $T(G)$ results in a cubical graph.

Suppose first that e is an edge of G. Then $G - e$ is a cubical graph and therefore is a subgraph of the cube $\mathcal{H}(X)$ where $X = \{z_1, \ldots, z_n\}$. We may assume that

$$a = \emptyset, \quad c = \{z_1\}, \quad x = \{z_2\}, \quad b = \{z_1, z_2\}.$$

For $z \notin X$, we define

$$v = x = \{z_2\}, \quad u = \{z\}, \quad w = \{z_1, z\}, \quad d = \{z_1, z_2, z\}.$$

It is clear that this assignment defines an embedding of $T(G) - e$ into the cube $\mathcal{H}(X \cup \{z\})$.

Suppose now that $e \notin E(G)$. Because G is a critical graph, the graph $G - x$ is a cubical subgraph of $T(G)$. By applying the construction from the previous paragraph, we obtain an embedding of $T(G) - e$ into a cube. □

Note that in our definition of transformation T, the orientation of the replacement G_0 was rather specific: vertex x of the edge ax was replaced by the lowest vertex of G_0 (cf. Figure 4.28) resulting in vertex d of $T(G)$. The result of Theorem 4.42 does not necessarily hold if an opposite orientation of G_0 is used. For instance, the right graph in Figure 4.31 is clearly cubical.

Figure 4.31. Obtaining a cubical graph from $K_{2,3}$.

By applying transformation T repeatedly, one can build up infinite families of critical graphs from a given critical graph. An example of such a family obtained from the graph G_8 is shown in Figure 4.32.

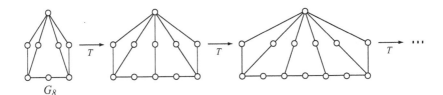

Figure 4.32. A sequence of critical graphs.

Let G be a critical graph with a unique 4-cycle u, v, w, x such that

$$\deg(u) > 3, \quad \deg(v) = \deg(x) = 2, \quad \deg(w) = 3 \tag{4.4}$$

For any such graph, the transformation T can be applied in two ways as shown in Figure 4.33, resulting in two critical graphs G_1 and G_2.

The graphs G_1 and G_2 are not isomorphic, because they differ in the number of vertices of degree three. Furthermore, G_1 and G_2 each have a unique 4-cycle and vertices of this cycle can be labeled in such a way that conditions (4.4) are satisfied.

Let $\mathcal{G}$ be a family of nonisomorphic critical graphs on n vertices, each having a unique 4-cycle satisfying (4.4). By applying T to each graph in $\mathcal{G}$, we obtain a new family $T(\mathcal{G})$ of graphs on $n + 3$ vertices. Note that any two graphs in the family $T(\mathcal{G})$ with different "predecessors" are nonisomorphic because their predecessors are not isomorphic. It follows that $|T(\mathcal{G})| = 2|\mathcal{G}|$.

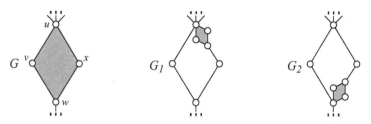

Figure 4.33. Producing two critical graphs from a single critical graph.

Let us apply this construction repeatedly starting with the critical graph on $n = 9$ vertices shown in Figure 4.34. It is easy to verify that this graph is indeed critical (cf. Exercise 4.42). At each stage, the number of new critical graphs is twice the number of graphs from the preceding stage. Hence, for each $n = 3k$ where $k \geq 3$, we obtain a family of $2^{n/3}$ critical graphs on n vertices.

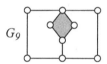

Figure 4.34. A critical graph on nine vertices.

To obtain exponentially many critical graphs on $n = 3k \pm 1$ vertices, we apply our construction to critical graphs G_{13} and G_{17} in Figure 4.35 (cf. Exercise 4.44).

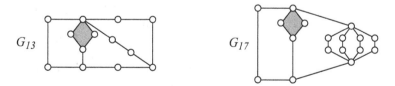

Figure 4.35. Critical graphs on 13 and 17 vertices.

Theorem 4.43. *For $n \geq 8$ there are at least $2^{(n-17)/3}$ nonisomorphic critical graphs on n vertices.*

Proof. We already established the result for $n \geq 17$. For the range $8 \leq n < 17$, it suffices to produce examples of critical graphs on n vertices. The graph G_8 in Figure 4.28 is a critical graph on 8 vertices. Critical graphs on 9, 12, and 15 vertices are obtained from the graph G_9 in Figure 4.34, and a critical graph on 16 vertices is obtained from the graph G_{13}. For an example of a critical graph on 10 vertices, see Exercise 4.46. $\square$

Notes

Apparently, the first study of cubical graphs appeared in Firsov (1965). In this short paper, V. V. Firsov established, in particular, the result of Theorem 4.28 and gave an example of a family of bipartite graphs that are not cubical. The paper was motivated by problems in mathematical linguistics.

It was conjectured by Garey and Graham (1975), and proved independently by Afrati et al. (1985) and Krumme et al. (1986), that the problem of deciding whether an arbitrary graph is cubical is $\mathcal{NP}$-complete. In other words, there is no computationally effective characterization of cubical graphs. However, the characterization of cubical graphs in terms of c-valuations (Theorem 4.8), which was established by Havel and Morávek (1972), is a valuable tool in studying cubical graphs, as evidenced by the proof of Theorem 4.11 (Havel and Liebl, 1972a) where the cubical dimension of a dichotomic tree is determined. Cubical dimensions for some other families of trees were determined by Havel and Liebl (1972b) and Nebeský (1974).

The notion of a critical graph was introduced independently in Havel and Morávek (1972), and Garey and Graham (1975). In both papers infinite families of critical graphs are constructed. The construction from Garey and Graham (1975) is presented in Section 4.7. Many more families of critical graphs are found in Havel (1983), and Gorbatov and Kazanskiy (1983).

Cubical graph theory is the subject of surveys by Harary et al. (1988) and Harary (1988), where the reader will find additional examples and references. Studies in the area are greatly motivated by problems in automata theory, interconnection networks, and parallel computing. For these applications and more recent references, see Xu (2001) and Hsu and Lin (2009).

Hamilton cycles are named after Sir William Rowan Hamilton (1805–1865), a renowned Irish astronomer, physicist, and mathematician, who made important contributions to classical mechanics, optics, and algebra. The reader will find an intriguing account of the history of the term in Biggs et al. (1986, Chapter 2). Note that the problem of deciding whether a given graph is hamiltonian is $\mathcal{NP}$-complete (Bondy and Murty, 2008, Chapter 8).

Various notions of connectivity and separation play an important role in graph theory and its applications. One example is *Steinitz' Theorem* which tells us that a graph is the graph of a 3-dimensional polytope if and only if it is planar and 3-connected (see Ziegler, 2006, Chapter 4). This theorem can be used to verify that graphs in Exercises 4.43 and 4.45 are indeed graphs of polytopes. For more results on connectivity and separation, the reader is referred to Bondy and Murty (2008) and other textbooks on graph theory.

Exercises

4.1. Show that the ray and the double ray are cubical graphs.

4.2. Show that the automorphism group of a cube is transitive.

4.3. Let H be a subgraph of a graph G, and let α be an automorphism of G. Show that the subgraph $\alpha(H)$ is isomorphic to H.

4.4. Show that all bipartite graphs on less than five vertices are cubical.

4.5. Show that the complete bipartite graph $K_{n,n}$ has $\frac{1}{2}(n-1)!n!$ Hamiltonian cycles.

4.6. Prove that G has a Hamiltonian path only if for every $S \subseteq V(G)$, the number of components of $G - S$ is at most $|S| + 1$.

4.7. Prove that every 5-vertex path in the dodecahedron graph (see Figure 3.17) extends to a Hamiltonian cycle.

4.8. Show that the Petersen graph is nonhamiltonian, but the deletion of any one vertex results in a hamiltonian graph.

4.9. Verify that the graphs labeled "cubical" in Figure 4.5 are indeed cubical graphs and that the remaining graphs are not cubical.

4.10. Show that $\dim_c(K_{1,n}) = n - 1$.

4.11. Let G be a cubical graph on n vertices. Show that

$$\lceil \log_2 n \rceil \leq \dim_c(G) \leq n - 1$$

(Garey and Graham, 1975).

4.12. What is the length of a longest induced path in Q_3? And the same question for a longest induced cycle in Q_3.

4.13. Show that the sum of two cubical graphs is a cubical graph.

4.14. Show that the colouring of the grid $P_5 \square P_3$ shown in Figure 4.13 is a c-valuation on the grid.

4.15. Prove that $\dim_c(P_5 \square P_3) = 5$.

4.16. Prove Theorem 4.12 and generalize its result to the weak Cartesian product of an infinite family of cubical graphs.

4.17. Show that the dichotomic tree DT_n has $2^{n+1} - 1$ vertices and $2^{n+1} - 2$ edges.

4.18. Show that the dichotomic tree DT_2 is not embeddable into the cube Q_3.

4.19. Let us connect the roots of two copies of the graph DT_2 by a path of length 3 (see Figure 4.36). Let DT_2^+ be the resulting graph.

a) Show that $\dim_c(DT_2^+) = \dim_c(DT_2) = 4$.
b) Show that DT_2^+ is a spanning tree of Q_4.
c) Do similar results hold for DT_3? For DT_n?

(Nebeský, 1974)

Figure 4.36. Exercise 4.19.

4.20. The cubical dimension of the grid $P_4 \square P_4$ is 4. Describe an embedding of this grid into the cube Q_4.

4.21. Show that equation (4.2) defines a proper edge-colouring of the cube $\mathcal{H}(X)$.

4.22. A positive integer n can be uniquely represented as $n = 2^m(2k + 1)$. Show that the function

$$f : x_{n-1}x_n \mapsto m + 1$$

defines a c-valuation on the ray with a vertex set $\{x_0, x_1, x_2, \ldots\}$.

4.23. Show that the two graphs in Figure 4.28 are critical.

4.24. Let uv and xy be edges of a graph G with an edge-transitive automorphism group. True or false: There exists an automorphism φ of G such that $\varphi(u) = x$ and $\varphi(v) = y$ (cf. Theorem 4.23).

4.25. Show that a graph obtained by vertex-pasting together two bipartite graphs is bipartite .

4.26. Let v be a vertex of degree k of a tree T. Prove that $T - v$ has precisely k components.

4.27. Show that:

a) A graph is even (cf. Exercise 2.22) if and only if each of its blocks is even.
b) A graph is bipartite if and only if each of its blocks is bipartite.

4.28. Prove Theorem 4.28.

4.29. Let uv be a cut-edge of a connected graph G with $|V(G)| \geq 3$. Show that at least one of the ends of uv is a cut-vertex.

4.30. Let G be a connected graph on at least three vertices. Prove that the following statements are equivalent.

a) G is a block.
b) Every two vertices of G belong to a common cycle.
c) Every vertex and edge of G belong to a common cycle.
d) Every two edges of G belong to a common cycle.
e) Given two vertices and an edge of G, there is a path joining the vertices that contains the edge.
f) For every three distinct vertices of G, there is a path joining any two of them that contains the third.
g) For every three distinct vertices of G, there is a path joining any two of them which does not contain the third

(Harary (1969), Theorem 3.3).

4.31. Let G be a connected graph with k blocks. Prove that:

a) $k \leq |V(G)| - 1$.
b) $k = |V(G)| - 1$ if and only if G is a tree.

4.32. A cubic graph has a cut-vertex if and only if it has a cut-edge.

4.33. True or false: A connected graph on at least three vertices is 2-connected if and only if given any two vertices and one edge, there is a path joining the vertices that does not contain the edge.

4.34. Prove that a connected graph with at least two edges is 2-connected if and only if any two adjacent vertices lie on a cycle.

4.35. Let G be a block on $n \geq 3$ vertices. Prove that

a) $\deg(v) > 1$ for all $v \in V(G)$.
b) G has at least n edges.
c) G has exactly n edges if and only if $G = C_n$.

4.36. Construct an ear decomposition of the Petersen graph.

4.37. Show that $d(G) \leq \lfloor |V(G)|/2 \rfloor$ for a 2-connected graph G, where $d(G)$ is the diameter of G (cf. Exercise 3.26).

4.38. For every $n \geq 4$ construct a 2-connected graph G on n vertices with $d(G) = 2$.

4.39. Prove Theorem 4.40.

4.40. Show that the graph G_8 in Figure 4.28 is critical.

4.41. Complete the proof of Theorem 4.42.

4.42. Show that the graph G_9 in Figure 4.34 is critical.

4.43. The polytope depicted in Figure 4.37 is called the *cubical octahedron*. Show that despite its name the graph of this polytope is not cubical. Find a critical subgraph of this graph.

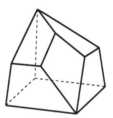

Figure 4.37. Exercise 4.43.

4.44. Show that the graphs G_{13} and G_{17} in Figure 4.35 are critical (Garey and Graham, 1975).

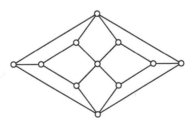

Figure 4.38. The Herschel graph.

4.45. The graph in Figure 4.38 is called the *Herschel graph*. Show that this graph is

a) bipartite
b) noncubical
c) noncritical
d) nonhamiltonian

It is, in fact, the graph of a polytope.

4.46. Show that the graph G_{10} in Figure 4.39 is critical.

Figure 4.39. Exercise 4.46.

4.47. Show that the graph in Figure 4.40 is critical.

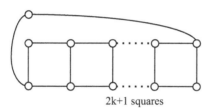

2k+1 squares

Figure 4.40. Exercise 4.47.

4.48. The graph in Figure 4.41 is called the *Heawood graph*. Show that this graph is not cubical.

Figure 4.41. The Heawood graph; Exercise 4.48.

5

Partial Cubes

This is the central chapter of the book. Unlike general cubical graphs, isometric subgraphs of cubes—partial cubes—can be effectively characterized in many different ways and possess much finer structural properties. In this chapter, we present what can be called a "micro" theory of partial cubes.

5.1 Definitions and Examples

As cubical graphs are subgraphs of cubes, partial cubes are *isometric* subgraphs of cubes (cf. Section 1.4).

Definition 5.1. A graph G is a *partial cube* if it can be isometrically embedded into a cube $\mathcal{H}(X)$ for some set X. We often identify G with its isometric image in $\mathcal{H}(X)$ and say that G is a *partial cube on the set X*.

Note that partial cubes are connected graphs. Clearly, they are cubical graphs. The converse is not true. The smallest counterexample is the graph G on seven vertices shown in Figure 5.1a.

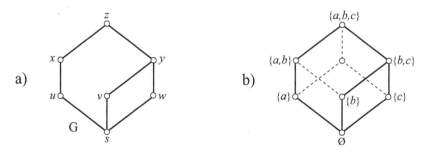

Figure 5.1. A cubical graph that is not a partial cube.

Indeed, suppose that $\varphi : G \to \mathcal{H}(X)$ is an embedding. We may assume that $\varphi(s) = \varnothing$. Then the images of vertices u, v, and w under the embedding φ are singletons in X, say, $\{a\}$, $\{b\}$, and $\{c\}$, respectively. It is easy to verify that we must have (cf. proof of Theorem 4.42)

$$\varphi(y) = \{b, c\}, \quad \varphi(z) = \{a, b, c\} \text{ and } [\varphi(x) = \{a, b\} \text{ or } \varphi(x) = \{a, c\}].$$

We assume that $\varphi(x) = \{a, b\}$ (see Figure 5.1b); the other case is treated similarly. The graph distance between vertices v and x in G is 3, whereas the distance between their images $\{b\}$ and $\{a, b\}$ in $\mathcal{H}(X)$ is 1. Therefore, φ is an embedding but not an isometric embedding.

In Section 4.1 we proved that paths, even cycles, and trees are cubical graphs (cf. Figure 4.1). In fact, they are partial cubes. It is not difficult to prove this assertion directly using constructions from Section 4.1 (cf. Exercise 5.1). It also follows from one of our characterization theorems (see Section 5.4). These theorems can be used to show that all cubical graphs in Figure 4.5 are partial cubes, proving that the graph in Figure 5.1a is indeed the smallest counterexample (cf. Exercise 5.2).

5.2 Well-Graded Families of Sets

Let G be a partial cube on X. Because G is an isometric subgraph of $\mathcal{H}(X)$, for any two vertices P and Q of G there is a sequence of vertices of G,

$$R_0 = P, R_1, \ldots, R_n = Q$$

such that $d(P, Q) = n$ and $d(R_i, R_{i+1}) = 1$ for $0 \le i < n$, where d is the Hamming distance on $\mathcal{H}(X)$. The sequence (R_i) is a shortest path connecting P to Q in G (and in $\mathcal{H}(X)$).

Definition 5.2. A family[1] $\mathcal{F}$ of distinct subsets of a set X is called a *well-graded family of sets* (*wg-family* for short) if, for any two distinct subsets $P, Q \in \mathcal{F}$, there is a sequence of sets in $\mathcal{F}$

$$R_0 = P, R_1, \ldots, R_n = Q$$

such that

$$|R_0 \triangle R_n| = n \text{ and } |R_i \triangle R_{i+1}| = 1, \quad \text{for all } 0 \le i < n. \tag{5.1}$$

According to this definition, the vertex set of a partial cube on X is a well-graded family of finite subsets of X.

Let $\mathfrak{P}(X) = K_2^X$ be the Cartesian power of K_2 (cf. Section 3.5). Any nonempty family $\mathcal{F}$ of subsets of X induces a subgraph $G_{\mathcal{F}} = (\mathcal{F}, E_{\mathcal{F}})$ of $\mathfrak{P}(X)$, where

$$E_{\mathcal{F}} = \{\{P, Q\} \in \mathcal{F} : |P \triangle Q| = 1\}.$$

---

[1] To avoid trivialities, we assume that families of sets under consideration contain more than one element.

Theorem 5.3. (i) *The graph $G_{\mathcal{F}}$ is an isometric subgraph of $\mathfrak{P}(X)$ if and only if the family $\mathcal{F}$ is well-graded.*

(ii) *If $\mathcal{F}$ is well-graded, then $G_{\mathcal{F}}$ is a partial cube.*

Proof. (i) (Necessity.) Suppose that $G_{\mathcal{F}}$ is an isometric subgraph of $\mathfrak{P}(X)$. Because the graph $G_{\mathcal{F}}$ is connected, it belongs to a connected component of the graph $\mathfrak{P}(X)$. Hence, by Lemma 3.5, $P \triangle Q$ is a finite set for any $P, Q \in \mathcal{F}$. Because $G_{\mathcal{F}}$ is an isometric subgraph, $\mathcal{F}$ is a wg-family.

(Sufficiency.) Let $\mathcal{F}$ be a wg-family and P, Q be two sets in $\mathcal{F}$. It is clear that a sequence (R_i) of vertices of $G_{\mathcal{F}}$ is a shortest PQ-path in $G_{\mathcal{F}}$ if and only if it satisfies condition (5.1). Therefore the graph $G_{\mathcal{F}}$ is an isometric subgraph of $\mathfrak{P}(X)$.

(ii) Let $\mathcal{F}$ be a wg-family of subsets of X. Then, by part (i), $G_{\mathcal{F}}$ is an isometric subgraph of $\mathfrak{P}(X)$. Let A be a vertex of $G_{\mathcal{F}}$. Because $G_{\mathcal{F}}$ is connected, it is a subgraph of the connected component $H(A)$ of $\mathfrak{P}(X)$. By Theorem 3.9, the graph $H(A)$ is isomorphic to the cube $\mathcal{H}(X)$. It follows that $G_{\mathcal{F}}$ is a partial cube. $\qquad\square$

The converse of claim (ii) does not hold as the following example demonstrates.

Example 5.4. Let $G_{\mathcal{F}}$ be a subgraph of the cube $Q_3 = \mathcal{H}(\{a, b, c\})$ induced by the family

$$\mathcal{F} = \{\varnothing, \{a\}, \{a, b\}, \{a, b, c\}, \{b, c\}\}$$

(see Figure 5.2). Clearly, $G_{\mathcal{F}}$ is not an isometric subgraph of Q_3. On the other hand, $G_{\mathcal{F}}$ is the path P_5 and therefore is a partial cube (Exercise 5.1). Thus, $G_{\mathcal{F}}$ is a partial cube, but not a partial cube on $X = \{a, b, c\}$.

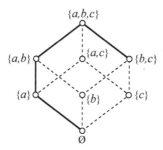

Figure 5.2. An induced subgraph of Q_3.

Example 5.5. Let $\mathcal{F}$ be the family of all subsets of $\mathbf{Z}$ in the form $(-\infty, n]$. It is not difficult to see that $\mathcal{F}$ is a wg-family of infinite subsets of $\mathbf{Z}$. The graph $G_{\mathcal{F}}$ is isomorphic to the double ray $\mathcal{Z}$. Therefore, by Theorem 5.3(ii), $\mathcal{Z}$ is an infinite partial cube (cf. Exercise 5.3).

Corollary 5.6. *The graph $G_{\mathcal{F}}$ is a partial cube on X if and only if $\mathcal{F}$ is a wg-family of finite subsets of X.*

Sequences of vertices (R_i) satisfying condition (5.1) are shortest paths in the graph $\mathfrak{P}(X)$. If $\mathcal{F}$ is a wg-family, then they are also shortest paths in the graph $G_{\mathcal{F}}$. The next two theorems establish some useful properties of these sequences.

Theorem 5.7. *Let $\mathcal{F}$ be a wg-family of subsets of a set X and let*

$$P = R_0, R_1, \ldots, R_n = Q$$

be a shortest path in $G_{\mathcal{F}}$. Then

(i) $d(R_i, R_j) = |j - i|$.
(ii) $R_i, \ldots, R_j$ *is a shortest $R_i R_j$-path in $G_{\mathcal{F}}$.*
(iii) $d(R_i, R_j) = d(R_i, R_k) + d(R_k, R_j)$ *for $i \le k \le j$.*
(iv) $R_i \cap R_j \subseteq R_k \subseteq R_i \cup R_j$ *for $i \le k \le j$,*

where d is the Hamming distance on $\mathfrak{P}(X)$.

Proof. (i) and (ii) are the results of Lemma 2.5. It remains to note that (iii) follows immediately from (i), and that (iii) and (iv) are equivalent conditions by Theorem 3.22. □

Theorem 5.8. *Let $\mathcal{F}$ be a wg-family of subsets of a set X and let*

$$P = R_0, R_1, \ldots, R_n = Q$$

be a shortest path in $G_{\mathcal{F}}$. Then

(i) *Either $R_{i+1} = R_i \cup \{x\}$ for some $x \in Q \setminus P$, or $R_{i+1} = R_i \setminus \{x\}$ for some $x \in P \setminus Q$. Equivalently, for any $0 \le i < n$ there is $x \in P \bigtriangleup Q$ such that $R_i \bigtriangleup R_{i+1} = \{x\}$.*
(ii) *For any $x \in P \bigtriangleup Q$ there is $0 \le i < n$ such that $R_i \bigtriangleup R_{i+1} = \{x\}$.*

Proof. (i) Inasmuch as $|R_i \bigtriangleup R_{i+1}| = d(R_i, R_{i+1}) = 1$, we have $R_i \bigtriangleup R_{i+1} = \{x\}$ for some $x \in X$. Hence, either $R_{i+1} = R_i \cup \{x\}$ and $x \notin R_i$, or $R_{i+1} = R_i \setminus \{x\}$ and $x \in R_i$.

Suppose that $R_{i+1} = R_i \cup \{x\}$, $x \notin R_i$. By Theorem 5.7(ii), R_{i+1} lies between R_i and $R_n = Q$. Therefore, by Theorem 3.22, $R_{i+1} \subseteq R_i \cup Q$. Hence, $x \in Q$.

Suppose now that $R_{i+1} = R_i \setminus \{x\}$, $x \in R_i$. Then, by Theorem 5.7(ii), R_i lies between R_{i+1} and $R_0 = P$. Therefore, by Theorem 3.22, $R_i \subseteq R_{i+1} \cup P$. It follows that $x \in P$.

(ii) Let $x \in P \bigtriangleup Q$. By symmetry, we may assume that $x \in Q \setminus P$. We prove claim (ii) by induction on $n = d(P, Q)$. For $n = 1$ the statement is trivial. For the induction step, suppose that $n > 1$ and assume that (ii) holds for paths of smaller length. If $x \in Q \setminus R_{n-1}$, we are done. Otherwise, $x \in R_{n-1}$ and the result follows from the induction hypothesis. □

We say that two distinct sets P and Q in a family of sets $\mathcal{F}$ are *adjacent in $\mathcal{F}$* if

$$P \cap Q \subseteq R \subseteq P \cup Q \text{ and } R \in \mathcal{F} \text{ implies } R = P \text{ or } R = Q.$$

In other words, P and Q are adjacent in $\mathcal{F}$ if they are the only sets in $\mathcal{F}$ that lie between P and Q.

One should distinguish two concepts of adjacency for a given family of sets $\mathcal{F}$: adjacency in $G_{\mathcal{F}}$ and adjacency in $\mathcal{F}$. Clearly, two vertices P and Q that are adjacent in $G_{\mathcal{F}}$ are also adjacent in $\mathcal{F}$. The converse is not true. For instance, the subsets $\varnothing$ and $\{b, c\}$ of the family $\mathcal{F}$ in Example 5.4 are adjacent in $\mathcal{F}$ but not in $G_{\mathcal{F}}$.

Theorem 5.9. *A family $\mathcal{F}$ of subsets of a set X is well-graded if and only if $d(P, Q) = 1$ for any two sets P, Q that are adjacent in $\mathcal{F}$.*

Proof. (Necessity.) Suppose that $\mathcal{F}$ is a wg-family and let P and Q be two sets that are adjacent in $\mathcal{F}$. Then there is a sequence (R_i) of sets in $\mathcal{F}$ satisfying condition (5.1). By Theorem 5.7(iv), $P \cap Q \subseteq R_i \subseteq P \cup Q$ for all $0 \leq i \leq n$. Because P and Q are adjacent in $\mathcal{F}$, we have $R_i \in \{P, Q\}$. Hence, $d(P, Q) = 1$.

(Sufficiency.) Let P, Q be two distinct sets in $\mathcal{F}$. We prove by induction on $n = d(P, Q)$ that there is a sequence (R_i) of sets in $\mathcal{F}$ satisfying condition (5.1).

The statement is trivial for $n = 1$. Suppose that $n > 1$ and that the statement is true for all $k < n$. Because $d(P, Q) > 1$, the sets P and Q are not adjacent in $\mathcal{F}$. Hence there is $R \in \mathcal{F}$ distinct from P and Q such that

$$P \cap Q \subseteq R \subseteq P \cup Q.$$

Then $d(P, R) + d(R, Q) = d(P, Q)$ and both distances $d(P, R)$ and $d(R, Q)$ are less than $n = d(P, Q)$. By the induction hypothesis, there is a sequence $(R_i) \in \mathcal{F}$ such that $P = R_0$, $R = R_j$, and $Q = R_n$, for some $0 < j < n$, satisfying condition (5.1). It follows that $\mathcal{F}$ is a wg-family. □

The last theorem in this section is an application of the criterion from Theorem 5.9.

Definition 5.10. A family of sets $\mathcal{F}$ is said to be an *independence system* (of sets) if, for any $P \in \mathcal{F}$ and any $Q \subseteq P$, the set Q also belongs to $\mathcal{F}$.

Theorem 5.11. *An independence system $\mathcal{F}$ is a wg-family. Accordingly, the graph $G_{\mathcal{F}}$ is a partial cube.*

Proof. Let P and Q be two adjacent sets in $\mathcal{F}$. Because $P \cap Q$ lies between P and Q, we have either $P \cap Q = P$ or $P \cap Q = Q$. By symmetry, we may assume that $P \subseteq Q$. Let x be an element of $Q \setminus R$. Because $\mathcal{F}$ is an independence system and the sets P and Q are adjacent in $\mathcal{F}$, we have $Q = P \cup \{x\}$; that is, $d(P, Q) = 1$. The result follows from Theorems 5.9 and 5.3(ii). □

5.3 Partial Orders

The main goal of this section is to show that the family of all partial orders on a finite set is well-graded and therefore defines a partial cube.

Let us recall (cf. Definition 3.3) that a *partial order* R on a set X is an irreflexive and transitive binary relation on X; that is,

(i) $(x, x) \notin R$, and
(ii) $(x, y) \in R$ and $(y, z) \in R$ implies $(x, z) \in R$,

for all $x, y, z \in X$. In this section we assume that X is a given nonempty finite set and denote the set of all partial orders on X by $\mathcal{PO}$.

Let Y be a nonempty subset of X. An element $y \in Y$ is said to be *maximal in Y with respect to a partial order R* if there is no element x in Y such that $(x, y) \in R$ (cf. Exercise 5.4). Likewise, an element $z \in Y$ is a *minimal element in Y with respect to R* if there is no element x in Y such that $(z, x) \in R$.

The following lemma is instrumental.

Lemma 5.12. *Suppose that a partial order P is a proper subset of a partial order Q. Then there exists $(a, b) \in Q$ such that $R = P \cup \{(a, b)\}$ is a partial order.*

Proof. Let us define

$$A = \{x \in X : (x, y) \in Q \setminus P \text{ for some } y \in X\}$$

and let a be a maximal element in A with respect to P. This means that $(x, a) \in P$ implies that $x \notin A$. Equivalently,

$$(x, a) \in P \text{ implies } (x, y) \notin Q \setminus P, \text{ for any } y \in X. \tag{5.2}$$

Furthermore, we define

$$B = \{y \in X : (a, y) \in Q \setminus P\}$$

and let b be a minimal element in B with respect to P. Then, for $z \in X$,

$$(b, z) \in P \text{ implies } (a, z) \notin Q \setminus P. \tag{5.3}$$

Clearly, $(a, b) \in Q \setminus P$ and $R = P \cup \{(a, b)\}$ is an irreflexive relation. Let us show that R is a transitive relation; that is, condition (ii) is satisfied for R. There are three possible cases:

1) $(x, y) \in P$ and $(y, z) \in P$. Then, by transitivity of P, $(x, z) \in P \subset R$.

2) $(x, y) = (a, b)$. We need to verify that $(a, b) \in R$, $(b, z) \in R$ together imply that $(a, z) \in R$. Clearly, $(b, z) \in P$. By (5.3), $(a, z) \notin Q \setminus P$. Because Q is a transitive relation, $(a, z) \in Q$. It follows that $(a, z) \in P \subset R$.

3) $(y, z) = (a, b)$. As in the previous case we need to show that $(x, b) \in R$, $(a, b) \in R$ imply $(x, b) \in R$. Obviously, $(x, a) \in P$. By (5.2), $(x, b) \notin Q \setminus P$; that is, $(x, b) \in P \subset R$, inasmuch as $(x, b) \in Q$, by transitivity of Q. $\square$

Theorem 5.13. *The family* $\mathcal{PO}$ *is well-graded.*

Proof. By Theorem 5.9, it suffices to show that $|P \triangle Q| = 1$ for any two adjacent elements P and Q of $\mathcal{PO}$. Note that $P \cap Q \in \mathcal{PO}$, for any two elements $P, Q \in \mathcal{PO}$ (cf. Exercise 5.5). Because $P \cap Q$ lies between P and Q in $\mathcal{PO}$ and P and Q are adjacent in $\mathcal{PO}$, we must have either $P \subseteq Q$ or $Q \subseteq P$. By Lemma 5.12, $P \triangle Q$ is a singleton. □

It follows that the graph $G_{\mathcal{PO}}$ is a partial cube (cf. Theorem 5.3). This graph is depicted in Figure 5.3 for $X = \{a, b, c\}$. The central vertex represents the empty partial order; $x < y$ stands for (y, x). For instance, the vertex labeled $b < a, b < c$ represents the partial order $\{(a, b), (c, b)\}$.

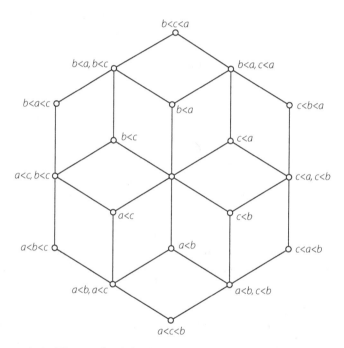

Figure 5.3. The graph of the family of partial orders on $X = \{a, b, c\}$.

A *linear order* on a set X is a complete partial order R on X; that is, it satisfies the completeness condition:

$$(x, y) \in R \text{ or } (y, x) \in R \text{ for all } x \neq y \text{ in } X.$$

Linear orders are represented by the vertices of the "big hexagon" in Figure 5.3.

The graph in Figure 5.3 is the Hasse diagram of the set of partial orders on $X = \{a, b, c\}$ ordered by the inclusion relation. Clearly, linear orders are maximal elements of this ordered set. In fact, this is true for any set X as the following argument demonstrates.

For a given partial order P on an arbitrary set X, we define the corresponding reflexive partial order P' by

$$(x, y) \in P' \text{ if and only if } (x, y) \in P \text{ or } x = y,$$

for all $x, y \in X$. It is easy to see that P' is transitive. Suppose that P is not a complete binary relation. Then there are two distinct elements $a, b \in X$ such that $(a, b) \notin P$ and $(b, a) \notin P$. Let us define a binary relation R on X by

$$(x, y) \in R \text{ if and only if } (x, y) \in P \text{ or } [(x, a) \in P' \text{ and } (b, y) \in P'],$$

for all $x, y \in X$, and show that this relation is a partial order on X properly containing P. Note that $(a, b) \in R$. It is clear that R is irreflexive. $(a, b) \in R$ and $(a, b) \notin P$, therefore we have $P \subset R$. Thus it suffices to establish the transitivity property of R.

Suppose that $(x, y) \in R$ and $(y, z) \in R$ for some $x, y, z \in X$. There are four possible cases:

(1) $(x, y) \in P$ and $(y, z) \in P$. Then $(x, z) \in P \subset R$, by transitivity of P.
(2) $(x, y) \in P$, $(y, a) \in P'$, and $(b, z) \in P'$. Then $(x, a) \in P'$, by transitivity of P'. Hence, $(x, z) \in R$.
(3) $(x, a) \in P'$, $(b, y) \in P'$, and $(y, z) \in P$. Then $(b, z) \in P'$, by transitivity of P'. Therefore, $(x, z) \in R$.
(4) $(x, a) \in P'$, $(b, y) \in P'$, and $(y, a) \in P'$, $(b, z) \in P'$. By transitivity of P', $(b, y) \in P'$, $(y, a) \in P'$ implies $(b, a) \in P'$, contradicting our assumption that a and b are distinct elements of X with $(b, a) \notin P$.

It follows that $(x, y) \in R$, $(y, z) \in R$ implies $(x, z) \in R$ for all $x, y, z \in X$; that is, R is transitive. We proved that a partial order which is not a linear order is properly contained in another partial order. Assuming that X is a finite set, we immediately obtain the following result.

Theorem 5.14. *A partial order on a finite set X is contained in a linear order. Equivalently, linear orders are the maximal elements of the poset $\mathcal{PO}$.*

Actually, the claim of the theorem also holds for infinite sets; see Notes at the end of the chapter.

5.4 Hereditary Structures

Let H be an isometric subgraph of a connected graph G. Then, by definition, we have

$$d_H(u, v) = d_G(u, v), \text{ for } u, v \in V(H).$$

Graph concepts that are defined in terms of the distance function on G are naturally inherited by the isometric subgraph H. In this section we apply this general observation to the notions of interval, semicube, the theta relation, and convexity. We use subscripts and superscripts H and G to distinguish subsets and relations on graphs H and G, respectively.

By the definition of an interval in a graph, we have

$$
\begin{aligned}
I_H(u, v) &= \{w \in V(H) : d_H(u, w) + d_H(w, v) = d_H(u, v)\} \\
&= \{w \in V(G) : d_G(u, w) + d_G(w, v) = d_G(u, v)\} \cap V(H) \\
&= I_G(u, v) \cap V(H).
\end{aligned}
$$

In words, the intervals in H are intersections of intervals in G with the vertex set $V(H)$. Likewise,

$$
\begin{aligned}
W_{ab}^H &= \{x \in V(H) : d_H(x, a) < d_H(x, b)\} \\
&= \{x \in V(G) : d_G(x, a) < d_G(x, b)\} \cap V(H) \\
&= W_{ab}^G \cap V(H),
\end{aligned}
$$

so the semicubes of H are intersections of semicubes of G with $V(H)$. Furthermore,

$$
\Theta_H = \{(xy, uv) \in E(H) \times E(H) : d_H(x, u) + d_H(y, v) \neq d_H(x, v) + d_H(y, u)\}
$$

and

$$
\Theta_G = \{(xy, uv) \in E(G) \times E(G) : d_G(x, u) + d_G(y, v) \neq d_G(x, v) + d_G(y, u)\}.
$$

It is clear that $\Theta_H = \Theta_G \cap (E(H) \times E(H))$, so the relation Θ_H is the restriction of Θ_G to the edge set $E(H)$.

We summarize these results in the following theorem.

Theorem 5.15. *Let H be an isometric subgraph of a connected graph G. Then*

(i) $I_H(u, v) = I_G(u, v) \cap V(H)$.
(ii) $W_{ab}^H = W_{ab}^G \cap V(H)$.
(iii) $\Theta_H = \Theta_G \cap (E(H) \times E(H))$.

Let S be a convex subset of a connected graph G and H be an isometric subgraph of G. By Theorem 5.15(i), for any two vertices $u, v \in S \cap V(H)$, we have

$$
I_H(u, v) = I_G(u, v) \cap V(H) \subseteq S \cap V(H).
$$

This proves the next theorem.

Theorem 5.16. *The intersection of a convex subset of the vertex set $V(G)$ of a graph G with the vertex set $V(H)$ of an isometric subgraph H of G is a convex subset of $V(H)$.*

We now apply the above results to establish two crucial properties of partial cubes.

Theorem 5.17. *Let G be a partial cube. Then*

(i) *The semicubes of G are convex subsets of the vertex set $V(G)$.*
(ii) *$\theta = \Theta$ is an equivalence relation on $E(G)$.*

Proof. By Theorem 2.22, $\theta = \Theta$, and by Theorem 2.25, conditions (i) and (ii) are equivalent. Therefore it suffices to prove (ii).

We may assume that G is an isometric subgraph of a cube $\mathcal{H}(X)$. By Theorem 3.28, $\Theta_{\mathcal{H}(X)}$ is an equivalence relation on $E(\mathcal{H}(X))$. By Theorem 5.15(iii), $\Theta = \Theta_G$ is an equivalence relation on $E(G)$. This proves (ii). $\qquad\square$

Furthermore, by combining the results of Theorems 5.15(ii) and 3.27, we obtain the following useful description of semicubes of a partial cube.

Theorem 5.18. *Let A and B be two adjacent vertices of a partial cube G on a set X with $A \triangle B = \{x\}$, $x \in X$. We may assume that $A = B \cup \{x\}$. Then*

$$W_{AB} = \{R \in V(G) : x \in R\} \ \text{and} \ W_{BA} = \{R \in V(G) : x \notin R\}.$$

5.5 Characterizations

The theta relations and fundamental sets associated with a given graph were our main tools in analyzing structural properties of bipartite graphs and characterizing them in Section 2.3. These structures play a central role in this section. Here we present the key results of this chapter, characterizations of partial cubes.

We say that a semicube W_{ab} *separates* two distinct vertices x and y of a graph G if either $x \in W_{ab}$, $y \in W_{ba}$ or $x \in W_{ba}$, $y \in W_{ab}$. In particular, both W_{ab} and W_{ba} separate a and b.

By Theorem 2.18, each pair of opposite semicubes $\{W_{ab}, W_{ba}\}$ in a bipartite graph G forms a partition of the vertex set $V(G)$. We orient this partition by calling, in an arbitrary way, one of the two opposite semicubes in each partition a *positive semicube*.

Theorem 5.19. *Let G be a connected graph. The following statements are equivalent:*

(i)   *G is a partial cube.*
(ii) *G is bipartite and all semicubes of G are convex.*
(iii) *G is bipartite and θ is an equivalence relation on $E(G)$.*

Proof. By Theorem 5.17, condition (i) implies conditions (ii) and (iii). Conditions (ii) and (iii) are equivalent by Theorem 2.25. Thus it suffices to prove that (ii) implies (i).

Let us assign to each vertex x of G the set $\mathcal{W}^+(x)$ of all positive semicubes containing x. In the next two paragraphs we show that $x \mapsto \mathcal{W}^+(x)$ is an isometry from $V(G)$ onto the well-graded family $\mathcal{F} = \{\mathcal{W}^+(x)\}_{x \in V(G)}$.

Let x and y be two distinct vertices of G and let P be a shortest path connecting x to y. By Theorem 2.12, no two distinct edges of P stand in relation θ (recall that $\theta = \Theta$ for bipartite graphs). The semicubes of G are assumed to be convex, thus it follows from Theorem 2.24 that distinct edges of P define distinct positive semicubes. By the same Theorem 2.24, each edge e of P defines a unique positive semicube separating ends of e. Clearly, this semicube also separates vertices x and y. On the other hand, any positive semicube separating x and y separates ends of some edge of P. Thus we have one-to-one correspondence between edges of the path P and positive semicubes separating its ends. It is clear that a positive semicube separates vertices x and y if and only if it belongs to $\mathcal{W}^+(x) \triangle \mathcal{W}^+(y)$. Therefore,

$$d(x, y) = |\mathcal{W}^+(x) \triangle \mathcal{W}^+(y)|; \tag{5.4}$$

that is, $x \mapsto \mathcal{W}^+(x)$ is an isometry from $V(G)$ onto $\mathcal{F}$.

For two distinct sets $\mathcal{W}^+(x), \mathcal{W}^+(y)$ in $\mathcal{F}$, let $x = x_0, x_1, \ldots, x_n = y$ be a shortest xy-path in G. By (5.4),

$$|\mathcal{W}^+(x_i) \triangle \mathcal{W}^+(x_{i+1})| = 1 \text{ and } |\mathcal{W}^+(x_0) \triangle \mathcal{W}^+(x_n)| = n.$$

Hence, $\mathcal{F}$ is a wg-family.

By Theorem 5.3, the graph $G_{\mathcal{F}}$ is an isometric subgraph of some Cartesian power of K_2. Therefore, by Theorem 3.9, $G_{\mathcal{F}}$ is a partial cube. Clearly, graphs G and $G_{\mathcal{F}}$ are isomorphic. Hence, G is a partial cube. □

Example 5.20. It was shown in Section 5.1 that the cubical graph G in Figure 5.1a is not a partial cube. This also follows immediately from Theorem 5.19. Indeed, the semicube $W_{sw} = \{s, u, v, x\}$ is not a convex set because vertices z and y lie between x and v but do not belong to W_{sw}. By using another equivalence from Theorem 5.19, we note that $(xz, sw) \in \Theta$ and $(sw, vy) \in \Theta$, but $(xz, vy) \notin \Theta$. Thus, Θ is not an equivalence relation, so G is not a partial cube.

Example 5.21. Semicubes of a finite path are initial and terminal segments of the path. They are clearly convex. This is also true for both rays. Therefore, all paths are partial cubes.

Example 5.22. It can be easily seen that semicubes of a cycle are convex sets. However, only even cycles are bipartite. By Theorem 5.19, even cycles are partial cubes.

Example 5.23. By Theorem 2.33, a graph G is a tree if and only if Θ is the identity relation on the edge set of G. By Theorem 5.19, trees are partial cubes. Note also that by Theorem 2.29 and Corollary 2.35, semicubes of a tree are convex sets (cf. Exercise 5.6).

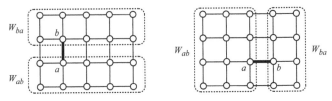

Figure 5.4. Semicubes in the (5×4)-grid.

Example 5.24. Typical semicubes in a 2-dimensional grid are depicted in Figure 5.4 for the grid $P_5 \square P_4$. It is clear that all semicubes of this grid are convex (cf. Exercise 5.7). By Theorem 5.19, $P_5 \square P_4$ is a partial cube.

Theorem 5.25. *Let G be a connected graph. The following statements are equivalent.*

(i) *G is a partial cube.*
(ii) *G is bipartite and, for all $xy, uv \in E(G)$,*

$$(xy, uv) \in \theta \ \text{ implies } \{W_{xy}, W_{yx}\} = \{W_{uv}, W_{vu}\}.$$

(iii) *G is bipartite and, for any pair of adjacent vertices of G, there is a unique pair of opposite semicubes separating these two vertices.*

Proof. (i) $\Rightarrow$ (ii). This implication follows from Theorems 5.19 and 2.24.
    (ii) $\Rightarrow$ (i). By (ii) and Theorem 2.23,

$$(xy, uv) \in \theta \ \text{ if and only if } \{W_{xy}, W_{yx}\} = \{W_{uv}, W_{vu}\}.$$

Therefore, θ is an equivalence relation. By Theorem 5.19, G is a partial cube.
    (ii) $\Rightarrow$ (iii). Suppose that semicubes W_{xy}, W_{yx} and W_{uv}, W_{vu} separate ends of an edge ab. Then $(ab, xy) \in \theta$ and $(ab, uv) \in \theta$. By condition (ii),

$$\{W_{xy}, W_{yx}\} = \{W_{ab}, W_{ba}\} = \{W_{uv}, W_{vu}\}.$$

(iii) $\Rightarrow$ (ii). If $(xy, uv) \in \theta$, then semicubes W_{xy}, W_{yx} separate vertices u and v. Clearly, semicubes W_{uv}, W_{vu} also separate u and v. By condition (iii), $\{W_{xy}, W_{yx}\} = \{W_{uv}, W_{vu}\}.$ □

Theorem 5.26. *Let G be a connected graph. The following statements are equivalent.*

(i) *G is a partial cube.*
(ii) *G is bipartite and*

$$d(x, u) = d(y, v), \quad d(x, v) = d(y, u), \tag{5.5}$$

for any edges $ab \in E(G)$ and $xy, uv \in F_{ab}$.

Proof. (i) $\Rightarrow$ (ii). Because $\theta = \Theta$ and $xy, uv \in F_{ab}$, we have $(xy, ab) \in \Theta$ and $(uv, ab) \in \Theta$. Therefore, $(xy, uv) \in \Theta$, by the transitivity property of the relation Θ (cf. Theorem 5.19). By Theorem 2.14, we have equations (5.5).

(ii) $\Rightarrow$ (i). Suppose that G is not a partial cube. Then, by Theorem 5.19, there is an edge ab such that the semicube W_{ba} is not convex. Therefore there are vertices $p, q \in W_{ba}$ such that there is a shortest pq-path that intersects the semicube W_{ab}. Let vu be the first edge of P that belongs to F_{ab} and xy be the last edge of P with the same property (cf. Figure 5.5). The vy-segment of P is a shortest path in G, therefore we have

$$d(v, y) = d(v, u) + d(u, x) + d(x, y) \neq d(u, x),$$

which contradicts (5.5). It follows that all semicubes of G are convex. By Theorem 5.19, G is a partial cube. $\qquad\square$

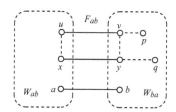

Figure 5.5. Proof of Theorem 5.26.

One can say that four vertices satisfying conditions (5.5) define a "rectangle" in G (the "opposite" sides and the two "diagonals" are equal). Then Theorem 5.26 states that a connected graph is a partial cube if and only if it is bipartite and for any edge ab, pairs of edges in F_{ab} define rectangles in the graph G (see the drawing in Figure 5.6).

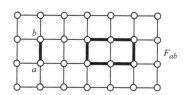

Figure 5.6. A rectangle in the (7×4)-grid.

Cubical graphs were characterized in Section 4.2 in terms of specific edge-colourings called c-valuations. We give a similar criterion for a connected graph to be a partial cube.

Theorem 5.27. *A connected graph $G = (V, E)$ is a partial cube if and only if there is a colouring $E \to X$ such that*

(i) *Edges of any shortest path in G are of different colours.*
(ii) *In each closed walk in G, every colour appears an even number of times.*

Proof. (Necessity.) We may assume that $G = G_{\mathcal{F}}$ where $\mathcal{F}$ is a wg-family of finite subsets of a set X. For any edge ST of G there is a unique element $x \in X$ such that $\{x\} = S \bigtriangleup T$. We prove below that the function $ST \mapsto x$ defines a required edge-colouring of G.

(i) Let $P = R_0, R_1, \ldots, R_n = Q$ be a shortest PQ-path in G. By Theorem 5.8, there is one-to-one correspondence between the edges of this path and the elements of $P \bigtriangleup Q$. Thus all edges of the path are of different colours.

(ii) Let $W = R_0, R_1, \ldots, R_n = R_0$ be a closed walk in G. If a colour x does not appear in W, we are done. Thus we assume that x occurs in W.

Let $R_i R_{i+1}$ be the first edge of W coloured by $x \in R_0$. Because $x \in R_0$, we must have $x \in R_i$, $x \notin R_{i+1}$. Because W is closed and $x \in R_0$, $x \notin R_{i+1}$, there must be another occurrence of x in W. Let $R_j R_{j+1}$ be the first edge after $R_i R_{i+1}$ coloured by x. Because $x \notin R_{i+1}$, we must have $x \in R_j$, $x \notin R_{j+1}$. By repeating this argument, we partition the occurrences of x in W into pairs, so the total number of these occurrences is even.

Essentially the same argument shows that the colour $x \notin R_0$ also appears an even number of times in W.

(Sufficiency.) Let $E \to X$ be a colouring satisfying conditions (i) and (ii), and let v_0 be a fixed vertex of G. For a vertex v of G and a shortest $v_0 v$-path P in G, we define

$$X_v = \{x \in X : x \text{ is a colour of an edge of } P\}$$

and set $X_{v_0} = \varnothing$. This set is well defined. Indeed, let Q be another shortest $v_0 v$-path. Then PQ^{-1} is a closed walk. By conditions (i) and (ii), the set X_v does not depend on the choice of P.

Let us show that the function $\varphi : v \mapsto X_v$ is an isometric embedding of G into the cube $\mathcal{H}(X)$. For $v, u \in V$, let P and R be shortest paths from v_0 to v and u, respectively, and let Q be a shortest vu-path in G. By condition (ii) applied to the closed walk PQR^{-1} and condition (i), $x \in X_v \bigtriangleup X_u$ if and only if x is a colour of an edge of Q. Hence, $d(v, u) = |X_v \bigtriangleup X_u|$, so φ is an isometric embedding. $\qquad\square$

Let us denote by $\vec{E} : \{(u, v) : uv \in E\}$ the set of all arcs of a graph $G = (V, E)$, and define the binary relation $\mathfrak{L}$ on $\vec{E}$ by

$$(u, v) \, \mathfrak{L} \, (x, y) \text{ if and only if } d(u, y) = d(v, x) = d(u, x) + 1 = d(v, y) + 1.$$

The relation $\mathfrak{L}$ is clearly reflexive and symmetric. Moreover, we have

$$(u, v) \, \mathfrak{L} \, (x, y) \text{ if and only if } (v, u) \, \mathfrak{L} \, (y, x), \tag{5.6}$$

for all $uv, xy \in E$. The relation $\mathfrak{L}$ and the theta relation Θ are closely related concepts as the following lemma asserts.

Lemma 5.28. *Let $\{u, v\}$ and $\{x, y\}$ be two pairs of adjacent vertices of a connected graph G. Then*

$$(u, v) \, \mathfrak{L} \, (x, y) \ \text{implies} \ uv \ominus xy.$$

If, in addition, the graph G is bipartite, then

$$uv \ominus xy \ \text{implies} \ [(u, v) \, \mathfrak{L} \, (x, y) \ \text{or} \ (u, v) \, \mathfrak{L} \, (y, x)].$$

Proof. The first implication follows immediately from the definition of $\mathfrak{L}$, because

$$d(u, y) + d(v, x) = d(u, x) + 1 + d(v, y) + 1 \neq d(u, x) + d(v, y).$$

The second implication is the result of Theorem 2.14. □

Lemma 5.29. *If G is a partial cube, then*

$$(u, v) \, \mathfrak{L} \, (x, y) \ \text{if and only if} \ W_{uv} = W_{xy}.$$

Proof. (Necessity.) By Lemma 5.28(i), $(u, v) \, \mathfrak{L} \, (x, y)$ implies $uv \ominus xy$, which in turn implies $\{W_{uv}, W_{vu}\} = \{W_{xy}, W_{yx}\}$, by Theorem 5.25. Because $(u, v) \, \mathfrak{L} \, (x, y)$, we have $d(v, x) = d(u, x) + 1$, which implies $x \in W_{uv}$. Hence, $W_{uv} = W_{xy}$.

  (Sufficiency.) If $W_{uv} = W_{xy}$, then $W_{vu} = W_{yx}$ and

$$x \in W_{uv}, \quad y \in W_{vu}, \quad u \in W_{xy}, \quad v \in W_{yx}.$$

Therefore, by Lemma 2.19,

$$d(u, y) = d(v, x) = d(u, x) + 1 = d(v, y) + 1,$$

that is, $(u, v) \, \mathfrak{L} \, (x, y)$ (see Figure 5.7). □

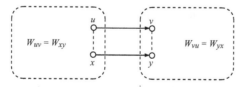

Figure 5.7. Proof of Lemma 5.29.

Theorem 5.30. *Let $G = (V, E)$ be a connected graph. The following statements are equivalent.*

(i) *G is a partial cube.*
(ii) *G is bipartite and $\mathfrak{L}$ is an equivalence relation on $\vec{E}$.*

Proof. (i) $\Rightarrow$ (ii). Let G be a partial cube. It suffices to prove that $\mathfrak{L}$ is transitive. Suppose that $(u, v) \mathfrak{L} (x, y)$ and $(x, y) \mathfrak{L} (w, z)$. Then, by Lemma 5.29, $W_{uv} = W_{xy} = W_{wz}$. By the same lemma, $(u, v) \mathfrak{L} (w, z)$, so $\mathfrak{L}$ is transitive.

(ii) $\Rightarrow$ (i). By Theorem 5.19, it suffices to prove that Θ is a transitive relation on E. Suppose that $uv \Theta xy$ and $xy \Theta wz$. By Lemma 5.28,

$$[(u, v) \mathfrak{L} (x, y) \text{ or } (u, v) \mathfrak{L} (y, x)] \text{ and } [(x, y) \mathfrak{L} (w, z) \text{ or } (x, y) \mathfrak{L} (z, w)].$$

There are four possibilities:

1) $(u, v) \mathfrak{L} (x, y)$ and $(x, y) \mathfrak{L} (w, z)$. Then, by transitivity of $\mathfrak{L}$, we have $(u, v) \mathfrak{L} (w, z)$, which implies $uv \Theta wz$, by Lemma 5.28.
2) $(u, v) \mathfrak{L} (x, y)$ and $(x, y) \mathfrak{L} (z, w)$. Then, as in the previous case, $uv \Theta zw$. Hence, $uv \Theta wz$.
3) $(u, v) \mathfrak{L} (y, x)$ and $(x, y) \mathfrak{L} (w, z)$. By (5.6), $(y, x) \mathfrak{L} (z, w)$. By transitivity of $\mathfrak{L}$, we have $(u, v) \mathfrak{L} (z, w)$, which implies $uv \Theta wz$, by Lemma 5.28.
4) $(u, v) \mathfrak{L} (y, x)$ and $(x, y) \mathfrak{L} (z, w)$. The argument from the previous case shows that we again have $uv \Theta wz$.

Thus in all four cases we have $uv \Theta wz$, so Θ is transitive. $\square$

5.6 Isometric Dimension

There are usually many ways in which a given partial cube can be isometrically embedded into a cube. For instance, the graph K_2 can be isometrically embedded in different ways into any cube $\mathcal{H}(X)$ with $|X| > 2$.

If a partial cube G is isometrically embeddable into a cube $\mathcal{H}(X)$, then it is also isometrically embeddable into any cube $\mathcal{H}(Y)$ with $|Y| \geq |X|$ (see Theorem 3.34 and discussion thereafter). Hence there is a cube of the minimum dimension in which G is isometrically embeddable.

Definition 5.31. The *isometric dimension* $\dim_I(G)$ of a partial cube G is the minimum dimension of a cube in which G is isometrically embeddable.

It is clear that

$$\dim_I(\mathcal{H}(X)) = \dim(\mathcal{H}(X)) = |X|$$

and that $\dim_c(G) \leq \dim_I(G)$, where $\dim_c(G)$ is the cubical dimension of G (see Section 4.1).

Example 5.32. The isometric and cubical dimensions of the n-dimensional cube are equal:

$$\dim_I(Q_n) = \dim_c(Q_n) = n.$$

On the other hand, it is easy to see that

$$\dim_c(P_4) = 2 < 3 = \dim_I(P_4),$$

for the path P_4.

By Theorem 3.28, the set of equivalence classes of the relation θ on the edge set E of a cube $\mathcal{H}(X)$ is in one-to-one correspondence with the set X. Therefore,

$$\dim_I(\mathcal{H}(X)) = |E/\theta|.$$

The main purpose of this section is to show that this result holds for arbitrary partial cubes.

Let $G = (V, E)$ be a partial cube on a set X. By Theorems 2.22 and 5.15(iii), the equivalence relation θ on E is the restriction of the equivalence relation $\Theta_{\mathcal{H}(X)}$ to the set E. Therefore, the equivalence classes of θ are nonempty intersections of the equivalence classes of $\Theta_{\mathcal{H}(X)}$ with the set E. By Theorem 3.28,

$$|E/\theta| \leq |X| = \dim(\mathcal{H}(X)).$$

Clearly, the above inequality holds for any partial cube $G = (V, E)$ that is isometrically embeddable into $\mathcal{H}(X)$. Accordingly, $|E/\theta| \leq \dim_I(G)$.

To prove that $\dim_I(G) = |E/\theta|$, it suffices now to construct a set X such that G is isometrically embeddable into $\mathcal{H}(X)$ and $|E/\theta| = |X|$.

Let $\mathcal{F}$ be a family of finite subsets of a set X. The *retract* of $\mathcal{F}$ is the family $\mathcal{F}'$ of intersections of the sets in $\mathcal{F}$ with the set $X' = \cup\mathcal{F} \setminus \cap\mathcal{F}$. Note that $\mathcal{F}'$ satisfies conditions

$$\cap\mathcal{F}' = \varnothing \text{ and } \cup\mathcal{F}' = X', \tag{5.7}$$

and that any family of sets satisfying these conditions is a retract of itself.

It is clear that $\alpha : P \mapsto P \cap X'$ where $P \in \mathcal{F}$ is a mapping from $\mathcal{F}$ onto $\mathcal{F}'$. For $R, S \in \mathcal{F}$, we have

$$\alpha(R) \bigtriangleup \alpha(S) = (R \cap X') \bigtriangleup (S \cap X') = (R \bigtriangleup S) \cap X' = R \bigtriangleup S.$$

Therefore, $\alpha : \mathcal{F} \to \mathcal{F}'$ is an isometry. It follows that the graphs $G_{\mathcal{F}}$ and $G_{\mathcal{F}'}$ are isomorphic. As a special case, we obtain the following result.

Theorem 5.33. *Partial cubes induced by a wg-family $\mathcal{F}$ and its retract $\mathcal{F}'$ are isomorphic.*

By Theorems 5.3 and 5.33, any partial cube $G = (V, E)$ is isomorphic to the partial cube $G_{\mathcal{F}}$ induced by a wg-family $\mathcal{F}$ satisfying conditions (5.7) for some set X. Let PQ be an edge of the cube $\mathcal{H}(X)$ with $P \bigtriangleup Q = \{x\}$. By (5.7), there are sets S and T in $\mathcal{F}$ such that $x \in S$ and $x \notin T$. Because $\mathcal{F}$ is well-graded, there is a sequence of sets $R_0 = S, R_1, \ldots, R_n = T$ satisfying conditions (5.1). Clearly, there is i such that $x \in R_i$ and $x \notin R_{i+1}$, so $R_i \triangle R_{i+1} = \{x\}$ (cf. Theorem 5.8). By Theorem 3.28, the edge $R_i R_{i+1}$ of $G_{\mathcal{F}}$ is in relation $\Theta_{\mathcal{H}(X)}$ to the edge PQ. It follows that the set of equivalence classes of the relation θ on the edge set of $G_{\mathcal{F}}$ is in one-to-one correspondence with the set of equivalence classes of $\Theta_{\mathcal{H}(X)}$. By Theorem 3.28, the cardinality of the latter set is $|X|$. Hence, $|E/\theta| = |X|$.

In summary, we have the following theorem.

Theorem 5.34. (i) Let $G = (V, E)$ be a partial cube. Then

$$\dim_I(G) = |E/\theta|,$$

where E/θ is the set of equivalence classes of the relation θ.

(ii) Let $G_{\mathcal{F}}$ be a partial cube induced by a wg-family $\mathcal{F}$ of finite subsets of a set X satisfying conditions (5.7). Then

$$\dim_I(G_{\mathcal{F}}) = |X|.$$

Example 5.35. Let $T = (V, E)$ be a finite tree. By Theorem 2.33, θ is the identity relation on E. Hence, by Theorem 5.19, T is a partial cube and the isometric dimension of T is $|V|$. In particular, the isometric dimension of the path P_n is $n - 1$. Let us recall (see Section 4.1) that the cubical dimension of P_n is $\lceil \log_2 n \rceil$. Therefore, $\dim_c(P_n) < \dim_I(P_n)$ for $n > 3$ (cf. Exercise 5.9).

Example 5.36. The relation Θ on the edge set of an even cycle C_{2n} is an equivalence relation with equivalence classes consisting of pairs of "opposite" edges of C_{2n} (cf. Exercise 5.10). Therefore, C_{2n} is a partial cube of isometric dimension n.

5.7 Cartesian Products of Partial Cubes

Let $\{(G_i, a_i)\}_{i \in I}$ and $\{(H_i, b_i)\}_{i \in I}$ be two families of connected rooted graphs. For each $i \in I$, let φ_i be an embedding $G_i \to H_i$ such that $\varphi_i(a_i) = b_i$. For the family $\varphi = \{\varphi_i\}_{i \in I}$, we define

$$\varphi(u) = \{\varphi_i(u_i)\}_{i \in I}, \text{ for every } u \in (G, a).$$

We want to show that φ is an embedding of the weak Cartesian product $(G, a) = \prod_{i \in I}(G_i, a_i)$ into the weak Cartesian product $(H, b) = \prod_{i \in I}(H_i, b_i)$. For $u \in (G, a)$, there is a finite subset J of I such that $u_i = a_i$ for $i \in I \setminus J$. Hence, $\varphi_i(u_i) = b_i$ for $i \in I \setminus J$, which implies that φ is a well-defined mapping of the vertex set of (G, a) into the vertex set of (H, b). Clearly, φ is one-to-one. Furthermore, for an edge uv of (G, a), there is $k \in I$ such that $u_k v_k$ is an edge of (G_k, a_k) and $u_i = v_i$ for $i \neq k$. Because φ_k is an embedding, $\varphi(u)\varphi(v)$ is an edge of (H, b). Therefore, φ is indeed an embedding of (G, a) into (H, b).

Suppose now that all embeddings in the family φ are isometric. Then (cf. Exercise 3.20)

$$d_H(\varphi(u), \varphi(v)) = \sum_{i \in I} d_{H_i}(\varphi_i(u_i), \varphi_i(v_i)) = \sum_{i \in I} d_{G_i}(u_i, v_i) = d_G(u, v).$$

for all vertices u, v of the graph (G, a). It follows that φ is an isometric embedding of (G, a) into (H, b). A weak Cartesian product of cubes is a cube (cf. Theorem 3.18), thus we have the following result.

Theorem 5.37. *A weak Cartesian product of a family of partial cubes is a partial cube.*

Example 5.38. Paths are partial cubes, therefore finite Cartesian products of these graphs (grids) are partial cubes. For instance, the $(m \times n \times q)$-grid $P_m \,\square\, P_n \,\square\, P_q$ is a partial cube.

Example 5.39. By Theorem 5.37, the Cartesian power $\mathcal{Z}^n$ (n-dimensional integer lattice) of the double ray $\mathcal{Z}$ is a partial cube.

Theorem 5.40. *Let $G = \prod_{i=1}^{n} G_i$ be the Cartesian product of a family $\{G_i\}_{1 \leq i \leq n}$ of finite partial cubes. Then*

$$\dim_I(G) = \sum_{i=1}^{n} \dim_I(G_i).$$

Proof. We may assume that $G_i = G_{\mathcal{F}_i}$, where $\mathcal{F}_i$ is a wg-family of subsets of a finite set X_i such that $\cap \mathcal{F}_i = \varnothing$, $\cup \mathcal{F}_i = X_i$, and the sets X_i are pairwise disjoint (cf. Section 5.6). By Theorem 5.34(ii), $\dim_I(G_i) = |X_i|$.

Let $\mathcal{F}$ be the family of subsets of $X = \cup_{i=1}^{n} X_i$ in the form $\cup_{i=1}^{n} R_i$, where $(R_1, \ldots, R_n)$ is a vertex of G. Clearly, the mapping $(R_1, \ldots, R_n) \mapsto \cup_{i=1}^{n} R_i$ is a bijection from the vertex set of G onto the family $\mathcal{F}$. It is also clear that two vertices are adjacent in G if and only if the corresponding subsets of X are adjacent in $G_{\mathcal{F}}$. Therefore the graphs G and $G_{\mathcal{F}}$ are isomorphic. By Theorem 5.34(ii), $\dim_I(G) = |X|$, because $\cap \mathcal{F} = \varnothing$ and $\cup \mathcal{F} = X$ (cf. Exercise 5.16). Because the sets X_i are pairwise disjoint, we have

$$\dim_I(G) = |X| = \sum_{i=1}^{n} |X_i| = \sum_{i=1}^{n} \dim_I(G_i).$$

The result follows. $\square$

5.8 Pasting Together Partial Cubes

Properties of vertex- and edge-pasting operations on cubical graphs were investigated in Section 4.5. In this section we apply these operations to partial cubes.

In what follows we use superscripts to distinguish subgraphs of two graphs G_1 and G_2. For instance, $W_{ab}^{(2)}$ stands for the semicube of G_2 defined by two adjacent vertices $a, b \in V_2$.

Theorem 5.41. *A graph $G = (V, E)$ obtained by vertex-pasting together partial cubes $G_1 = (V_1, E_1)$ and $G_2 = (V_2, E_2)$ is a partial cube.*

Proof. (Cf. proof of Theorem 4.18.) Let $a = \{a_1, a_2\}$ be the vertex of G obtained by identifying vertices $a_1 \in V_1$ and $a_2 \in V_2$. Clearly, G is a bipartite graph. Let xy be an edge of G. Without loss of generality we may assume that $xy \in E_1$ and $a \in W_{xy}$. Note that any path between vertices in V_1 and V_2 must go through a. Because $a \in W_{xy}$, we have, for any $v \in V_2$,

$$d(v, x) = d(v, a) + d(a, x) < d(v, a) + d(a, y) = d(v, y),$$

which implies $V_2 \subseteq W_{xy}$ and $W_{yx} \subseteq V_1$. It follows that $W_{xy} = W_{xy}^{(1)} \cup V_2$ and $W_{yx} = W_{yx}^{(1)}$. The sets $W_{xy}^{(1)}$, $W_{yx}^{(1)}$ and V_2 are convex subsets of V. Inasmuch as $W_{xy}^{(1)} \cap V_2 = \{a\}$, the set $W_{xy} = W_{xy}^{(1)} \cup V_2$ is also convex. By Theorem 5.19, the graph G is a partial cube. $\qquad\square$

Theorem 5.42. *Let $G = (V, E)$ be a partial cube obtained by vertex-pasting together finite partial cubes $G_1 = (V_1, E_1)$ and $G_2 = (V_2, E_2)$. Then*

$$\dim_I(G) = \dim_I(G_1) + \dim_I(G_2).$$

Proof. The isometric dimension of a partial cube is the cardinality of the quotient set of the relation θ, therefore it suffices to prove that there are no edges $xy \in E_1$, $uv \in E_2$ such that $(xy, uv) \in \theta$. Suppose that G_1 and G_2 are pasted together along vertices $a_1 \in V_1$ and $a_2 \in V_2$, and let $a = \{a_1, a_2\} \in E$. Let $xy \in E_1$, $uv \in E_2$ be two edges in E. We may assume that $u \in W_{xy}$. Because a is a cut-vertex of G and $u \in W_{xy}$, we have

$$d(u, a) + d(a, x) = d(u, x) < d(u, a) + d(a, y).$$

Hence, $d(a, x) < d(a, y)$, which implies

$$d(v, x) = d(v, a) + d(a, x) < d(v, a) + d(a, y) = d(v, y).$$

It follows that $v \in W_{xy}$. Therefore, $(xy, uv) \notin \theta$. $\qquad\square$

Note that blocks of a partial cube are clearly partial cubes themselves.

Corollary 5.43. *Let G be a finite partial cube and $\{G_1, \ldots, G_n\}$ be the family of its blocks. Then*

$$\dim_I(G) = \sum_{i=1}^{n} \dim_I(G_i).$$

By Theorem 4.25, the graph G obtained by edge-pasting together two partial cubes G_1 and G_2 is cubical (and therefore is bipartite). We show that in fact it is a partial cube. The following lemma is instrumental; it describes the semicubes of G in terms of semicubes of graphs G_1 and G_2.

Lemma 5.44. *Let uv be an edge of G. Then*

(i)   *For $uv \in E_1$, $a, b \in W_{uv}$ implies $W_{uv} = W_{uv}^{(1)} \cup V_2$, $W_{vu} = W_{vu}^{(1)}$.*
(ii) *For $uv \in E_2$, $a, b \in W_{uv}$ implies $W_{uv} = W_{uv}^{(2)} \cup V_1$, $W_{vu} = W_{vu}^{(2)}$.*
(iii) *$a \in W_{uv}$, $b \in W_{vu}$ implies $W_{uv} = W_{ab}$.*

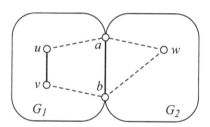

Figure 5.8. Proof of Lemma 5.44.

Proof. We prove parts (i) and (iii) (cf. Figure 5.8); proof of (ii) is left to the reader (cf. Exercise 5.17).

(i) Any path from $w \in V_2$ to u or v contains a or b and $a, b \in W_{uv}$, thus we have $w \in W_{uv}$. Hence, $W_{uv} = W_{uv}^{(1)} \cup V_2$ and $W_{vu} = W_{vu}^{(1)}$.

(iii) $(ab, uv) \in \theta$ in G_1, thus we have $W_{uv}^{(1)} = W_{ab}^{(1)}$, by Theorem 5.25(ii). Let w be a vertex in $W_{uv}^{(2)}$. Then, by the triangle inequality,

$$d(w, u) < d(w, v) \le d(w, b) + d(b, v) < d(w, b) + d(b, u).$$

Any shortest path from w to u contains a or b, therefore we have

$$d(w, a) + d(a, u) = d(w, u).$$

Hence,

$$d(w, a) + d(a, u) < d(w, b) + d(b, u).$$

Inasmuch as $(ab, uv) \in \theta$ in G_1, we have $d(a, u) = d(b, v)$, by Theorem 5.26(ii). It follows that $d(w, a) < d(w, b)$; that is, $w \in W_{ab}^{(2)}$. We proved that $W_{uv}^{(2)} \subseteq$

$W_{ab}^{(2)}$. By symmetry, $W_{vu}^{(2)} \subseteq W_{ba}^{(2)}$. Because two opposite semicubes form a partition of V_2, we have $W_{uv}^{(2)} = W_{ab}^{(2)}$. The result follows.   □

Theorem 5.45. *A graph G obtained by edge-pasting together partial cubes G_1 and G_2 is a partial cube.*

Proof. By Theorem 5.19(ii), we need to show that for any edge uv of G the semicube W_{uv} is a convex subset of V. There are two possible cases.

(i) $uv = ab$. The semicube W_{ab} is the union of semicubes $W_{ab}^{(1)}$ and $W_{ab}^{(2)}$ that are convex subsets of V_1 and V_2, respectively. It is clear that any shortest path connecting a vertex in $W_{ab}^{(1)}$ with a vertex in $W_{ab}^{(2)}$ contains vertex a and therefore is contained in W_{ab}. Hence, W_{ab} is a convex set. A similar argument proves that the set W_{ba} is convex.

(ii) $uv \neq ab$. We may assume that $uv \in E_1$. To prove that the semicube W_{uv} is a convex set, we consider two cases.

(a) $a, b \in W_{uv}$. (The case when $a, b \in W_{vu}$ is treated similarly.) By Lemma 5.44(i), the semicube W_{uv} is the union of the semicube $W_{uv}^{(1)}$ and the set V_2 which are both convex sets. Any shortest path P from a vertex in V_2 to a vertex in $W_{uv}^{(1)}$ contains either a or b. It follows that $P \subseteq W_{uv}^{(1)} \cup V_2 = W_{uv}$. Therefore the semicube W_{uv} is convex.

(b) $a \in W_{uv}$, $b \in W_{vu}$. (The case when $b \in W_{uv}$, $a \in W_{vu}$ is treated similarly.) By Lemma 5.44(ii), $W_{uv} = W_{ab}$. The result follows from part (i) of the proof.   □

Theorem 5.46. *Let G be a graph obtained by edge-pasting together finite partial cubes G_1 and G_2. Then*

$$\dim_I(G) = \dim_I(G_1) + \dim_I(G_2) - 1.$$

Proof. Let θ, θ_1, and θ_2 be the theta relations on E, E_1, and E_2, respectively. By Lemma 5.44, for $uv, xy \in E_i$ ($i \in \{1, 2\}$) we have

$$(uv, xy) \in \theta \text{ if and only if } (uv, xy) \in \theta_i.$$

Let $uv \in E_1$, $xy \in E_2$, and $(uv, xy) \in \theta$. Suppose that $(uv, ab) \notin \theta$. We may assume that $a, b \in W_{uv}$. By Lemma 5.44(i), $V_2 \subset W_{uv}$, a contradiction, inasmuch as $xy \in E_2$. Hence, $uv \, \theta \, xy \, \theta \, ab$. It follows that each equivalence class of the relation θ is either an equivalence class of θ_1, an equivalence class of θ_2, or the class containing the edge ab. Therefore

$$|E/\theta| = |E_1/\theta_1| + |E_2/\theta_2| - 1.$$

The result follows from Theorem 5.34(i).   □

In conclusion, we outline another proof of Theorem 5.45 given in terms of wg-families.

Let X_1 and X_2 be two sets such that their intersection $X_1 \cap X_2$ is a singleton $\{z\}$, and let $\mathcal{F}_1$ and $\mathcal{F}_2$ be wg-families of finite subsets of X_1 and X_2, respectively, both containing the empty set $\varnothing$ and the set $\{z\}$ (cf. proof of Theorem 4.25). We want to show that the family $\mathcal{F} = \mathcal{F}_1 \cup \mathcal{F}_2$ is well-graded. Let P and Q be distinct sets in $\mathcal{F}$. If $P, Q \in \mathcal{F}_1$ or $P, Q \in \mathcal{F}_2$, then there is a sequence (R_i) satisfying conditions (5.1), for the families $\mathcal{F}_1$ and $\mathcal{F}_2$ are well-graded. Suppose now that, say, $P \in \mathcal{F}_1$ and $Q \in \mathcal{F}_2$. Let us consider two possible cases:

(i) $P \cap Q = \varnothing$. The families $\mathcal{F}_1$ and $\mathcal{F}_2$ are well-graded, therefore there are two sequences of sets satisfying conditions (5.1) connecting P to $\varnothing$ in $\mathcal{F}_1$ and $\varnothing$ to Q in $\mathcal{F}_2$. The lengths of these sequences are $|P|$ and $|Q|$, respectively. By concatenating these sequences at $\varnothing$, we obtain a sequence of sets in $\mathcal{F}$ satisfying conditions (5.1), because $d(P, Q) = |P \triangle Q| = |P| + |Q|$.

(ii) $P \cap Q = \{z\}$. As in the previous case, there are two sequences of sets satisfying conditions (5.1) connecting P to $\{z\}$ in $\mathcal{F}_1$ and $\{z\}$ to Q in $\mathcal{F}_2$. The lengths of these sequences are $|P| - 1$ and $|Q| - 1$, respectively. By concatenating these sequences at $\{z\}$, we obtain a sequence of sets in $\mathcal{F}$ satisfying conditions (5.1), because $d(P, Q) = |P \triangle Q| = |P| + |Q| - 2$.

It follows that the graph $G_{\mathcal{F}}$ is a partial cube. This graph is obtained by edge-pasting together the partial cubes $G_{\mathcal{F}_1}$ and $G_{\mathcal{F}_2}$ along the edges with ends $\varnothing$ and $\{z\}$.

A similar argument can be used to prove all other statements in this section (cf. Exercise 5.18).

5.9 Expansions and Contractions of Partial Cubes

In this section we investigate properties of (isometric) expansion and contraction operations on graphs and, in particular, prove in two different ways that a graph is a partial cube if and only if it can be obtained from the trivial graph K_1 by a sequence of expansions.

A remark about notations is in order. In the product $\{1, 2\} \times (V_1 \cup V_2)$, we denote $V_i' = \{i\} \times V_i$ and $x^i = (i, x)$ for $x \in V_i$, where $i \in \{1, 2\}$. Let us also recall that the disjoint union operation is denoted by $+$.

Definition 5.47. Let $G = (V, E)$ be a connected graph, and let $G_1 = (V_1, E_1)$ and $G_2 = (V_2, E_2)$ be two isometric subgraphs of G such that $G = G_1 \cup G_2$. The *expansion* of G with respect to G_1 and G_2 is the graph $G' = (V', E')$ constructed as follows from G (see Figure 5.9).

(i) $V' = V_1 + V_2 = V_1' \cup V_2'$.
(ii) $E' = E_1 + E_2 + M$, where M is the matching $\bigcup_{x \in V_1 \cap V_2} \{x^1 x^2\}$.

In this case, we also say that G is a *contraction* of G'.

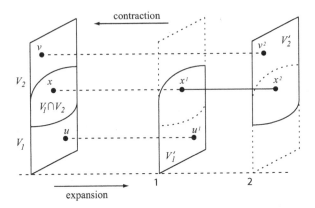

Figure 5.9. Expansion/contraction processes.

Example 5.48. Suppose that $G_1 = G_2 = G$. It is clear that $G' = G \,\square\, K_2$. By repeatedly applying repeatedly the expansion operation to the trivial graph K_1, we obtain the finite cubes $Q_1 = K_2, Q_2, Q_3, \dots$.

The following two examples give geometric illustrations for the expansion and contraction procedures.

Example 5.49. Let a and b be two opposite vertices of the graph $G = C_4$. Clearly, the two distinct paths P_1 and P_2 from a to b are isometric subgraphs of G defining an expansion $G' = C_6$ of G (see Figure 5.10).

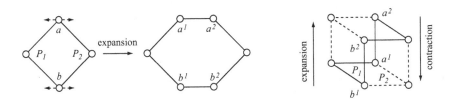

Figure 5.10. An expansion of the cycle C_4.

Example 5.50. Another isometric expansion of the graph $G = C_4$ is shown in Figure 5.11. Here, the path P_1 is the same as in Example 5.49 and $G_2 = G$.

We define a *projection* $p : V' \to V$ by $p(x^i) = x$ for $x \in V$. Clearly, the restriction of p to V_1' is a bijection $p_1 : V_1' \to V_1$ and its restriction to V_2' is a bijection $p_2 : V_2' \to V_2$. These bijections define isomorphisms $G'[V_1'] \to G_1$ and $G'[V_2'] \to G_2$.

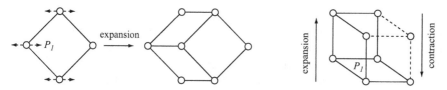

Figure 5.11. Another isometric expansion of the cycle C_4.

Let P' be a path in G'. The vertices of G obtained from the vertices in P' under the projection p define a walk P in G; we call this walk P the *projection* of the path P'. It is clear that

$$\ell(P) = \ell(P'), \text{ if } P' \subseteq G'[V_1'] \text{ or } P' \subseteq G'[V_2'], \tag{5.8}$$

where $\ell(P)$ stands for the length of the path P. In this case, P is a path in G and either $P = p_1(P')$ or $P = p_2(P')$. On the other hand,

$$\ell(P) < \ell(P'), \text{ if } P' \cap G'[V_1'] \neq \varnothing \text{ and } P' \cap G'[V_2'] \neq \varnothing \tag{5.9}$$

and P is not necessarily a path.

We frequently use the results of the next lemma in this section.

Lemma 5.51. (i) For $u^1, v^1 \in V_1'$, any shortest path $P_{u^1v^1}$ in G' belongs to $G'[V_1']$ and its projection $P_{uv} = p_1(P_{u^1v^1})$ is a shortest path in G. Accordingly,

$$d_{G'}(u^1, v^1) = d_G(u, v)$$

and $G'[V_1']$ is a convex subgraph of G'. A similar statement holds for vertices $u^2, v^2 \in V_2'$.

(ii) For $u^1 \in V_1'$ and $v^2 \in V_2'$,

$$d_{G'}(u^1, v^2) = d_G(u, v) + 1.$$

Let $P_{u^1v^2}$ be a shortest path in G'. There is a unique edge $x^1x^2 \in M$ such that $x^1, x^2 \in P_{u^1v^2}$ and the segments $P_{u^1x^1}$ and $P_{x^2v^2}$ of the path $P_{u^1v^2}$ are shortest paths in $G'[V_1']$ and $G'[V_2']$, respectively. The projection P_{uv} of $P_{u^1v^2}$ in G' is a shortest path in G.

Proof. (i) Let $P_{u^1v^1}$ be a path in G' that intersects V_2'. Because $G[V_1]$ is an isometric subgraph of G, there is a path P_{uv} in G that belongs to $G[V_1]$. Then $p_1^{-1}(P_{uv})$ is a path in $G'[V_1']$ of the same length as P_{uv}. By (5.8) and (5.9),

$$\ell(p_1^{-1}(P_{uv})) < \ell(P_{u^1v^1}).$$

Therefore any shortest path $P_{u^1v^1}$ in G' belongs to $G'[V_1']$. The result follows.

(ii) Let $P_{u^1v^2}$ be a shortest path in G' and P_{uv} be its projection to V. By (5.9),

$$d_{G'}(u^1, v^2) = \ell(P_{u^1v^2}) > \ell(P_{uv}) \geq d_G(u, v).$$

Inasmuch as there is no edge of G joining vertices in $V_1 \setminus V_2$ and $V_2 \setminus V_1$, a shortest path in G from u to v must contain a vertex $x \in V_1 \cap V_2$. Because G_1 and G_2 are isometric subgraphs, there are shortest paths P_{ux} in G_1 and P_{xv} in G_2 such that their union is a shortest path from u to v. Then, by the triangle inequality and part (i) of the proof, we have (cf. Figure 5.9)

$$d_{G'}(u^1, v^2) \le d_{G'}(u^1, x^1) + d_{G'}(x^1, x^2) + d_{G'}(x^2, v^2) = d_G(u, v) + 1.$$

The last two displayed formulas imply $d_{G'}(u^1, v^2) = d_G(u, v) + 1$.

$u^1 \in V_1'$ and $v^2 \in V_2'$, therefore the path $P_{u^1v^2}$ must contain an edge, say x^1x^2, in M. This path is a shortest path in G', thus this edge is unique. Then the segments $P_{u^1x^1}$ and $P_{x^2v^2}$ of $P_{u^1v^2}$ are shortest paths in $G'[V_1']$ and $G'[V_2']$, respectively. Clearly, P_{uv} is a shortest path in G.   □

Example 5.52. Lemma 5.51 claims, in particular, that the projection of a shortest path in an extension G' of a graph G is a shortest path in G. In general, the converse is not true. Consider the graph G shown in Figure 5.12 and two paths in G:

$$V_1 = abcef \quad \text{and} \quad V_2 = bde.$$

The graph G' in Figure 5.12 is an expansion of G with respect to V_1 and V_2. The path $abdef$ is a shortest path in G; it is not a projection of a shortest path in G'.

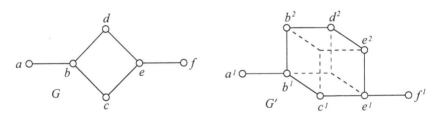

Figure 5.12. A shortest path that is not a projection of a shortest path.

Let a^1a^2 be an edge in the matching $M = \cup_{x \in V_1 \cap V_2}\{x^1x^2\}$. This edge defines five fundamental sets (cf. Section 2.3): the semicubes $W_{a^1a^2}$ and $W_{a^2a^1}$, the sets of vertices $U_{a^1a^2}$ and $U_{a^2a^1}$, and the set of edges $F_{a^1a^2}$. The next theorem follows immediately from Lemma 5.51. It gives a hint to a connection between the expansion process and partial cubes.

Theorem 5.53. *Let G' be an expansion of a connected graph G and notations are chosen as above. Then*

(i) $W_{a^1a^2} = V_1'$ *and* $W_{a^2a^1} = V_2'$ *are convex semicubes of G'.*

(ii) $F_{a^1a^2} = M$ defines an isomorphism between induced subgraphs $G'[U_{a^1a^2}]$ and $G'[U_{a^2a^1}]$, each of which is isomorphic to the subgraph $G_1 \cap G_2$.

The result of Theorem 5.53 justifies the following constructive definition of the contraction operation.

Definition 5.54. Let ab be an edge of a connected graph $G' = (V', E')$ such that

(i) Semicubes W_{ab} and W_{ba} are convex and form a partition of V'.
(ii) The set F_{ab} is a matching and defines an isomorphism between subgraphs $G'[U_{ab}]$ and $G'[U_{ba}]$.

The graph G obtained from the graphs W_{ab} and W_{ba} by pasting them along subgraphs $G'[U_{ab}]$ and $G'[U_{ba}]$ is said to be a *contraction* of the graph G'.

One can say that, in the case of finite partial cubes, the contraction operation is defined by an orthogonal projection of a cube onto one of its facets (cf. Figures 5.10 and 5.11).

By Theorem 5.53, the sets V_1' and V_2' are opposite semicubes of G'. These semicubes are defined by edges of M. Their projections are the sets V_1 and V_2 that are not necessarily semicubes of G. For other semicubes in G' we have the following result.

Lemma 5.55. For any two adjacent vertices $u, v \in V_i$,

$$W_{u^i v^i} = p^{-1}(W_{uv}),$$

where $i \in \{1, 2\}$.

Proof. By Lemma 5.51,

$$d_{G'}(x^j, u^i) < d_{G'}(x^j, v^i) \text{ if and only if } d_G(x, u) < d_G(x, v),$$

for $x \in V$ and $i, j \in \{1, 2\}$. The result follows.   □

Corollary 5.56. If uv is an edge of $G_1 \cap G_2$, then $W_{u^1v^1} = W_{u^2v^2}$.

The following lemma is an immediate consequence of Lemma 5.51. We use it implicitly in our arguments later.

Lemma 5.57. Let $u, v \in V_1$ and $x \in V_1 \cap V_2$. Then

$$x^1 \in W_{u^1v^1} \quad \text{if and only if} \quad x^2 \in W_{u^1v^1}.$$

The same result holds for semicubes in the form $W_{u^2v^2}$.

In general, the projection of a convex subgraph of G' is not a convex subgraph of G. For instance, the projection of the convex path $b^2d^2e^2$ in Figure 5.12 is the path bde that is not a convex subgraph of G. On the other hand, we have the following result.

Theorem 5.58. Let $G' = (V', E')$ be an expansion of a graph $G = (V, E)$ with respect to subgraphs $G_1 = (V_1, E_1)$ and $G_2 = (V_2, E_2)$. The projection of a convex semicube of G' different from V_1' and V_2' is a convex semicube of the graph G.

Proof. It suffices to consider the case when $W_{uv} = p(W_{u^1 v^1})$ for $u, v \in V_1$ (cf. Lemma 5.55). Let $x, y \in W_{uv}$ and $z \in V$ be a vertex such that

$$d_G(x, z) + d_G(z, y) = d_G(x, y).$$

We need to show that $z \in W_{uv}$.

(i) $x, y \in V_1$ (the case when $x, y \in V_2$ is treated similarly). Suppose that $z \in V_1$. Then $x^1, y^1, z^1 \in V_1'$ and, by Lemma 5.51,

$$d_{G'}(x^1, z^1) + d_{G'}(z^1, y^1) = d_{G'}(z^1, y^1).$$

$x^1, y^1 \in W_{u^1 v^1}$ and $W_{u^1 v^1}$ is convex, thus $z^1 \in W_{u^1 v^1}$. Hence, $z \in W_{uv}$.

Suppose now that $z \in V_2 \setminus V_1$. Consider a shortest path P_{xy} in G from x to y containing z. This path contains vertices $x', y' \in V_1 \cap V_2$ such that (cf. Figure 5.13)

$$d_G(x, x') + d_G(x', z) = d_G(x, z) \text{ and } d_G(y, y') + d_G(y', z) = d_G(y, z).$$

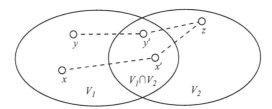

Figure 5.13. Proof of Theorem 5.58.

P_{xy} is a shortest path in G, therefore we have

$$d_G(x, x') + d_G(x', y) = d_G(x, y), \quad d_G(x, y') + d_G(y', y) = d_G(x, y),$$

and

$$d_G(x', z) + d_G(z, y') = d_G(x', y').$$

Because $x, x', y \in V_1$, we have $x^1, x'^1, y^1 \in V_1'$. Because $x^1, y^1 \in W_{u^1 v^1}$ and $W_{u^1 v^1}$ is convex, $x'^1 \in W_{u^1 v^1}$. Hence, $x' \in W_{uv}$ and, similarly, $y' \in W_{uv}$. Because $x'^2, y'^2, z^2 \in V_2'$ and $W_{u^1 v^1}$ is convex, $z^2 \in W_{u^1 v^1}$. Hence, $z \in W_{uv}$.

(ii) $x \in V_1 \setminus V_2$ and $y \in V_2 \setminus V_1$. We may assume that $z \in V_1$. By Lemma 5.51,

$$d_{G'}(x^1, y^2) = d_G(x, y) + 1 = d_G(x, z) + d_G(z, y) + 1$$
$$= d_{G'}(x^1, z^1) + d_{G'}(z^1, y^2).$$

$x^1, y^2 \in W_{u^1v^1}$ and $W_{u^1v^1}$ is convex, thus $z^1 \in W_{u^1v^1}$. Hence, $z \in W_{uv}$. $\qquad\square$

By using the results of Lemma 5.51, it is not difficult to show that the class of connected bipartite graphs is closed under the expansion and contraction operations. The next theorem establishes this result for the class of partial cubes.

Theorem 5.59. (i) *An expansion of a partial cube is a partial cube.*
(ii) *A contraction of a partial cube is a partial cube.*

Proof. (i) Let $G = (V, E)$ be a partial cube and $G' = (V', E')$ be its expansion with respect to isometric subgraphs $G_1 = (V_1, E_1)$ and $G_2 = (V_2, E_2)$. By Theorem 5.19(ii), it suffices to show that the semicubes of G' are convex.

By Theorem 5.53, the semicubes V_1' and V_2' are convex, so we consider a semicube in the form $W_{u^1v^1}$ where $uv \in E_1$ (the other case is treated similarly). Let $P_{x'y'}$ be a shortest path connecting two vertices in $W_{u^1v^1}$ and P_{xy} be its projection to G. By Lemma 5.55, $x, y \in W_{uv}$ and, by Lemma 5.51, P_{xy} is a shortest path in G. Because W_{uv} is convex, P_{xy} belongs to W_{uv}. Let z' be a vertex in $P_{x'y'}$ and $z = p(z') \in P_{xy}$. By Lemma 5.51,

$$d_G(z, u) < d_G(z, v) \text{ implies } d_{G'}(z', u^1) \leq d_{G'}(z', v^1).$$

G' is a bipartite graph, thus $d_{G'}(z', u^1) < d_{G'}(z', v^1)$. Hence, $P_{x'y'} \subseteq W_{u^1v^1}$, so $W_{u^1v^1}$ is convex.

(ii) Let $G = (V, E)$ be a contraction of a partial cube $G' = (V', E')$. By Theorem 5.19(ii), we need to show that the semicubes of G are convex. By Lemma 5.55, all semicubes of G are projections of semicubes of G' distinct from V_1' and V_2'. By Theorem 5.58, the semicubes of G are convex. $\qquad\square$

Theorem 5.60. *A finite connected graph is a partial cube if and only if it can be obtained from K_1 by a sequence of expansions. Furthermore, the number of expansions needed to produce a partial cube G from K_1 is $\dim_I(G)$.*

Proof. By Theorem 5.59(i), a sequence of expansions produces a partial cube from K_1. Conversely, by Theorems 5.19(ii) and 2.26, a partial cube admits a contraction that produces, by Theorem 5.59(ii), a partial cube. By applying this procedure repeatedly, we obtain the graph K_1. It is clear that the contraction operation reduces the number of equivalence classes of the theta relation by one. By Theorem 5.34, there are exactly $\dim_I(G)$ expansions needed to produce the partial cube G from K_1. $\qquad\square$

As shown in Sections 5.2 and 5.6, a finite partial cube is isomorphic to the partial cube $G_{\mathcal{F}}$ where $\mathcal{F}$ is a wg-family of subsets of some finite set X satisfying conditions $\cap \mathcal{F} = \varnothing$ and $\cup \mathcal{F} = X$. In what follows we present proofs of Theorems 5.59 and 5.60 given in terms of wg-families of sets.

The expansion process for a partial cube $G_{\mathcal{F}}$ on X can be described as follows. Let $\mathcal{F}_1$ and $\mathcal{F}_2$ be wg-families of finite subsets of X such that

$$\mathcal{F}_1 \cap \mathcal{F}_2 \neq \varnothing, \quad \mathcal{F}_1 \cup \mathcal{F}_2 = \mathcal{F}$$

and the distance between any two sets $P \in \mathcal{F}_1 \setminus \mathcal{F}_2$ and $Q \in \mathcal{F}_2 \setminus \mathcal{F}_1$ is greater than one. Note that $G_{\mathcal{F}_1}$ and $G_{\mathcal{F}_2}$ are partial cubes, $G_{\mathcal{F}_1} \cap G_{\mathcal{F}_2} \neq \varnothing$, and $G_{\mathcal{F}_1} \cup G_{\mathcal{F}_2} = G_{\mathcal{F}}$. Let $X' = X + \{p\}$, where $p \notin X$, and

$$\mathcal{F}_2' = \{Q + \{p\} : Q \in \mathcal{F}_2\}, \quad \mathcal{F}' = \mathcal{F}_1 \cup \mathcal{F}_2'.$$

It is quite clear that the graphs $G_{\mathcal{F}_2'}$ and $G_{\mathcal{F}_2}$ are isomorphic and the graph $G_{\mathcal{F}}'$ is an expansion of the graph $G_{\mathcal{F}}$.

Theorem 5.61. *An expansion of a partial cube is a partial cube.*

Proof. We need to verify that $\mathcal{F}'$ is a wg-family of finite subsets of X'. By Theorem 5.9, it suffices to show that the distance between any two adjacent sets in $\mathcal{F}'$ is one. It is obvious if each of these two sets belongs to one of the families $\mathcal{F}_1$ or $\mathcal{F}_2'$. Suppose that $P \in \mathcal{F}_1$ and $Q + \{p\} \in \mathcal{F}_2'$ are adjacent; that is, for any $S \in \mathcal{F}'$ we have

$$P \cap (Q + \{p\}) \subseteq S \subseteq P \cup (Q + \{p\}) \text{ implies } S = P \text{ or } S = Q + \{p\}. \quad (5.10)$$

If $Q \in \mathcal{F}_1$, then

$$P \cap (Q + \{p\}) \subseteq Q \subseteq P \cup (Q + \{p\}),$$

inasmuch as $p \notin P$. By (5.10), $Q = P$ implying $d(P, Q + \{p\}) = 1$.

If $Q \in \mathcal{F}_2 \setminus \mathcal{F}_1$, there is $R \in \mathcal{F}_1 \cap \mathcal{F}_2$ such that

$$d(P, R) + d(R, Q) = d(P, Q),$$

because $\mathcal{F}$ is well graded. By Theorem 3.22,

$$P \cap Q \subseteq R \subseteq P \cup Q,$$

which implies

$$P \cap (Q + \{p\}) \subseteq R + \{p\} \subseteq P \cup (Q + \{p\}).$$

By (5.10), $R + \{p\} = Q + \{p\}$, a contradiction. The result follows. $\qquad\square$

It is easy to recognize the fundamental sets (cf. Section 2.3) in an isometric expansion $G_{\mathcal{F}'}$ of a partial cube $G_{\mathcal{F}}$. Let $P \in \mathcal{F}_1 \cap \mathcal{F}_2$ and $Q = P + \{p\} \in \mathcal{F}_2'$ be two vertices defining an edge in $G_{\mathcal{F}'}$ according to Definition 5.47(ii). Clearly, the families $\mathcal{F}_1$ and $\mathcal{F}_2'$ are the semicubes W_{PQ} and W_{QP} of the graph $G_{\mathcal{F}'}$ (cf. Theorem 5.18) and therefore are convex subsets of $\mathcal{F}'$. The set F_{PQ} is the set of edges defined by p as in Theorem 5.18. In addition, $U_{PQ} = \mathcal{F}_1 \cap \mathcal{F}_2$ and $U_{QP} = \{R + \{p\} : R \in \mathcal{F}_1 \cap \mathcal{F}_2\}$.

Let G be a partial cube induced by a wg-family $\mathcal{F}$ of finite subsets of a set X. As before, we assume that $\cap \mathcal{F} = \varnothing$ and $\cup \mathcal{F} = X$. Let PQ be an edge of G. We may assume that $Q = P + \{p\}$ for some $p \notin P$. Then (see Theorem 5.18)

$$W_{PQ} = \{R \in \mathcal{F} : p \notin R\} \text{ and } W_{QP} = \{R \in \mathcal{F} : p \in R\}.$$

Let $X' = X \setminus \{p\}$ and $\mathcal{F}' = \{R \setminus \{p\} : R \in \mathcal{F}\}$. It is clear that the graph $G_{\mathcal{F}'}$ induced by the family $\mathcal{F}'$ is isomorphic to the contraction of $G_{\mathcal{F}}$ defined by the edge PQ. Geometrically, the graph $G_{\mathcal{F}'}$ is the orthogonal projection of the graph $G_{\mathcal{F}}$ along the edge PQ (cf. Figures 5.10 and 5.11).

Theorem 5.62. (i) *A contraction $G_{\mathcal{F}'}$ of a partial cube $G_{\mathcal{F}}$ is a partial cube.*
(ii) *If $G_{\mathcal{F}}$ is finite, then $\dim_I(G_{\mathcal{F}'}) = \dim_I(G_{\mathcal{F}}) - 1$.*

Proof. (i) For $p \in X$ we define $\mathcal{F}_1 = \{R \in \mathcal{F} : p \notin R\}$, $\mathcal{F}_2 = \{R \in \mathcal{F} : p \in R\}$, and $\mathcal{F}_2' = \{R \setminus \{p\} \in \mathcal{F} : p \in R\}$. Note that $\mathcal{F}_1$ and $\mathcal{F}_2$ are semicubes of $G_{\mathcal{F}}$ and $\mathcal{F}_2'$ is isometric to $\mathcal{F}_2$. Hence, $\mathcal{F}_1$ and $\mathcal{F}_2'$ are wg-families of finite subsets of $X' = X \setminus \{p\}$. We need to prove that $\mathcal{F}' = \mathcal{F}_1 \cup \mathcal{F}_2'$ is a wg-family. By Theorem 5.9, it suffices to show that $d(P, Q) = 1$ for any two adjacent sets $P, Q \in \mathcal{F}'$. This is true if $P, Q \in \mathcal{F}_1$ or $P, Q \in \mathcal{F}_2'$, because these two families are well graded. For $P \in \mathcal{F}_1 \setminus \mathcal{F}_2'$ and $Q \in \mathcal{F}_2' \setminus \mathcal{F}_1$, the sets P and $Q + \{p\}$ are not adjacent in $\mathcal{F}$, because $\mathcal{F}$ is well graded and $Q \notin \mathcal{F}$. Hence there is $R \in \mathcal{F}_1$ such that

$$P \cap (Q + \{p\}) \subseteq R \subseteq P \cup (Q + \{p\})$$

and $R \neq P$. Inasmuch as $p \notin R$, we have

$$P \cap Q \subseteq R \subseteq P \cup Q.$$

$R \neq P$ and $R \neq Q$, thus the sets P and Q are not adjacent in $\mathcal{F}'$. The result follows.

(ii) If G is a finite partial cube, then

$$\dim_I(G') = |X'| = |X| - 1 = \dim_I(G) - 1,$$

by Theorem 5.34(ii).   □

5.10 Uniqueness of Isometric Embeddings

Let G be the partial cube depicted in Figure 5.14, left. There are different ways of embedding this graph into the cube $\mathcal{H}(\{a, b, c\}) = Q_3$. (This cube is shown in Figure 5.14, right).

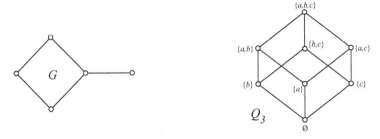

Figure 5.14. Graph G and the cube Q_3.

Two images G_1 and G_2 of G under isometric embeddings are shown in Figure 5.15. Note that as geometric figures the graphs G_1 and G_2 are congruent; that is, by moving one of them it is possible to superimpose it onto the other so that the two images become identified with each other in all their parts. More formally, there is an automorphism α of the cube Q_3 such that $\alpha(G_1) = G_2$. Indeed, the automorphism $\alpha = \hat{\sigma} \circ \alpha_{\{ac\}}$, where σ is the permutation of $\{a, b, c\}$ transposing elements a and b (for notation, see Section 3.8), maps G_1 onto G_2.

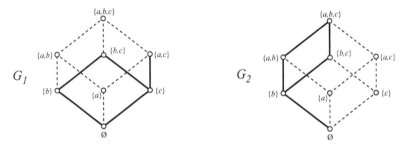

Figure 5.15. Graph G and the cube Q_3.

In this section we show that this claim holds for arbitrary finite-dimensional partial cubes: the image of a finite graph under its isometric embedding into a cube is unique up to an automorphism of the cube.

Note that the claim does not hold for infinite graphs. Consider, for instance, an infinite set X and let X' be a proper subset of X of the same cardinality. The cube $\mathcal{H}(X')$ is a proper isometric subgraph of the cube $\mathcal{H}(X)$ and isomorphic to $\mathcal{H}(X)$. However, it is evident that there is no automorphism of $\mathcal{H}(X)$ that maps the "smaller" cube onto the "larger" one.

It is convenient to formulate the main result of this section in terms of wg–families of X.

Theorem 5.63. *Let* $\mathcal{F}$ *and* $\mathcal{G}$ *be wg-families of subsets of a finite set* X, *and let* α *be an isometry from* $\mathcal{F}$ *onto* $\mathcal{G}$. *There is an isometry* φ *of* $\mathcal{P}(X)$ *onto* *itself such that* $\varphi(\mathcal{F}) = \mathcal{G}$. *Moreover,* φ *can be chosen so it is an extension of* α; *that is,* $\varphi|_{\mathcal{F}} = \alpha$.

Some general remarks are in order. A metric space is said to be *homogeneous* if for any two points of the space there is an isometry of the space onto itself mapping one of the points into another. Let Y be a homogeneous metric space, A and B be two subspaces of Y, and α be an isometry from A onto B. The isometry α is said to be *extendable* to an isometry φ of Y onto itself if $\varphi|_A = \alpha$. Let c be a fixed point in Y. For a given $a \in A$, let $b = \alpha(a) \in B$. Inasmuch as Y is homogeneous, there are isometries β and γ from Y onto itself such that $\beta(a) = c$ and $\gamma(b) = c$. Then $\lambda = \gamma\alpha\beta^{-1}$ is an isometry from $\beta(A)$ onto $\gamma(B)$ such that $\lambda(c) = c$. Clearly, α is extendable to an isometry of Y if and only if λ is extendable.

The argument in the foregoing paragraph shows that in Theorem 5.63 we may assume that the wg-families $\mathcal{F}$ and $\mathcal{G}$ both contain the empty set and $\varphi(\varnothing) = \varnothing$. We obtain the result of Theorem 5.63 by proving several lemmas.

For a given family $\mathcal{F} \subseteq \mathcal{P}_f(X)$ we denote by $D(\mathcal{F})$ the union $\cup \mathcal{F}$ and define a function $r_{\mathcal{F}} : D(\mathcal{F}) \to \mathbf{N}$ by

$$r_{\mathcal{F}}(x) = \min\{|R| : x \in R, \, R \in \mathcal{F}\}.$$

For $k \in \mathbf{N}$ a subset $X_k^{\mathcal{F}}$ of X is defined by

$$X_k^{\mathcal{F}} = \{x \in X : r_{\mathcal{F}}(x) = k\}.$$

Clearly, $X_i^{\mathcal{F}} \cap X_j^{\mathcal{F}} = \varnothing$ for $i \neq j$, and $\cup_k X_k^{\mathcal{F}} = D(\mathcal{F})$. Note that some of the sets $X_k^{\mathcal{F}}$ could be empty for $k > 1$.

Example 5.64. Let $X = \{a, b, c\}$ and $\mathcal{F} = \{\varnothing, \{a\}, \{b\}, \{a, b\}, \{a, b, c\}\}$. We have $r_{\mathcal{F}}(a) = r_{\mathcal{F}}(b) = 1$, $r_{\mathcal{F}}(c) = 3$, and

$$X_1^{\mathcal{F}} = \{a, b\}, \quad X_2^{\mathcal{F}} = \varnothing, \quad X_3^{\mathcal{F}} = \{c\}.$$

Lemma 5.65. *The set* $X_1^{\mathcal{F}}$ *is not empty and, for any nonempty set* $P \in \mathcal{F}$, *there is* $x \in P$ *such that* $P \setminus \{x\} \in \mathcal{F}$.

Proof. $\mathcal{F}$ is well-graded and contains the empty set, thus there is a nested sequence $\varnothing, R_1, \ldots, R_k = P$ of distinct sets in $\mathcal{F}$ such that $|R_{i+1} \setminus R_i| = 1$. Because R_1 is a singleton, we have $X_1^{\mathcal{F}} \neq \varnothing$. Clearly, $R_{k-1} = P \setminus \{x\}$ for some $x \in P$. Thus $P \setminus \{x\} \in \mathcal{F}$. $\square$

Lemma 5.66. *For* $P \in \mathcal{F}$ *and* $x \in P$,

$$r_{\mathcal{F}}(x) = |P| \text{ implies } P \setminus \{x\} \in \mathcal{F}.$$

Proof. By Lemma 5.65, there is $y \in P$ such that $P \setminus \{y\} \in \mathcal{F}$.

$$|P \setminus \{y\}| = |P| - 1 < r_{\mathcal{F}}(x),$$

thus we have $x \notin P \setminus \{y\}$. Therefore, $y = x$ and the result follows. $\qquad \square$

Let $\mathcal{F}$ and $\mathcal{G}$ be two wg-families each containing $\varnothing$, and $\alpha : \mathcal{F} \to \mathcal{G}$ be an isometry such that $\alpha(\varnothing) = \varnothing$. We show that there is a permutation $\sigma : X \to X$ such that $\alpha = \hat{\sigma}|_{\mathcal{F}}$, where $\hat{\sigma}$ is the automorphism of $\mathcal{H}(X)$ defined by σ (see Section 3.8).

Because α is an isometry, it preserves the betweenness relation. For any two sets $P, Q \in \mathcal{F}$, we have

$$P \subseteq Q \text{ if and only if } \alpha(P) \subseteq \alpha(Q), \tag{5.11}$$

inasmuch as P lies between $\varnothing$ and Q and $\alpha(\varnothing) = \varnothing$. We also have

$$|\alpha(P)| = |P|, \text{ for } P \in \mathcal{F}, \tag{5.12}$$

because $|P| = d(\varnothing, P) = d(\varnothing, \alpha(P)) = |\alpha(P)|$.

Lemma 5.67. *If $x \in P \in \mathcal{F}$ and $P \setminus \{x\} \in \mathcal{F}$, then there is $y \in \alpha(P)$ such that $\alpha(P) \setminus \{y\} = \alpha(P \setminus \{x\})$.*

Proof. By property (5.11), $P \setminus \{x\} \subset P$ implies $\alpha(P \setminus \{x\}) \subset \alpha(P)$. Because $d(P \setminus \{x\}, P) = 1$, we have $d(\alpha(P \setminus \{x\}), \alpha(P)) = 1$. The result follows. $\qquad \square$

Let us define a relation $\sigma \subseteq D(\mathcal{F}) \times D(\mathcal{G})$ as follows: $(x, y) \in \sigma$ if and only if $x \in D(\mathcal{F})$ and $y \in D(\mathcal{G})$ satisfy conditions of Lemma 5.67 for some $P \in \mathcal{F}$. By Lemmas 5.66 and 5.67, for any $x \in D(\mathcal{F})$ there is $y \in D(\mathcal{G})$ such that $(x, y) \in \sigma$. Conversely, for any $y \in D(\mathcal{G})$ there is $x \in D(\mathcal{F})$ such that $(x, y) \in \sigma$. Indeed, it suffices to apply the results of Lemmas 5.66 and 5.67 to the family $\mathcal{G}$ and the inverse isometry α^{-1}. We show that the relation σ is a bijection.

Lemma 5.68. *If $x \in X_k^{\mathcal{F}}$ and $(x, y) \in \sigma$, then $y \in X_k^{\mathcal{G}}$. Conversely, if $y \in X_k^{\mathcal{G}}$ and $(x, y) \in \sigma$, then $x \in X_k^{\mathcal{F}}$.*

Proof. Let $P \in \mathcal{F}$ be a set of cardinality k defining $r_{\mathcal{F}}(x) = k$. Then $r_{\mathcal{G}}(y) \leq k$, because $y \in \alpha(P)$ and, by (5.12), $|\alpha(P)| = k$.

Suppose that $m = r_{\mathcal{G}}(y) < k$. Then there is $Q \in \mathcal{G}$ such that $y \in Q$ and $|Q| = m$. By Lemma 5.66, $Q \setminus \{y\} \in \mathcal{G}$. By Lemma 5.67,

$$\alpha(P \setminus \{x\}) \cap Q \subseteq \alpha(P) \subseteq \alpha(P \setminus \{x\}) \cup Q.$$

The isometry α^{-1} preserves the betweenness relation, thus we have

$$(P \setminus \{x\}) \cap \alpha^{-1}(Q) \subseteq P \subseteq (P \setminus \{x\}) \cup \alpha^{-1}(Q).$$

Thus, $x \in \alpha^{-1}(Q)$, a contradiction, because $r_{\mathcal{F}}(x) = k$ and, by (5.12),

$$|\alpha^{-1}(Q)| = |Q| = m < k.$$

It follows that $r_{\mathcal{G}}(y) = k$; that is, $y \in X_k^{\mathcal{G}}$.

We prove the converse statement by applying the above argument to the inverse isometry α^{-1}. $\qquad \square$

We proved that for every $k \geq 1$ the restriction of σ to $X_k^{\mathcal{F}}$ is a relation $\sigma_k \subseteq X_k^{\mathcal{F}} \times X_k^{\mathcal{G}}$.

Lemma 5.69. *The relation σ_k is a bijection for every $k \geq 1$.*

Proof. First we prove that σ_k is a function. Suppose that there are $z \neq y$ such that $(x, y) \in \sigma_k$ and $(x, z) \in \sigma_k$. Then there are two distinct sets $P, Q \in \mathcal{F}$ defining y and z, respectively, such that

$$x \in P \cap Q, \quad k = r_{\mathcal{F}}(x) = |P| = |Q|, \quad P \setminus \{x\} \in \mathcal{F}, \quad Q \setminus \{x\} \in \mathcal{F}.$$

By Lemma 5.67,

$$\alpha(P) \setminus \{y\} = \alpha(P \setminus \{x\}), \quad \alpha(Q) \setminus \{z\} = \alpha(Q \setminus \{x\}),$$

for some $y \in \alpha(P)$ and $z \in \alpha(Q)$. We have

$$d\left(\alpha(P), \alpha(Q)\right) = d\left(P, Q\right) = d\left(P \setminus \{x\}, Q \setminus \{x\}\right)$$
$$= d\left(\alpha(P) \setminus \{y\}, \alpha(Q) \setminus \{z\}\right).$$

Thus, $y, z \in \alpha(P) \cap \alpha(Q)$. In particular, $z \in \alpha(P) \setminus \{y\}$, a contradiction, because $|\alpha(P) \setminus \{y\}| = k - 1$ but, by Lemma 5.68, $r_{\mathcal{G}}(z) = k$.

By applying the above argument to α^{-1}, we prove that for any $y \in X_k^{\mathcal{G}}$ there is a unique $x \in X_k^{\mathcal{F}}$ such that $(x, y) \in \sigma_k$. Hence, σ_k is a bijection. $\qquad \square$

Let us recall that nonempty sets $X_k^{\mathcal{F}}$ and $X_k^{\mathcal{G}}$ form partitions of the sets $D(\mathcal{F})$ and $D(\mathcal{G})$, respectively. Therefore, we established the following result:

Corollary 5.70. *The relation σ is a bijection from $D(\mathcal{F})$ onto $D(\mathcal{G})$.*

$\mathcal{F}$ and $\mathcal{G}$ are finite families of finite sets, therefore the bijection σ can be extended to a permutation of the set X. We denote this permutation by the same symbol σ.

Lemma 5.71. $\alpha(P) = \hat{\sigma}(P)$ *for $P \in \mathcal{F}$.*

Proof. The proof is by induction on $k = |P|$. The case $k = 1$ is trivial, inasmuch as $\alpha(\{x\}) = \{\sigma_1(x)\}$ for $\{x\} \in \mathcal{F}$.

Suppose that $\alpha(R) = \hat{\sigma}(P)$ for all $R \in \mathcal{F}$ such that $|R| < k$. Let P be a set in $\mathcal{F}$ of cardinality k. By Lemma 5.65, $P = R \cup \{x\}$ for some $R \in \mathcal{F}$ and $x \notin R$. It is clear that $m = r_{\mathcal{F}}(x) \leq k$ and $|R| = k - 1$.

If $m = k$, then $\alpha(P) = \alpha(R) \cup \{\sigma(x)\} = \hat{\sigma}(P)$, by the definition of σ and the induction hypothesis.

Suppose that $m < k$. Because $m = r_{\mathcal{F}}(x)$, there is a set $Q \in \mathcal{F}$ containing x such that $|Q| = m$. By Lemma 5.66, there is $S \in \mathcal{F}$ such that $S = Q \setminus \{x\}$. Because $x \in P$, we have

$$S \cap P \subseteq Q \subseteq S \cup P,$$

which implies

$$\alpha(S) \cap \alpha(P) \subseteq \alpha(Q) \subseteq \alpha(S) \cup \alpha(P)$$

Thus, by the induction hypothesis,

$$\sigma(S) \cup \{\sigma(x)\} = \sigma(Q) \subseteq \sigma(S) \cup \alpha(P).$$

$\sigma(x) \notin \sigma(S)$, thus we have $\sigma(x) \in \alpha(P)$. Because $\alpha(P) = \sigma(R) \cup \{y\}$ for $y \notin \sigma(R)$, and $x \notin R$, we have $y = \sigma(x)$; that is, $\alpha(P) = \sigma(P)$.   □

The claim of Theorem 5.63 follows from the last lemma. The result of Theorem 5.63 holds for some infinite wg-families; see Exercise 5.19. The theorem can be also reformulated as follows.

Theorem 5.72. *For any two finite isomorphic partial cubes G_1 and G_2 on a set X, there is an automorphism of the cube $\mathcal{H}(X)$ that maps one of the partial cubes onto the other. Moreover, for any isomorphism $\alpha : G_1 \to G_2$, there is an automorphism $\sigma : \mathcal{H}(X) \to \mathcal{H}(X)$ such that $\sigma|_{G_1} = \alpha$.*

In conclusion, we present a geometric interpretation of Theorems 5.63 and 5.72. Let $\mathcal{M}$ be a nonempty family of subsets of a metric space Y. We say that X is $\mathcal{M}$-*homogeneous* if for any two sets $A, B \in \mathcal{M}$ and an isometry from A onto B, this isometry can be extended to an isometry of the space Y onto itself. By Theorem 5.72, the cube $\mathcal{H}(X)$ is $\mathcal{M}$-homogeneous with respect to the family $\mathcal{M}$ of all finite partial cubes on X.

When $\mathcal{M}$ is the set of all singletons of Y, the space Y is simply called *homogeneous* (see the text after Theorem 5.63). A metric space Y is said to be *fully homogeneous* if it is $\mathfrak{P}(Y)$-homogeneous. Cubes are homogeneous metric spaces. However, even finite cubes Q_n are not fully homogeneous if $n \geq 4$, as the following example illustrates.

Example 5.73. Let $X = \{a, b, c, d\}$. Consider two families of subsets of X:

$$\mathcal{A} = \{\varnothing, \{a, d\}, \{b, d\}, \{c, d\}\} \text{ and } \mathcal{B} = \{\varnothing, \{a, b\}, \{a, c\}, \{b, c\}\}.$$

Clearly, $\mathcal{A}$ and $\mathcal{B}$ are isometric. The distance from the set $\{d\}$ to all sets in $\mathcal{A}$ is one. On the other hand, it is easy to verify that there is no subset of X which is on distance one from all sets in $\mathcal{B}$ (see Figure 5.16). Thus an isometry from $\mathcal{A}$ onto $\mathcal{B}$ cannot be extended to an isometry from $\mathcal{H}(X)$ onto itself.

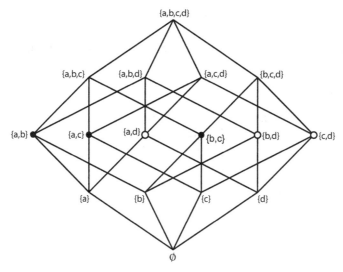

Figure 5.16. Families $\mathcal{A}$ and $\mathcal{B}$ in the hypercube Q_4 ($\varnothing \in \mathcal{A} \cap \mathcal{B}$).

5.11 Median Graphs

Let us recall (cf. Exercise 2.33) that a *median* of a triple of vertices $\{u, v, w\}$ of a connected graph G is a vertex in $I(u, v) \cap I(u, w) \cap I(v, w)$. A graph G is a *median graph* if every triple of vertices of G has a unique median. We denote the median of a triple $\{u, v, w\}$ in a median graph by $\langle u, v, w \rangle$.

The goal of this section is to show that the class of median graphs is a proper subclass of the class of partial cubes.

· Two remarks are in order. First, for any edge uv of a graph, $I(u, v) = \{u, v\}$. Therefore, $\langle u, v, w \rangle \in \{u, v\}$ for any vertex w of a median graph, provided that uv is an edge of the graph. Second, a median graph is bipartite. Indeed, suppose that it is not. Then, by Theorem 2.3, there is an edge uv and a vertex w of the graph such that $d(w, u) = d(w, v)$. It is clear that $I(u, v) \cap I(u, w) \cap I(v, w) = \varnothing$, a contradiction.

To prove that a median graph is a partial cube, we use the characterization of partial cubes established in Theorem 5.25. Specifically, we prove that two edges of a median graph that stand in the relation θ define equal pairs of opposite semicubes (see Theorem 5.25(ii)):

$$xy\,\theta\,uv \text{ implies } \{W_{xy}, W_{yx}\} = \{W_{uv}, W_{vu}\}. \tag{5.13}$$

We begin by proving a special case of the claim.

Lemma 5.74. *Let uv and xy be two edges of a median graph G such that $xy\,\theta\,uv$ with $x \in W_{uv}$, $y \in W_{vu}$, and $d(x, u) = 1$. Then $W_{xy} = W_{uv}$ and $W_{yx} = W_{vu}$.*

Note that the vertices u, x and v, y belong to the fundamental sets U_{uv} and U_{vu}, respectively (see Figure 5.17), and induce a square in G (cf. Exercise 2.16). In what follows, we use the result of Corollary 2.4 implicitly.

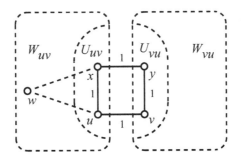

Figure 5.17. Proof of Lemma 5.74.

Proof. Opposite semicubes are complements of each other in the vertex set, therefore it suffices to show that $W_{uv} \subseteq W_{xy}$.

Let $w \in W_{uv}$. Because $d(u, x) = 1$, we have either (i) $d(w, u) = d(w, x) + 1$ or (ii) $d(w, x) = d(w, u) + 1$. We consider these two cases separately.

(i) By applying the triangle inequality to the triples $\{w, v, y\}$ and $\{w, y, x\}$, we have

$$d(w, v) \le d(w, y) + 1 \le d(w, x) + 2.$$

On the other hand,

$$d(w, v) = d(w, u) + 1 = d(w, x) + 2.$$

It follows that $d(w, y) = d(w, x) + 1$; that is, $w \in W_{xy}$.

(ii) $d(w, x) = d(w, u) + 1$, thus we have $u \in I(w, x)$. Clearly, $u \in I(w, u)$ and $u \in I(x, v)$ (see Figure 5.17). Thus, $u = \langle w, v, x \rangle$.

Because $d(x, y) = 1$, we have

$$\text{either} \quad d(w, y) = d(w, x) + 1 \quad \text{or} \quad d(w, x) = d(w, y) + 1.$$

Suppose that $d(w, x) = d(w, y) + 1$. Then $y \in I(w, x)$.

$$d(w, v) = d(w, u) + 1 = d(w, x) = d(w, y) + 1 = d(w, y) + d(y, v),$$

therefore we have $y \in I(w, v)$. Clearly, $y \in I(x, v)$ (see Figure 5.17). It follows that y is a median of the triple $\{w, v, x\}$. Because G is a median graph and $y \ne u = \langle w, v, x \rangle$, we have a contradiction. Therefore we must have $d(w, y) = d(w, x) + 1$, which means that $w \in W_{xy}$.

We proved that any vertex in W_{uv} is also in W_{xy}. Hence, $W_{uv} \subseteq W_{xy}$ and the result follows. ∎

Suppose now that $d(x, u) > 1$ in (5.13) (as in Lemma 5.74, we assume that $x \in W_{uv}$, $y \in W_{vu}$). Let $u = u_0, u_1, \ldots, u_n = x$ be a shortest ux-path. By the result of Exercise 2.15, this path is a subset of W_{uv}. The median $v_1 = \langle u_1, v, y \rangle$ is in W_{vu}, inasmuch as $I(v, y) \subseteq W_{vy}$ (by the same Exercise 2.15). Because $v_1 \in I(v, u_1)$ and $d(u_1, v) = 2$, the quadruple $\{u, u_1, v_1, v\}$ induces a square in G. By Lemma 5.74, we have $W_{u_1, v_1} = W_{uv}$ and $W_{v_1, u_1} = W_{vu}$. Clearly, $u_1 v_1 \,\theta\, uv$.

We apply the construction from the previous paragraph to edges $u_1 v_1$ and xy resulting in an edge $u_2 v_2$ with $W_{u_2, v_2} = W_{u_1 v_1}$ and $W_{v_2, u_2} = W_{v_1 u_1}$ (see Figure 5.18).

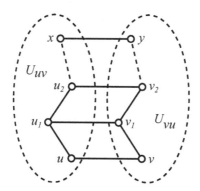

Figure 5.18. Sequence of squares.

An obvious inductive argument shows that $W_{xy} = W_{uv}$ and $W_{yx} = W_{vu}$. Thus, by Theorem 5.25, G is a partial cube. On the other hand, the graph obtained from the cube Q_3 by deleting one vertex is a partial cube but not a median graph (cf. Exercise 5.21). This completes the proof of the following theorem.

Theorem 5.75. *A median graph is a partial cube. Furthermore, the class of median graphs is a proper subclass of the class of partial cubes.*

5.12 Average Length and the Wiener Index

Let $G = (V, E)$ be a finite graph on n vertices. We recall the definitions of the *total distance* $\mathrm{td}(G)$ and the *average length* $\ell_{av}(G)$ of G (cf. Exercise 3.26):

$$\mathrm{td}(G) = \frac{1}{2} \sum_{u, v \in V} d(u, v), \quad \ell_{av}(G) = \frac{\sum_{u, v \in V} d(u, v)}{n(n-1)}.$$

Note that there are $n(n-1)$ nonzero terms in the sum $\sum_{u, v \in V} d(u, v)$.

In this section we compute these functions for a partial cube G on n vertices. First, we assume that $G = G_{\mathcal{F}}$ for a wg-family $\mathcal{F}$ of subsets of a finite set X such that $\cup \mathcal{F} = X$ and $\cap \mathcal{F} = \varnothing$.

By Theorem 5.18, we can label pairs of opposite semicubes by elements of the set X as follows:

$$W_x = \{R \in \mathcal{F} : x \in R\}, \quad \overline{W}_x = \{R \in \mathcal{F} : x \notin R\}, \quad x \in X.$$

$\chi_{A \triangle B} = \chi_A + \chi_B - 2\chi_A \cdot \chi_B$ (cf. Exercise 5.29), thus we have the following formula for the Hamming distance:

$$d(A, B) = \sum_{x \in X} (\chi_A(x) + \chi_B(x) - 2\chi_A(x) \cdot \chi_B(x)).$$

Therefore,

$$\sum_{A, B \in \mathcal{F}} d(A, B) = \sum_{A, B \in \mathcal{F}} \sum_{x \in X} (\chi_A(x) + \chi_B(x) - 2\chi_A(x) \cdot \chi_B(x))$$

$$= \sum_{x \in X} \sum_{A, B \in \mathcal{F}} (\chi_A(x) + \chi_B(x) - 2\chi_A(x) \cdot \chi_B(x)).$$

For a given x, the term in the last sum is nonzero if and only if $A \triangle B = \{x\}$. Note also that this nonzero term is 1. Either $A \in W_x$, $B \in \overline{W}_x$ or $B \in W_x$, $A \in \overline{W}_x$, therefore we have

$$\sum_{A, B \in \mathcal{F}} (\chi_A(x) + \chi_B(x) - 2\chi_A(x) \cdot \chi_B(x)) = 2|W_x||\overline{W}_x|.$$

Hence,

$$d(A, B) = 2 \sum_{x \in X} |W_x||\overline{W}_x|.$$

We obtained the following result.

Theorem 5.76. *Let $\mathcal{F}$ be a finite wg-family of subsets of a finite set X such that $\cup \mathcal{F} = X$ and $\cap \mathcal{F} = \varnothing$. Then*

$$td(G_{\mathcal{F}}) = \sum_{x \in X} |W_x||\overline{W}_x| \quad \text{and} \quad \ell_{av}(G_{\mathcal{F}}) = \frac{2}{n(n-1)} \sum_{x \in X} |W_x||\overline{W}_x|,$$

where $n = |\mathcal{F}|$.

Any finite partial cube G is isomorphic to a partial cube $G_{\mathcal{F}}$ for some wg-family $\mathcal{F}$ of subsets of a finite set X satisfying conditions $\cup \mathcal{F} = X$ and $\cap \mathcal{F} = \varnothing$. By Theorem 5.34, $\dim_I(G_{\mathcal{F}}) = |X|$, so the above theorem can be reformulated as follows.

Theorem 5.77. *Let G be a finite partial cube. Then*

$$td(G) = \sum_{i=1}^{m} |W_i||\overline{W}_i| \text{ and } \ell_{av}(G) = \frac{2}{n(n-1)} \sum_{i=1}^{m} |W_i||\overline{W}_i|,$$

where $n = |V(G)|$, $m = \dim_I(G)$, and $\{W_i, \overline{W}_i\}_{1 \le i \le m}$ is the family of mutually opposite semicubes of G.

Finite subgraphs of the hexagonal lattice (see Figure 1.15) play an important role in chemical graph theory. In the framework of this theory, the total distance of a graph G is called the *Wiener index* and denoted by $W(G)$. Thus, by definition,

$$W(G) = \frac{1}{2} \sum_{u,v \in V(G)} d(u,v).$$

Let C be a cycle of the hexagonal lattice. A *benzenoid graph* is formed by the vertices and edges of this lattice lying on and in the interior of C. An example of a benzenoid graph is shown in Figure 5.19 (interior edges of the graph are indicated by dashed lines).

Note that the hexagonal lattice is a bipartite graph (Exercise 2.3). Moreover, we show in Chapter 7 that this lattice is a partial cube. It is clear from the example in Figure 5.19 that a benzenoid graph need not be an isometric subgraph of the hexagonal lattice.

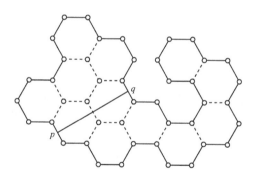

Figure 5.19. A benzenoid graph.

However, a benzenoid graph is a partial cube. We prove this claim by a geometric argument, assuming that the cells of the underlying lattice are regular hexagons. Let G be a benzenoid graph. A straight line segment S with ends p and q is said to be a *cut segment* if (i) S is perpendicular to one of the three edge directions, (ii) both p and q are the center of an edge, and (iii) the graph obtained from G by removing all edges intersected by S has exactly two components G_S and G'_S (see Figure 5.19). Let uv be an edge of G and S be a cut segment intersecting this edge. It is clear that the components G_S and

G'_S are the semicubes W_{uv} and W_{vu}, and that these subgraphs are convex. By Theorem 5.19, G is a partial cube.

Because the components G_S and G'_S are opposite semicubes of G, the edges that are intersected by the cut segments S form an equivalence class of the theta relation Θ on $E(G)$. Thus there is one-to-one correspondence between the set of cut segments and the quotient set $E(G)/\Theta$. By Theorem 5.77, we have the following result.

Theorem 5.78. *Let G be a benzenoid graph on n vertices and $E_1, \ldots, E_m$ be the equivalence classes of Θ. For $i = 1, \ldots, m$, let $u_i v_i \in E_i$ and $n_i = |W_{u_i v_i}|$. Then*

$$W(G) = \sum_{i=1}^{m} n_i (n - n_i).$$

Theorem 5.78 is particularly useful for finding closed formulas for the Wiener index of some families of benzenoid graphs (cf. Exercise 5.31).

5.13 Linear and Weak Orders

We denote by $\mathcal{LO}$ the family of all linear orders on a fixed set X of cardinality $n \geq 2$. It is clear that this family is not well-graded if $n > 2$ (cf. Figure 5.3). Therefore, $G_{\mathcal{LO}}$ is not a partial cube. However, there is a graph with the vertex set $\mathcal{LO}$ which is a partial cube. Let us denote by $G(\mathcal{LO})$ the graph on the vertex set $\mathcal{LO}$ in which a pair LL' of linear orders is an edge whenever L and L' are adjacent in the family $\mathcal{LO}$. This graph is a partial cube. We begin by outlining a proof of this claim.

Let L_0 be a fixed linear order on X and let $\mathcal{F} = \{L \cap L_0 : L \in \mathcal{LO}\}$ be the family of all intersections of linear orders in $\mathcal{LO}$ with L_0. First, we show that the assignment $L \mapsto L \cap L_0$ defines a one-to-one correspondence between families $\mathcal{LO}$ and $\mathcal{F}$. Second, we prove that $\mathcal{F}$ is a wg-family. Finally, we prove that the assignment $L \mapsto L \cap L_0$ defines a graph isomorphism from $G(\mathcal{LO})$ onto $G_{\mathcal{F}}$.

Lemma 5.79. *Let L_0, L, and L' be linear orders. If $L \cap L_0 = L' \cap L_0$, then $L = L'$.*

Proof. The proof is by contradiction. Suppose that $L \neq L'$. Then there is an ordered pair (x, y) such that $(x, y) \in L$ and $(x, y) \notin L'$. Accordingly, $(y, x) \notin L$ and $(y, x) \in L'$. Note that either $(x, y) \in L_0$ or $(y, x) \in L_0$.

If $(x, y) \in L_0$, then $(x, y) \in L \cap L_0 = L' \cap L_0$, which implies $(x, y) \in L'$, a contradiction.

If $(y, x) \in L_0$, then $(y, x) \in L' \cap L_0 = L \cap L_0$ implying $(y, x) \in L$, again a contradiction. $\qquad\square$

By Lemma 5.79, the assignment $L \mapsto L \cap L_0$ defines a bijection from $\mathcal{LO}$ onto $\mathcal{F}$.

Let L be a linear order on X. Then there is a sequence called *permutation* $x_1, x_2, \ldots, x_n$ of elements of X such that, for any $x \neq y$ in X, $(x, y) \in L$ if $x = x_i$, $y = x_j$ with $i < j$ (cf. Exercise 5.33c). This property justifies the name "linear order" for L. We say that x covers y in L if $(x, y) \in L$ and there is no z such that $(x, z) \in L$ and $(z, y) \in L$. In other words, x covers y in L if $x = x_k$, $y = x_{k+1}$ for some $1 \leq k < n$ in the permutation defined by L. (One can think of the permutation of X defined by L as of a list of elements of X in a "decreasing order" with respect to L in which each element covers the next one.)

If L and L' are distinct linear orders, then there is $(x, y) \in L$ such that $(x, y) \notin L'$, implying, by the completeness property of L', $(y, x) \in L'$. The next lemma is a stronger version of this simple observation.

Lemma 5.80. *Let L and L' be two distinct linear orders on X. There are elements $x, y \in X$ such that x covers y in L and $(y, x) \in L'$.*

Proof. Let $X = \{x_1, \ldots, x_n\}$ be the enumeration of X defined by L. We cannot have $(x_k, x_{k+1}) \in L'$ for all $1 \leq k < n$, for otherwise we would have $L' = L$, because L' is a linear order. Hence, there is k such that $(x_k, x_{k+1}) \notin L'$. The result follows for $x = x_k$, $y = x_{k+1}$, because x_k covers x_{k+1} in L and $(x_{k+1}, x_k) \in L'$. $\qquad\square$

Lemma 5.81. *Let L be a linear order on X. For $(x, y) \in L$, the relation*

$$L' = (L \setminus \{(x, y)\}) \cup \{(y, x)\}$$

is a linear order if and only if x covers y in L. Furthermore, in this case y covers x in L'.

Proof. (Necessity.) Suppose that L' is a linear order and there is $z \in X$ such that $(x, z) \in L$ and $(z, y) \in L$. Then $(x, z) \in L'$ and $(z, y) \in L'$, because $z \neq x$ and $z \neq y$. By transitivity of L', we have $(x, y) \in L'$, a contradiction. Hence, x covers y in L.

(Sufficiency.) Suppose that x covers y in L. Then, for some k, we have $x = x_k$, $y = x_{k+1}$ in the permutation $x_1, \ldots, x_n$ defined by L. Let L' be a linear order defined by the permutation $x_1, \ldots, x_{k+1}, x_k, \ldots, x_n$. For any x_i and x_j with $i < j$ and $i \neq k$, $j \neq k+1$, the pair (x_i, x_j) belongs to both L and L'. Therefore, $L' = (L \setminus \{(x_k, x_{k+1})\}) \cup \{(x_{k+1}, x_k)\}$ and x_{k+1} covers x_k in L'. $\qquad\square$

Now we are ready to prove the second claim in our outline.

Theorem 5.82. *The family $\mathcal{F} = \{L \cap L_0 : L \in \mathcal{L}0\}$ is well-graded.*

Proof. We use the criterion from Theorem 5.9. Let

$$P = L \cap L_0 \quad \text{and} \quad P' = L' \cap L_0$$

be two adjacent elements of $\mathcal{F}$. By Lemma 5.80, there are $x, y \in X$ such that x covers y in L and $(y, x) \in L'$. By Lemma 5.81, the relation

$$L'' = (L \setminus \{(x, y)\}) \cup \{(y, x)\}$$

is a linear order. Clearly, $L \cap L' \subseteq L'' \subseteq L \cup L'$. Therefore,

$$P \cap P' = L \cap L' \cup L_0 \subseteq L'' \cap L_0 \subseteq (L \cup L') \cap L_0 = P \cup P';$$

that is, $P'' = L'' \cap L_0$ lies between P and P' in $\mathcal{F}$.

We have

$$P'' = L'' \cap L_0 = [(L \cap L_0) \setminus (\{(x, y)\} \cap L_0)] \cup ([\{(y, x)\} \cap L_0]$$

$$= \begin{cases} P \setminus \{(x, y)\}, & \text{if } (x, y) \in L_0, \\ P \cup \{(y, x)\}, & \text{if } (x, y) \notin L_0. \end{cases}$$

Thus, the Hamming distance $d(P, P'') = 1$. Inasmuch as $P'' \neq P$ and the relations P and P' are adjacent in $\mathcal{F}$, we must have $P'' = P'$. By Theorem 5.9, the family $\mathcal{F}$ is well-graded. $\qquad\square$

The next lemma describes edges of the graph $G(\mathcal{LO})$.

Lemma 5.83. *Two linear orders L and L' are adjacent in $\mathcal{LO}$ if and only if there are $x, y \in X$ such that x covers y in L and*

$$L' = (L \setminus \{(x, y)\}) \cup \{(y, x)\}.$$

Proof. (Necessity.) Let L and L' be two linear orders that are adjacent in $\mathcal{LO}$. By Lemma 5.80, there are $x, y \in X$ such that x covers y in L and $(y, x) \in L'$. By Lemma 5.81, the relation

$$L'' = (L \setminus \{(x, y)\}) \cup \{(y, x)\}$$

is a linear order. Clearly, L'' lies between L and L' and is distinct from L. Hence, $L'' = L'$.

(Sufficiency.) We need to show that L and $L' = (L \setminus \{(x, y)\}) \cup \{(y, x)\}$ are adjacent in $\mathcal{LO}$. Let L'' be a relation lying between L and L'; that is,

$$L \cap L' = L \setminus \{(x, y)\} \subseteq L'' \subseteq L \cup \{(y, x)\} = L \cup L'.$$

$L \setminus \{(x, y)\}$ and $L \cup \{(y, x)\}$ are not linear orders, therefore we must have either $L'' = L$ or $L'' = L'$. The result follows. $\qquad\square$

Finally, we prove that graphs $G(\mathcal{LO})$ and $G_{\mathcal{F}}$ are isomorphic.

Theorem 5.84. *The assignment $L \to L \cap L_0$ defines a graph isomorphism from $G(\mathcal{LO})$ onto $G_{\mathcal{F}}$.*

Proof. By Lemma 5.79, $L \to L \cap L_0$ is a one-to-one correspondence between $\mathcal{LO}$ and $\mathcal{F}$. Thus we need to prove two claims: (1) for any two linear orders L and L' adjacent in $\mathcal{LO}$, the relations $L \cap L_0$ and $L' \cap L_0$ are adjacent in $\mathcal{F}$, and (2) for any two relations $L \cap L_0$ and $L' \cap L_0$ that are adjacent in $\mathcal{F}$, the linear orders L and L' are adjacent in $\mathcal{LO}$.

(1) Let L and L' be two linear orders adjacent in $\mathcal{LO}$. By Lemma 5.83, there are $x, y \in X$ such that x covers y in L and $L' = (L \setminus \{(x, y)\}) \cup \{(y, x)\}$. Then

$$L' \cap L_0 = [(L \cap L_0) \setminus (\{(x, y)\} \cap L_0)] \cup ([\{(y, x)\} \cap L_0]$$
$$= \begin{cases} (L \cap L_0) \setminus \{(x, y)\}, & \text{if } (x, y) \in L_0, \\ (L \cap L_0) \cup \{(y, x)\}, & \text{if } (x, y) \notin L_0. \end{cases}$$

Therefore, the relations $L \cap L_0$ and $L' \cap L_0$ are adjacent in $\mathcal{F}$.

(2) Let $L \cap L_0$ and $L' \cap L_0$ be two relations adjacent in $\mathcal{F}$. By symmetry, we may assume that

$$L' \cap L_0 = (L \cap L_0) \setminus \{(x, y)\}, \tag{5.14}$$

for some $(x, y) \in L \cap L_0$. Let (u, v) be an ordered pair in L different from (x, y). There are two possible cases:

i) $(u, v) \in L_0$. Then, by (5.14), $(u, v) \in L'$, inasmuch as $(u, v) \neq (x, y)$.

ii) $(u, v) \notin L_0$. Suppose that $(u, v) \notin L'$. Then $(v, u) \in L' \cap L_0$. By (5.14), $(v, u) \in L$. This contradicts our assumption that $(u, v) \in L$. Therefore, (u, v) is in L'.

We proved that any ordered pair (u, v) in L distinct from (x, y) (which is also in L) belongs to L'. Because L' is distinct from L and these relations are linear orders, we conclude that $L' = (L \setminus \{(x, y)\}) \cup \{(y, x)\}$. It follows that L and L' are adjacent in $\mathcal{LO}$. $\square$

Now we consider another class of binary relations that is of importance in applications. A partial order W on a set X is called a *weak order* if it is negatively transitive; that is, $(x, y) \in W$ implies $(x, z) \in W$ or $(z, y) \in W$, for any $x, y, z \in X$. As before, we assume that X is a given set of cardinality $n \geq 2$.

We denote by $\mathcal{WO}$ the set of all weak orders on X. This set is partially ordered by the inclusion relation. The graph in Figure 5.20 is the Hasse diagram of $\mathcal{WO}$ for $X = \{a, b, c\}$ (cf. the graph in Figure 5.3). It is clear that the family $\mathcal{WO}$ is not well-graded. For instance, there is no weak order on Hamming distance one from the empty weak order.

The result of the following lemma can be used as a constructive definition of a weak order on X. The proof is left as an exercise (cf. Exercise 5.35).

Lemma 5.85. *A binary relation W on X is a weak order if and only if there is a partition $X = X_1 \cup \cdots \cup X_k$, $1 \leq k \leq n$, such that $(x, y) \in W$ if and only if $x \in X_i$, $y \in X_j$ for some $i < j$.*

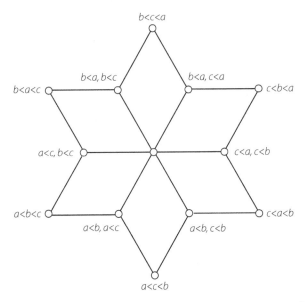

Figure 5.20. The graph of the family of weak orders on $X = \{a, b, c\}$.

Example 5.86. The linear order $c < b < a$ in Figure 5.20 is defined by the partition $\{a\} \cup \{b\} \cup \{c\}$, and the weak order $a < b$, $c < b$ in the same figure is defined by the partition $\{b\} \cup \{a, c\}$.

If $X = X_1 \cup \cdots \cup X_k$ is a partition defining a weak order W, we say that W is a weak k-order, denote it by $W = \langle X_1, \ldots, X_k \rangle$, and call it an *ordered partition*. For instance, weak n-orders are linear orders, and the only weak 1-order is the empty weak order. The set of all weak k-orders on X is denoted by $\mathcal{WO}(k)$.

As in the case of linear orders, we introduce a graph on the vertex set $\mathcal{WO}$ by defining its edges to be pairs WW' of weak orders that are adjacent to each other in $\mathcal{WO}$, and denote this graph by $G(\mathcal{WO})$.

Our goal is to show that this graph is a partial cube. For this we need some structural properties of weak orders. The proofs of the next two lemmas and corollary are omitted (cf. Exercise 5.36).

Lemma 5.87. *Two weak orders W and W' are adjacent in $\mathcal{WO}$ if and only if either*

$$W = \langle X_1, \ldots, X_k \rangle \text{ and } W' = \langle X_1, \ldots, X_i \cup X_{i+1}, \ldots, X_k \rangle$$

or

$$W' = \langle X_1, \ldots, X_k \rangle \text{ and } W = \langle X_1, \ldots, X_i \cup X_{i+1}, \ldots, X_k \rangle,$$

for some $1 \leq i < k$.

In words, one of two weak orders that are adjacent in $\mathcal{WO}$ is obtained from the other by joining two consecutive elements of its ordered partition. Note that, for $i = k - 1$, we have $\langle X_1, \ldots, X_{k-1} \cup X_k \rangle$ in the above formulas.

Lemma 5.88. *A weak order $W = \langle X_1, \ldots, X_k \rangle$ contains a weak order W' if and only if*

$$W' = \left\langle \bigcup_{i=1}^{j_1} X_i, \ \bigcup_{i=j_1+1}^{j_2} X_i, \ldots, \ \bigcup_{i=j_m+1}^{k} X_i \right\rangle,$$

for some sequence of indices $1 \le j_1 < \cdots < j_m < k$.

One can say that $W' \subseteq W$ if elements of the ordered partition W' are "enlargements" of the consecutive elements of the ordered partition W.

Corollary 5.89. *Two weak orders are adjacent in $\mathcal{WO}$ if and only if they are adjacent in the Hasse diagram of the poset $\mathcal{WO}$. Accordingly, the graph $G(\mathcal{WO})$ is the Hasse diagram of $\mathcal{WO}$.*

For a weak order $W \in \mathcal{WO}$, we denote by J_W the set of all weak 2-orders that are contained in W:

$$J_W = \{U \in \mathcal{WO}(2) : U \subseteq W\}$$

and denote by $\mathcal{F}$ the family of all such subsets of $\mathcal{WO}(2)$:

$$\mathcal{F} = \{J_W : W \in \mathcal{WO}\}.$$

To prove that $G(\mathcal{WO})$ is a partial cube, we first show that posets $G(\mathcal{WO})$ and $\mathcal{F}$ are isomorphic. Then we prove that $\mathcal{F}$ is a wg-family.

Theorem 5.90. *A weak order admits a unique representation as a union of weak 2-orders; that is, for any $W \in \mathcal{WO}$ there is a unique set $J \subseteq \mathcal{WO}(2)$ such that*

$$W = \bigcup_{U \in J} U. \tag{5.15}$$

Furthermore, $J = J_W$ in (5.15).

Proof. If W is the empty weak order then it has a unique representation in the form (5.15) for $J = \varnothing$. Thus we may assume that $W = \langle X_1 \ldots, X_k \rangle$ where $k \ge 2$. By Lemma 5.88, each weak order in J_W is in the form

$$W_i = \left\langle \bigcup_{j=1}^{i} X_j, \bigcup_{i+1}^{k} X_j \right\rangle, \quad 1 \le i < k.$$

Let us prove that $W = \bigcup_{i=1}^{k-1} W_i$. If $(x, y) \in W$, then $x \in X_p$, $y \in X_q$ with $p < q$. Hence, $(x, y) \in W_p$. Therefore, $(x, y) \in \bigcup_{i=1}^{k-1} W_i$. On the other hand, if (x, y) belongs to the union of W_i, then it belongs to one of them. Clearly, this implies $(x, y) \in W$. We established (5.15) for $J = J_W$.

Now let $W = \langle X_1, \ldots, X_k \rangle$ be a weak order representable in the form (5.15) for some $J \subseteq \mathcal{WO}(2)$. We prove that $J = J_W$. Clearly,

$$J \subseteq J_W = \{W_1, \ldots, W_{k-1}\}.$$

Suppose that there is $W_p \notin J$. For any $x \in X_p$, $y \in X_{p+1}$, we have $(x, y) \in W$ and $(x, y) \notin W_i$ for $i \neq p$. This contradicts our assumption that (5.15) holds for W. It follows that $J = J_W$. □

By Theorem 5.90, the correspondence $W \mapsto J_W$ establishes an isomorphism between the posets $\mathcal{WO}$ and $\mathcal{F}$.

Theorem 5.91. *The family $\mathcal{F}$ is an independence system. Accordingly, the graph $G_{\mathcal{F}}$ is a partial cube (cf. Theorem 5.11).*

Proof. We need to show that the family $\mathcal{F}$ is closed under taking subsets. Let $W' = \bigcup_{U \in J'} U$ for some subset J' of $J_W \in \mathcal{F}$. As the union of negatively transitive relations, the relation W' itself is negatively transitive (cf. Exercise 5.37). It is a partial order, because $W' \subseteq W$. Therefore, W' is a weak order. By Theorem 5.90, $J' = J_{W'} \in \mathcal{F}$. □

$\mathcal{F}$ is a wg-family (cf. Theorem 5.11), therefore the graph $G_{\mathcal{F}}$ is the Hasse diagram of $\mathcal{F}$. Because posets $\mathcal{WO}$ and $\mathcal{F}$ are isomorphic, their Hasse diagrams are also isomorphic. By Corollary 5.89 and Theorem 5.91, we have the following result.

Theorem 5.92. *The graph $G(\mathcal{WO})$ is a partial cube.*

Notes

Isometric embeddings of graphs into cubes were first studied by Firsov (1965) (cf. Notes to Chapter 4). The term "partial cube" was coined by Hans-Jürgen Bandelt and appeared for the first time in Imrich and Klavžar (1998) (see also Imrich and Klavžar, 2000).

The concept of a well-graded family of sets was introduced by Falmagne and Doignon (1997) in connection with studies in the area of stochastic evolution of preference structures (see also Falmagne, 1997). It plays a key role in media theory (see Eppstein et al. (2008) and Chapter 8 of this book). Under a different name, wg-families were introduced earlier in Kuzmin and Ovchinnikov (1975) and Ovchinnikov (1980).

The well-gradedness property of the family of partial orders on a finite set (Theorem 5.13) was established by Kenneth P. Bogart (1973). The result of Theorem 5.14 for arbitrary partial orders is known as Szpilrajn's Extension

Theorem. It was established by the Polish mathematician Edward Szpilrajn (later known as Edward Marczewski). In Szpilrajn (1930) he attributes it to S. Banach, M. Kuratowski, and A. Tarski.

As isometric subgraphs of cubes, partial cubes inherit many fine metric properties of cubes (cf. Section 5.4). The metric structures of partial cubes expressed in terms of fundamental sets are the main tools in studying these graphs. Unlike cubical graphs, partial cubes can be effectively characterized. Historically, the first characterization (Theorem 5.19 in Section 5.5) was obtained by Dragomir Djoković (1973) who also introduced the relation θ. Part (iii) of Theorem 5.19 for the relation Θ is due to Peter Winkler (1984). The results of Theorems 5.25 and 5.26 are found in Ovchinnikov (2008c). For more characterizations of partial cubes and related problems see Avis (1981), Roth and Winkler (1986), and Chepoi (1988, 1994). The relation $\mathcal{L}$ is called the *like relation of a graph* in Eppstein et al. (2008), where it appears in the context of media theory (see Chapter 8). Metric structures of partial cubes allow for effective recognition algorithms for these graphs; see Imrich and Klavžar (2000) and Eppstein (2008). Theorem 5.34 in Section 5.6 was established in Djoković (1973).

Pasting (gluing) together two spaces is a standard technique in topology; see, for instance, Bourbaki (1966) and Munkres (2000). In the context of graph theory, this concept was adopted in Ovchinnikov (2008c).

Mulder (1980) introduced graph expansions in his studies of median graphs. Specifically, he proved that finite median graphs can be obtained from the graph K_1 by a sequence of convex expansions. For partial cubes, this result (cf. Theorem 5.60) was obtained by Chepoi (1988) (see also Chepoi, 1994) who used isometric expansions (cf. Definition 5.47). In the context of the oriented matroid theory, the result of Theorem 5.60 was obtained by Fukuda and Handa (1993).

An isometric embedding of a graph into a cube is unique up to an automorphism of the cube (Theorem 5.72). This property of partial cubes is also known as ℓ_1-rigidity of partial cubes; see Deza and Laurent (1997).

Median graphs have a long history and an extensive list of applications. The term "median graph" was coined by Ladislav Nebeský (1971). The reader is referred to the books by Mulder (1980) and Imrich and Klavžar (2000), and the survey by Bandelt and Chepoi (2008), where additional references to the pertinent literature on median graphs are found.

Benzenoid graphs are of importance in chemistry where they represent benzenoid hydrocarbons. For more information about this applied area see Gutman and Cyvin (1989). Klavžar et al. (1995) proved that benzenoid graphs are partial cubes.

The family $\mathcal{PO}$ of all partial orders on a given set is well-graded (Theorem 5.13) and therefore can be modeled as a partial cube. Some proper subfamilies of $\mathcal{PO}$ are also well-graded. They are studied in Chapter 7. The families $\mathcal{LO}$ and $\mathcal{WO}$ of linear and weak orders, respectively, are examples of families of sets that are not well-graded, but still can be modeled as partial cubes. These constructions require many fine structural properties of these orders. In Section 5.13, some of these properties are presented as exercises. The reader who is inclined to learn more about these relations is referred to the books by Fishburn (1985), Mirkin (1979), Roberts (1979), and Trotter (1992).

Exercises

5.1. Use embeddings from Section 4.1 to show that paths, even cycles, and trees are partial cubes.

5.2. Let G be a cubical graph on seven vertices that is not a partial cube. Show that G is isomorphic to the graph in Figure 5.1a.

5.3. Represent the double ray $\mathbb{Z}$ as a partial cube on some set X.

5.4. Let R be a partial order on a finite set X. Show that there is at least one maximal element in X with respect to R. Give an example of a partial order for which there is more than one maximal element in X.

5.5. Show that the intersection of two partial orders is a partial order.

5.6. Show that semicubes of a tree are convex sets.

5.7. Let G be the $(m \times n)$-grid.

a) Show that the semicubes of G are convex.
b) Describe the relation θ and show that it is an equivalence relation on the edge set of G.
c) Show that $\dim_I(G) = m + n - 2$.

5.8. Give an example of a partial cube G different from Q_3 for which $\dim_c(G) = \dim_I(G) = 3$.

5.9. Prove that $\lceil \log_2 n \rceil < n - 1$ for $n > 3$.

5.10. Let $C_{2n} = v_0, v_1, \ldots, v_{2n-1}, v_{2n} = v_0$ be an even cycle. Show that $(v_i v_{i+1}, v_j, v_{j+1}) \in \Theta$ for $0 \le i < j < 2n$ if and only if $j = i + n$. Deduce that Θ is an equivalence relation on the edge set of C_{2n} with equivalence classes consisting of pairs $\{v_i, v_{i+1}, v_{i+n}v_{i+n+1}\}$, $0 \le i < n$, of "opposite" edges of the cycle.

5.11. Prove that an isometric subgraph of a partial cube is a partial cube.

5.12. Prove that a semicube of a partial cube is a partial cube.

5.13. Show that the colouring in Theorem 5.27 is proper.

5.14. Give an example of a connected graph G that has two edges uv, xy such that $(uv, xy) \in \Theta$, but $((u, v), (x, y)) \notin \mathfrak{L}$ and $((u, v), (y, x)) \notin \mathfrak{L}$ (cf. Lemma 5.28).

5.15. Show that a connected bipartite graph in which every edge is contained in at most one cycle is a partial cube.

5.16. Let $\{X_i\}_{i \in I}$ be a family of pairwise disjoint sets and, for every $i \in I$, let $\mathfrak{F}_i$ be a family of subsets of X_i such that $\cap \mathfrak{F}_i = \varnothing$ and $\cup \mathfrak{F}_i = X_i$. We denote by $\mathfrak{F}$ the family of all subsets R of the set $X = \cup_{i \in I} X_i$ such that $R \cap X_i \in \mathfrak{F}_i$ for all $i \in I$. Show that $\cap \mathfrak{F} = \varnothing$ and $\cup \mathfrak{F} = X$.

5.17. Prove Lemma 5.44(ii).

5.18. Prove theorems in Section 5.8 using wg-families of sets.

5.19. Let $\mathfrak{F}$ and $\mathfrak{G}$ be isometric wg-families of finite subsets of a set X such that
$$|\cup \mathfrak{F}| = |\cup \mathfrak{G}| \quad \text{and} \quad |X \setminus \cup \mathfrak{F}| = |X \setminus \cup \mathfrak{G}|$$
and let α be an isometry from $\mathfrak{F}$ onto $\mathfrak{G}$. Show that there is an isometry $\varphi : \mathfrak{F} \to \mathfrak{G}$ such that $\varphi|_{\mathfrak{F}} = \mathfrak{G}$ (cf. Theorem 5.63).

5.20. Let X be a metric space whose points are the six vertices of the truncated equilateral triangle shown in Figure 5.21 and distances between the points are the Euclidean distances in the plane.

a) Show that for any two isometric subsets of X there is an isometry of X that maps one of the subsets onto another.
b) Show that X is not fully homogeneous. (Hint: Let $Y = \{a, b\}$ in Figure 5.21. The isometry $\varphi : Y \to Y$ defined by $\varphi(a) = b$, $\varphi(b) = a$ is not expandable to an isometry of X onto itself.)

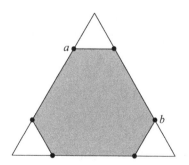

Figure 5.21. Exercise 5.20.

5.21. Let $X = \{a, b, c\}$ and $\mathcal{F} = \mathfrak{P}(X) \setminus \{\varnothing\}$. Show that $G_{\mathcal{F}}$ is a partial cube but not a median graph.

5.22. Prove that trees are median graphs.

5.23. Show that cycles are not median graphs.

5.24. Show that fundamental sets U_{ab} in a median graph induce convex subgraphs. Is this true for partial cubes?

5.25. Let A, B, and C be vertices of a cube. Show that a unique median of the triple $\{A, B, C\}$ is given by

$$\langle A, B, C \rangle = (A \cap B) \cup (A \cap C) \cup (B \cap C).$$

5.26. Let $u = (u_1, \ldots, u_n)$, $v = (v_1, \ldots, v_n)$, and $w = (w_1, \ldots, w_n)$ be three vertices of the cube Q^n with the vertex set $\{0, 1\}^n$. Show that the triple $\{u, v, w\}$ has a unique median $c = (c_1, \ldots, c_n)$ where

$$c_i = (u_i \wedge v_i) \vee (v_i \wedge w_i) \vee (w_i \wedge u_i), \quad i \in \{1, \ldots, n\}$$

(cf. Exercise 5.25). For notations $\wedge$ and $\vee$, see Section 3.3.

5.27. Let $u = (u_1, u_2)$, $v = (v_1, v_2)$, and $w = (w_1, w_2)$ be three vertices of the lattice $\mathcal{Z}^2$ with the vertex set $\mathbf{Z}^2$. Show that the triple $\{u, v, w\}$ has a unique median $c = (c_1, c_2)$, where

$$c_i = (u_i \wedge v_i) \vee (v_i \wedge w_i) \vee (w_i \wedge u_i), \quad i \in \{1, 2\}$$

(cf. Exercise 5.26).

5.28. Let c be a median of a triple of vertices $\{u, v, w\}$ of a graph. Show that the sum
$$d(x, u) + d(x, v) + d(x, w)$$
attains its minimum value at $x = c$.

5.29. Let X be a set. Prove the following properties of characteristic functions:

a) $\chi_{A \cup B} = \chi_A + \chi_B - \chi_A \cdot \chi_B$,
b) $\chi_{A \cap B} = \chi_A \cdot \chi_B$,
c) $\chi_{A \triangle B} = \chi_A + \chi_B - 2\chi_A \cdot \chi_B$,
d) $\chi_{\bar{A}} = 1 - \chi_A$,

where $A, B \subseteq X$ and $\bar{A} = X \setminus A$.

5.30. Let G be the graph shown in Figure 5.22.

a) Show that G is a partial cube.
b) Find the isometric dimension of G.
c) Find $\mathrm{td}(G)$ and $\ell_{av}(G)$.

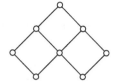

Figure 5.22. Exercise 5.30.

5.31. Let H_k be the kth benzenoid graph from the so-called coronene/circumcoronere series. The first three graphs of this series are displayed in Figure 5.23. Show that the Wiener index of H_k is given by

$$W(H_k) = \frac{164k^5 - 30k^3 + k}{5}$$

(Gutman and Klavžar, 1996; Imrich and Klavžar, 2000).

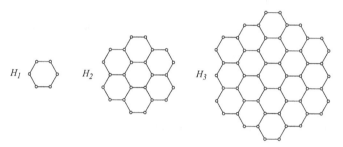

Figure 5.23. Exercise 5.31.

5.32. Let C be a cycle of the square lattice $\mathcal{Z}^2$ and G be the graph formed by the vertices and edges of the lattice lying on and in the interior of C (see Figure 5.24). Prove that G is a partial cube.

The geometric figure formed by the squares of G is called a *polyomino* (Golomb, 1994).

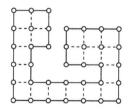

Figure 5.24. Exercise 5.32.

5.33. Let X be a finite set of cardinality $n \geq 2$, and let L be a linear order on X. An element $u \in X$ is a *minimum* element with respect to L if $(x, u) \in L$ for all $x \in X \setminus \{u\}$.

a) Show that there is a unique minimum element in X with respect to L.
b) Show that the restriction of L to a nonempty subset Y of X is a linear order on Y.
c) Show that there is an enumeration $X = \{x_1, \ldots, x_n\}$ such that $(x, y) \in L$ if and only if $x = x_i$ and $x = x_j$ for $i < j$.

5.34. Let L be a linear order on X and let $P \subseteq L$ be a partial order. Prove that $P = L \cap L'$ for some linear order L' if and only if $L \setminus P$ is a partial order (Eppstein et al., 2008, Theorem 3.5.8).

5.35. Prove Lemma 5.85. (Hint: Show that, for a given weak order W, the relation I_W defined by

$$(x, y) \in I_W \text{ if and only if } (x, y) \notin W \text{ and } (y, x) \notin W$$

is an equivalence relation on X. Then use the equivalence classes of I_W.)

5.36. Prove Lemmas 5.87, 5.88, and Corollary 5.89.

5.37. Show that the union of a family of negatively transitive relations is itself negatively transitive.

5.38. Let X be a set of cardinality n. Prove that

a) $|W\mathcal{O}(2)| = 2^n - 2$.
b) $|W\mathcal{O}(n-1)| = n!(n-2)/2$.
c) $|W\mathcal{O}| = \sum_{k=1}^{n} S(n, k)k!$, where $S(n, k)$ is the Stirling number of the second kind (cf. Stanley, 1997).

5.39. Let X be a set of cardinality n. Show that

$$\ell_{av}(\mathcal{LO}) = \frac{n!n(n-1)}{4(n!-1)} \sim 0.25(n^2 - 1),$$

where ℓ_{av} is the average length function (Section 5.12) and $\sim$ stands for "asymptotically equivalent".

5.40. Let X be a finite set, $n = |X|$, and $f(n) = |W\mathcal{O}|$ (cf. Exercise 5.38).

a) Show that, for $n = 3$, $\ell_{av}(W\mathcal{O}) = 30/13$.
b) Show that

$$\ell_{av}(W\mathcal{O}) = \frac{2\sum_{k=1}^{n-1} \binom{n}{k} f(k)f(n-k)[f(n) - f(k)f(n-k)]}{f(n)(f(n) - 1)}.$$

5.41. Let G be a partial cube on $\{1, \ldots, m\}$. We denote by $D(G)$ the distance matrix $(d_G(i,j))_{1 \leq i,j \leq m}$, and by $n_+(G)$ and $n_-(G)$ the number of positive and negative eigenvalues of $D(G)$, respectively. Prove that

a) $n_+(G) = 1$.
b) $n_-(G) = \dim_I(G)$.
c) $\det(D(G)) \neq 0$ if and only if G is a tree

(Graham and Winkler, 1985).

6

Lattice Embeddings

Any partial cube is isomorphic to an isometric subgraph of an integer lattice (Theorem 6.4). The dimension of this lattice may be much lower than the isometric dimension of the partial cube making lattice representations a valuable tool in visualizing large and infinite partial cubes.

6.1 Integer Lattices

If I is an infinite set and G is a connected graph, then the Cartesian power G^I is a disconnected graph (cf. Section 3.6). As we demonstrated in Example 3.13, the graph P_3^ω has two nonisomorphic components (weak Cartesian powers of P_3). On the other hand, for any I, the components of P_2^I are pairwise isomorphic (each is isomorphic to the cube on I). We begin this section by proving this claim for the Cartesian power $\mathcal{Z}^I$, where $\mathcal{Z}$ stands for the double ray (cf. Section 1.8).

Let us recall (cf. Section 3.5) that vertices of the Cartesian power $\mathcal{Z}^I$ are functions from I to the set of integers $\mathbf{Z}$. Two vertices $u, v : I \to \mathbf{Z}$ are adjacent in $\mathcal{Z}^I$ if there is $i \in I$ such that $|u(i) - v(i)| = 1$ and $u(j) = v(j)$ for all $j \neq i$. Let us denote by $\mathcal{Z}^I(a)$ the connected component of $\mathcal{Z}^I$ containing vertex a, and by $\mathcal{Z}(I)$ the component containing the zero function on I.

Theorem 6.1. *For any $a \in \mathcal{Z}^I$, the graphs $\mathcal{Z}^I(a)$ and $\mathcal{Z}(I)$ are isomorphic.*

Proof. We show that the mapping $\varphi : V(\mathcal{Z}^I(a)) \to V(\mathcal{Z}(I))$ defined by $\varphi(u) = u - a$ (a translation of $\mathcal{Z}^I$) is an isomorphism of $\mathcal{Z}^I(a)$ onto $\mathcal{Z}(I)$.

By definition, u is a vertex of $\mathcal{Z}^I(a)$ if and only if $u(i) - a(i) \neq 0$ for at most finitely many $i \in I$. Clearly, the former condition is equivalent to the claim that $\varphi(u)$ is a vertex of $\mathcal{Z}(I)$. Thus, φ is a bijection from $V(\mathcal{Z}^I(a))$ onto $V(\mathcal{Z}(I))$.

A pair uv is an edge of $\mathcal{Z}^I(a)$ if and only if there is $i \in I$ such that $|u(i) - v(i)| = 1$ and $u(j) = v(j)$ for all $j \neq i$, which is equivalent to

$$|\varphi(u)(i) - \varphi(v)(i)| = 1 \text{ and } \varphi(u)(j) = \varphi(v)(j), \text{ for all } j \neq i.$$

Hence, uv is an edge of $\mathcal{Z}^I(a)$ if and only if $\varphi(u)\varphi(v)$ is an edge of $\mathcal{Z}(I)$. The result follows. $\square$

As in the case of cubes, we use the component of an infinite Cartesian power of $\mathcal{Z}$ containing the zero function in our definition of an integer lattice.

Definition 6.2. Let $\mathbf{Z}_X$ be the set of all functions $X \to \mathbf{Z}$ that take nonzero value for at most finitely many $x \in X$. The *integer lattice* on X is the graph $\mathcal{Z}(X)$ having $\mathbf{Z}_X$ as its set of vertices; two vertices u, v are adjacent in $\mathcal{Z}(X)$ if

$$\begin{cases} |u(x) - v(x)| = 1, & \text{for some } x \in X, \text{ and} \\ u(y) = v(y), & \text{for all } y \in X \setminus \{x\}. \end{cases}$$

The *dimension* of the integer lattice $\mathcal{Z}(X)$ is the cardinality of the set X. If $|X| = n$, then $\mathcal{Z}(X) = \mathcal{Z}^n$.

The graphs of the double ray $\mathcal{Z}$ and the integer lattice $\mathcal{Z}^2$ are depicted in Figures 1.14 and 1.15, respectively. The graph distance on $\mathcal{Z}(X)$ is given by

$$d(u, v) = \sum_{x \in X} |u(x) - v(x)|$$

(cf. Exercise 3.20).

By Theorem 5.37, for any set X, the integer lattice $\mathcal{Z}(X)$ is a partial cube. We give a more concrete proof of this fact in the next theorem.

Theorem 6.3. *For any set X, the cube $\mathcal{H}(X)$ can be isometrically embedded into $\mathcal{Z}(X)$, and the integer lattice $\mathcal{Z}(X)$ can be isometrically embedded into the cube $\mathcal{H}(X \times \mathbf{Z})$.*

Proof. The function $\alpha : A \to \chi_A$, where χ_A is the characteristic function of $A \subseteq X$, is clearly an isometric embedding of the cube $\mathcal{H}(X)$ into the integer lattice $\mathcal{Z}(X)$.

In the other direction, for $u \in \mathbf{Z}_X$ define the set

$$B_u = \{(x, k) \in X \times \mathbf{Z} : k \leq u(x)\}.$$

We have

$$B_u \triangle B_v = \{(x, k) : u(x) < k \leq v(x) \text{ or } v(x) < k \leq u(x)\}.$$

Let $A_u = B_u \triangle B_0$, where 0 stands for the zero function on X. Clearly, each A_u is a finite set and $A_u \triangle A_v = B_u \triangle B_v$. Hence,

$$d(A_u, A_v) = |A_u \triangle A_v| = |B_u \triangle B_v| = \sum_{x \in X} |u(x) - v(x)| = d(u, v),$$

so $\beta : u \to A_u$ is an isometric embedding of the integer lattice $\mathcal{Z}(X)$ into the cube $\mathcal{H}(X \times \mathcal{Z})$. $\square$

Theorem 6.4. *A graph is a partial cube if and only if it is isometrically embeddable into an integer lattice.*

Proof. If a graph is partial cube, it can be isometrically embedded into some cube $\mathcal{H}(X)$, and by Theorem 6.3, into the lattice $\mathcal{Z}(X)$. On the other hand, if a graph is isometrically embeddable into an integer lattice $\mathcal{Z}(X)$, then by the same theorem it is also isometrically embeddable into $\mathcal{H}(X \times \mathbf{Z})$, and therefore is a partial cube.   □

As the isometric dimension of a partial cube G is the minimum dimension of a cube in which G is isometrically embeddable, its "lattice dimension" can be defined in terms of lattice embeddings.

Definition 6.5. The *lattice dimension* $\dim_{\mathcal{Z}}(G)$ of a partial cube G is the minimum dimension of an integer lattice in which G is isometrically embeddable.

It follows immediately from the first claim of Theorem 6.3 that the lattice dimension of a partial cube is bounded by its isometric dimension:

$$\dim_{\mathcal{Z}}(G) \leq \dim_I(G).$$

In fact, the lattice dimension of a partial cube may be much lower than its isometric dimension, making the lattice representation an invaluable tool in visualizing large and infinite partial cubes.

Example 6.6. Clearly, the lattice dimension of the path P_n is 1, whereas its isometric dimension is $n - 1$. Note also that $\dim_{\mathcal{Z}}(\mathcal{Z}) = 1$ and

$$\dim_I(\mathcal{Z}) = |\mathbf{Z}| = \aleph_0,$$

where $\aleph_0$ is the first infinite cardinal number.

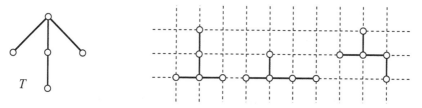

Figure 6.1. Three embeddings of the tree T into $\mathcal{Z}^2$.

Example 6.7. Let T be the tree depicted in Figure 6.1, left, together with its three isometric embeddings into $\mathcal{Z}^2$. It is clear that

$$2 = \dim_{\mathcal{Z}}(T) < \dim_I(T) = 4.$$

6.2 Tree Embeddings

As we know from Example 5.35, any tree T is a partial cube of isometric dimension $|E(T)|$. In this section, we show that the lattice dimension of a finite tree with n leaves is $\lceil n/2 \rceil$. All embeddings are assumed to be isometric.

A finite tree with exactly two leaves is a path (Exercise 6.1), thus it is embeddable into $\mathbb{Z}$ and therefore its lattice dimension is 1. In what follows, T is a finite tree with more than two leaves.

Lemma 6.8. *The lattice dimension of the star $K_{1,n}$ is $\lceil n/2 \rceil$.*

Proof. Suppose that $K_{1,n}$ is embedded into $\mathbb{Z}^d$. We may assume that the center of $K_{1,n}$ is represented by the zero vertex in $\mathbf{Z}^d$. Then each leaf in $K_{1,n}$ is represented by the head of a coordinate unit vector e_k or by the head of the vector $-e_k$. It follows that $2d \geq n$. Because d is an integer, we have $d \geq \lceil n/2 \rceil$.

To prove the claim it suffices to construct an embedding of $K_{1,n}$ into $\mathbb{Z}^{\lceil n/2 \rceil}$. Let $v_1, \ldots, v_n$ be the leaves and c be the center of $K_{1,n}$. We map c into the zero vertex of $\mathbb{Z}^{\lceil n/2 \rceil}$, v_1 and v_2 into the heads of e_1 and $-e_1$, respectively, and so on. It is clear that we obtain an embedding of $K_{1,n}$ into $\mathbb{Z}^{\lceil n/2 \rceil}$ (cf. Figure 6.2). □

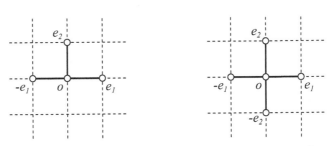

Figure 6.2. Embeddings of stars $K_{1,3}$ and $K_{1,4}$ into $\mathbb{Z}^2$.

Lemma 6.9. *Let T be a tree with n leaves. Suppose that T is embedded into $\mathbb{Z}^d$. Then*

$$d \geq \lceil n/2 \rceil.$$

Proof. Let uv be an inner edge of T, that is, an edge with ends that are not leaves. (If T has no inner edges, then it is a star and the result follows from Lemma 6.8.) Vertices u and v are represented by vertices of $\mathbb{Z}^d$ that are different only in one coordinate, say,

$$u = (x_1, x_2, \ldots, x_d), \quad v = (x'_1, x_2, \ldots, x_d)$$

with $|x_1 - x'_1| = 1$. Let us select all edges in $\mathbb{Z}^d$ with vertices that have the same first coordinates as u and v. Inasmuch as T is a tree and the embedding is isometric, the edge uv is the only edge among those selected that belongs to T. Now we remove all the selected edges in $\mathbf{Z}^d$ and, for each selected edge, identify vertices defining this edge (cf. Figure 6.3).

We obtained a tree that is a contraction of T having the same number n of leaves and embedded into another copy of $\mathcal{Z}^d$. By repeating this process, we end up with the star $K_{1,n}$ embedded into $\mathcal{Z}^d$. By Lemma 6.8, we have $d \geq \lceil n/2 \rceil$. $\square$

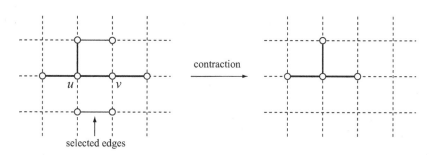

contraction

selected edges

Figure 6.3. Proof of Lemma 6.9.

We say that a tree T is a *quasi-star* if it has a unique vertex of degree greater than two. For example, the tree T in Figure 6.1 is a quasi-star with three leaves. By applying Lemma 6.9 and modifying the construction from the proof of Lemma 6.8, we obtain the following result (cf. Exercise 6.3).

Lemma 6.10. *The lattice dimension of a quasi-star with n leaves is $\lceil n/2 \rceil$.*

Lemma 6.11. *A tree $T = (V, E)$ with precisely three leaves is a quasi-star.*

Proof. By formula (1.1) and Theorem 2.32, we have

$$\sum_{v \in V} \deg_T(v) = 2|E| = 2|V| - 2.$$

Therefore,

$$\sum_{v \in V} (\deg_T(v) - 1) = |V| - 2.$$

There are exactly $|V| - 3$ positive terms in the sum on the left side of this equality. It follows that one of the terms is 2 and the remaining positive terms equal 1. Hence, there is a unique vertex of degree three. $\square$

For a given leaf $v \in T$ we denote by v_T the closest vertex in T that has a degree greater than two and call the unique $v_T v$-path the *hanging path* to v (cf. Exercise 6.2). The following theorem is the main result of this section.

Theorem 6.12. *The lattice dimension of a tree T with n leaves is $\lceil n/2 \rceil$.*

Proof. By Lemma 6.9, it suffices to show that there is an embedding of T into $\mathcal{Z}^{\lceil n/2 \rceil}$. We construct this embedding recursively.

By Lemmas 6.10 and 6.11, a tree with exactly three leaves is embeddable into $\mathcal{Z}^2$.

Let us assume that, for a given $n > 3$, any tree with $k < n$ leaves is embeddable into $\mathcal{Z}^{\lceil k/2 \rceil}$. By Lemma 6.10, we may also assume that T is not a quasi-star. Then there are leaves u and v with $u_T \neq v_T$. By deleting all vertices of the hanging $u_T u$- and $v_T v$-paths different from u_T and v_T, we obtain a new tree T' with $(n-2)$ leaves. Under our assumption, T' can be embedded into $\mathcal{Z}^d$ with

$$d = \left\lceil \frac{n-2}{2} \right\rceil = \left\lceil \frac{n}{2} \right\rceil - 1.$$

Now we embed, in a natural way, $\mathcal{Z}^d$ into $\mathcal{Z}^{d+1}$ and add the u_T, u- and $v_T v$-paths to the image of T' in the positive and negative directions of the new dimension, respectively. It is clear that we constructed an isometric embedding of T into $\mathcal{Z}^{\lceil n/2 \rceil}$. $\qquad\square$

Suppose that a tree T with n leaves is embedded into $\mathcal{Z}^d$ with $d = \lceil n/2 \rceil$. The projection of T into the kth factor is a path P_k of length $l_k > 0$. Thus T is actually embedded into the $(l_1 \times \cdots \times l_d)$-grid

$$P = P_1 \,\square\, P_2 \,\square \cdots \square\, P_d.$$

Note that the grid P may not be uniquely determined by the tree T. For instance, embeddings of the tree T in Figure 6.1 are into (3×3)- and (4×2)-grids. However, the sequence $(l_1, \ldots, l_d)$ determines both the lattice and isometric dimensions of T. Namely,

$$\dim_Z(T) = d \quad \text{and} \quad \dim_I(T) = l_1 + \cdots + l_d.$$

Let us recall that two embeddings of a partial cube into a cube are congruent (see Section 5.10; in particular, Theorem 5.63). This is not necessarily true for the lattice embeddings; the three embeddings in Figure 6.1 are pairwise noncongruent.

6.3 The Automorphism Group of a Lattice

Let X be an arbitrary set. There are three basic types of automorphisms of the lattice $\mathcal{Z}(X)$:

1) *Translations.* For a given $a \in \mathbf{Z}_X$, the translation $\alpha_a : \mathbf{Z}_X \to \mathbf{Z}_X$ is defined by $\alpha_a(u) = u + a$.
2) *Permutations.* A bijection $\sigma : X \to X$ defines a transformation $\hat{\sigma}$ of $\mathcal{Z}(X)$ by $\hat{\sigma}(u)(x) = u(\sigma(x))$ for all $u \in \mathbf{Z}_X$, $x \in X$.

3) *Reflections.* For a subset $Y \subseteq X$ we define a transformation ρ_Y of $\mathcal{Z}(X)$ by

$$\rho_Y(u)(x) = \begin{cases} -u(x), & \text{if } x \in Y, \\ u(x), & \text{if } x \in X \setminus Y, \end{cases}$$

and call it a *reflection through the origin* (it fixes the zero function).

The reader is encouraged to verify that these transformations are indeed automorphisms (cf. Exercise 6.4).

The set of translations is a subgroup K of the group $\mathrm{Aut}(\mathcal{Z}(X))$. The group K is isomorphic to the additive group $\mathbf{Z}_X$ (cf. Definition 6.2) and is called the *translation group*.

As in the case of cubes, the automorphisms $\hat{\sigma}$ form a subgroup of the group $\mathrm{Aut}(\mathcal{Z}(X))$ isomorphic to the symmetric group $S(X)$ (cf. Section 3.8). We denote this subgroup by H_p.

It can be readily verified that $\rho_Y \circ \rho_Z = \rho_{Y \triangle Z}$ and $\rho_Y^{-1} = \rho_Y$. Hence reflections form a subgroup of $\mathrm{Aut}(\mathcal{Z}(X))$ isomorphic to the group $\mathcal{P}_f(X)$ (cf. Section 3.8). This subgroup is denoted by H_r.

Let α be an automorphism of $\mathcal{Z}(X)$ and let $a = \alpha(0)$, where 0 stands for the zero function on X. Then $\beta = \alpha_a^{-1} \circ \alpha = \alpha_{-a} \circ \alpha$ is an automorphism fixing 0; that is, $\beta(0) = 0$.

Let us consider a family of layers (cf. Section 3.6) of $\mathcal{Z}(X)$

$$L_x = L_x(0) = \{u \in \mathcal{Z}(X) : u(y) = 0 \quad \text{for } y \neq x\}$$

(these layers can be regarded as "coordinate axes" in $\mathcal{Z}(X)$). By Lemma 3.14, the automorphism β maps each layer L_x onto another layer L_y preserving the zero function. (Although we proved Lemma 3.14 assuming that X is a countable set and P is a finite path, the proof remains valid for an arbitrary set X and an infinite ray.) β is an automorphism, thus the equation $\beta(L_x) = L_y$ defines a permutation σ of the set X:

$$y = \sigma(x) \text{ if and only if } \beta(L_x) = L_y.$$

Consider the automorphism $\gamma = \hat{\sigma}^{-1} \circ \beta$. Clearly, for any $x \in X$, the restriction $\gamma|_{L_x}$ is an isomorphism of a layer L_x onto itself with $\gamma(0) = 0$. Inasmuch as each layer L_x is isomorphic to $\mathcal{Z}$, the mapping $\gamma|_{L_x}$ is either the identity mapping of $\mathcal{Z}$ onto itself, or a reflection through the origin. Let Y be the set of all $x \in X$ for which $\gamma|_{L_x}$ is a reflection. Then ρ_Y is a reflection of $\mathcal{Z}$. It follows that the automorphism $\iota = \rho_Y^{-1} \circ \gamma$ fixes all layers L_x.

We want to show that ι is the identity of $\mathrm{Aut}(\mathcal{Z}(X))$. For a given $u \in \mathcal{Z}(X)$ and $x \in X$, we define a vertex u_x of $\mathcal{Z}(X)$ by

$$u_x(y) = \begin{cases} u(x), & \text{if } y = x, \\ 0, & \text{if } y \neq x, \end{cases} \quad \text{for } y \in X.$$

(This is the "x-coordinate" of u in $\mathcal{Z}(X)$.) Clearly, $u_x \in L_x$ and $\iota(u_x) = u_x$. Let $v = \iota(u)$. Because ι is an automorphism preserving zero, we must have $d(v, 0) = d(u, 0)$, that is,

$$\sum_{z \in X} |v(z)| = \sum_{z \in X} |u(z)|. \tag{6.1}$$

Because ι fixes u_x, we have $d(v, u_x) = d(u, u_x)$, so

$$\sum_{z \in X} |v(z) - u_x(z)| = \sum_{z \in X} |u(z) - u_x(z)|,$$

or, simplifying,

$$|v(x) - u(x)| + \sum_{z \neq x} |v(z)| = \sum_{z \neq x} |u(z)|. \tag{6.2}$$

By subtracting equation (6.1) from equation (6.2), we obtain

$$|v(x)| - |u(x)| = |v(x) - u(x)| \geq 0, \quad \text{for all } x \in X. \tag{6.3}$$

Hence, $|v(x)| \geq |u(x)|$ for all $x \in X$. By (6.1), we must have $|v(x)| = |u(x)|$ for all x. It follows from (6.3) that $v = u$.

We proved that $\iota = \rho_Y^{-1} \circ \widehat{\sigma}^{-1} \circ \alpha_a^{-1} \circ \alpha$ is the identity element of the group $\mathrm{Aut}(\mathcal{Z}(X))$. Therefore,

$$\alpha = \alpha_a \circ \widehat{\sigma} \circ \rho_Y. \tag{6.4}$$

In summary, we obtained the following result.

Theorem 6.13. $\mathrm{Aut}(\mathcal{Z}(X)) = K H_p H_r$

Note that subgroups K and H_r are commutative. It is convenient to represent an automorphism $\rho_Y \in H_r$ as the mapping $u \to \mu_Y u$, where μ_Y is a function on X defined by

$$\mu_Y(x) = \begin{cases} -1, & \text{if } x \in Y, \\ 1, & \text{if } x \in X \setminus Y. \end{cases}$$

We use decomposition (6.4) to prove the next theorem.

Theorem 6.14. K is a normal subgroup of $\mathrm{Aut}(\mathcal{Z}(X))$.

Proof. We need to show that $\alpha \circ \alpha_b \circ \alpha^{-1} \in K$ for all $\alpha \in \mathrm{Aut}(\mathcal{Z}(X))$ and $b \in \mathbf{Z}_X$. This is done below step by step using (6.4).

First we have

$$(\rho_Y \circ \alpha_b \circ \rho_Y^{-1})(u)(x) = (\mu_Y(x)(u)(x) + b(x))\mu_Y(x) = u(x) + \mu_Y(x)b(x).$$

Hence,

$$\rho_Y \circ \alpha_b \circ \rho_Y^{-1} = \alpha_c, \quad \text{where } c = \rho_Y(b).$$

Next,

$$(\widehat{\sigma} \circ \alpha_c \circ \widehat{\sigma}^{-1})(u)(x) = u(x) + c(\sigma(x)) = \alpha_{\widehat{\sigma}(c)}(u)(x).$$

Therefore,

$$\widehat{\sigma} \circ \alpha_c \circ \widehat{\sigma}^{-1} = \alpha_d, \text{ where } d = \widehat{\sigma}(c).$$

Finally, $\alpha_a \circ \alpha_d \circ \alpha_a^{-1} = \alpha_d$, because K is a commutative group.
We proved that

$$\alpha \circ \alpha_b \circ \alpha^{-1} = (\alpha_a \circ \widehat{\sigma} \circ \rho_Y) \circ \alpha_b \circ (\alpha_a \circ \widehat{\sigma} \circ \rho_Y)^{-1}$$
$$= \alpha_a \circ (\widehat{\sigma} \circ (\rho_Y \circ \alpha_b \circ \rho_Y^{-1}) \circ \widehat{\sigma}^{-1}) \circ \alpha_a^{-1} = \alpha_d.$$

$\alpha_d \in K$, therefore the result follows.   □

Let $H = H_p H_r$, so $\mathrm{Aut}(\mathcal{Z}(X)) = KH$. Because $H \cap K = \{\iota\}$ (cf. Exercise 6.6), the group $\mathrm{Aut}(\mathcal{Z}(X))$ is the semidirect product of the subgroups K and H (cf. Theorem 3.31).

Theorem 6.15. *The automorphism group of the lattice $\mathcal{H}(X)$ is the semidirect product of the subgroup K and the subgroup H.*

Note that the group $\mathrm{Aut}(\mathcal{Z}(X))$ is vertex-transitive because the translation group K is vertex-transitive.

6.4 Lattice Dimension of Finite Partial Cubes

Let us first consider a simple geometric example. Let G be the graph depicted in Figure 6.4a.

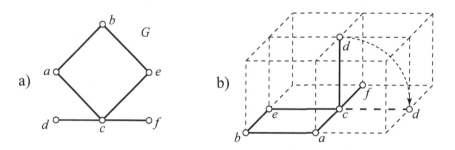

Figure 6.4. Unfolding graph G.

This graph is the partial cube B18 from Figure 4.5. It is easy to verify that $\dim_I(G) = 4$. An embedding of G into the lattice $\mathcal{Z}^3$ is shown in Figure 6.4b. We can "unfold" this embedding along the path acf, resulting in the isometric embedding of G into $\mathcal{Z}^2$ (see Figure 6.4b). Note that the path acf is the intersection of two semicubes

$$W_{ab} = \{a, c, d, f\} \quad \text{and} \quad W_{cd} = \{a, b, c, e, f\}.$$

These semicubes satisfy the following condition.

$$W_{ab} \cup W_{cd} = V(G) \quad \text{and} \quad W_{ab} \cap W_{cd} \neq \varnothing.$$

In fact, whenever we have two semicubes with a nonempty intersection and covering the entire vertex set, and the graph is embedded into a lattice, we can "unfold" this embedding along the intersection of semicubes, obtaining an embedding into a lattice of a lower dimension.

Definition 6.16. Let $G = (V, E)$ be a partial cube. The *semicube graph* $\mathrm{Sc}(G)$ has all semicubes of G as the set of its vertices. Two semicubes W_{ab} and W_{cd} are adjacent in $\mathrm{Sc}(G)$ if

$$W_{ab} \cup W_{cd} = V \quad \text{and} \quad W_{ab} \cap W_{cd} \neq \varnothing. \tag{6.5}$$

An example of a semicube graph is depicted in Figure 6.5. In this graph, the edge with ends $W_{ab} = \{a, c, d, f\}$ and $W_{cd} = \{a, b, c, e, f\}$ corresponds to the unfolding shown in Figure 6.4. Other edges of this graph correspond to unfoldings of different embeddings of the graph G.

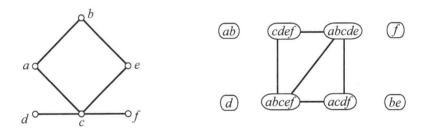

Figure 6.5. A partial cube (left) and its semicube graph (right).

Lemma 6.17. *Condition (6.5) is equivalent to either of two conditions:*

$$W_{ba} \subset W_{cd} \quad \text{or} \quad W_{dc} \subset W_{ab}, \tag{6.6}$$

where $\subset$ stands for the proper inclusion.

Proof. We prove that condition (6.5) is equivalent to the first condition in (6.6). Suppose that condition (6.5) holds. $W_{ab} \cup W_{cd} = V$, thus we have

$$W_{ba} = V \setminus W_{ab} = (W_{ab} \cup W_{cd}) \setminus W_{ab} = W_{cd} \setminus W_{ab}.$$

Because $W_{ab} \cap W_{cd} \neq \varnothing$, we have $W_{cd} \setminus W_{ab} \subset W_{cd}$. Hence, $W_{ba} \subset W_{cd}$. Conversely, suppose that $W_{ba} \subset W_{cd}$. Then

$$V = W_{ab} \cup W_{ba} \subseteq W_{ab} \cup W_{cd},$$

implying $W_{ab} \cup W_{cd} = V$. Furthermore,

$$W_{cd} \setminus W_{ba} = W_{cd} \setminus (V \setminus W_{ab}) = W_{cd} \cap W_{ab} \neq \varnothing,$$

because $W_{ba} \subset W_{cd}$. Hence, (6.5) holds. $\square$

Let $G = (V, E)$ be a finite partial cube isometrically embedded into the integer lattice $\mathbb{Z}^d$. As in the case of trees (cf. Section 6.2), we may assume that G is an isometric subgraph of a grid

$$R = R_1 \,\square\, R_2 \,\square\, \cdots \,\square\, R_d,$$

where R_k is a path of length $l_k > 0$. Each path R_k is a projection of G into the kth factor of $\mathbb{Z}^d$. The equivalence classes of the theta relation on $E(G)$ are in one-to-one correspondence with edges of the paths R_k. Therefore,

$$\dim_I(G) = \sum_{k=1}^{d} l_k.$$

Without loss of generality, we may assume that each R_k is in the form $(0, 1, \ldots, l_k)$. It can be easily seen that all semicubes in G are intersections of G with grids $R'_{k,j}$ and $R''_{k,j}$, where (cf. Figure 6.6)

$$R'_{k,j} = R_1 \,\square\, \cdots \,\square\, Q_k^j \,\square\, \cdots \,\square\, R_d, \quad R''_{k,j} = R_1 \,\square\, \cdots \,\square\, S_k^j \,\square\, \cdots \,\square\, R_d,$$
$$Q_k^j = (0, 1, \ldots, j), \ 0 \leq j < l_k \ \text{ and } \ S_k^j = (j, \ldots, l_k), \ 0 < j \leq l_k.$$

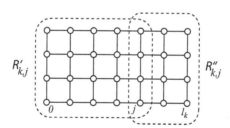

Figure 6.6. Semicubes $R'_{k,j}$ and $R''_{k,j}$.

We denote

$$L_{k,j} = R'_{k,j} \cap V \ (0 \leq j < l_k) \quad \text{and} \quad U_{k,j} = R''_{k,j} \cap V \ (0 < j \leq l_k)$$

the *lower* and *upper semicubes* of G, respectively. Clearly,

$$L_{k,j} \cup U_{k,j} = V \quad \text{and} \quad L_{k,j} \cap U_{k,j} \neq \varnothing, \quad \text{for} \ \ 0 < j < l_k, \ 1 \leq k \leq d,$$

so these semicubes are adjacent in the semicube graph $\mathrm{Sc}(G)$. Note that all vertices $L_{k,0}$, U_{k,l_k} $(1 \le k \le d)$ have degree zero in $\mathrm{Sc}(G)$.

Let M be the set of edges in $\mathrm{Sc}(G)$ joining semicubes $L_{k,j}$ and $U_{k,j}$. It is clear that the set M is a matching in $\mathrm{Sc}(G)$ and

$$|M| = \sum_{k=1}^{d}(l_k - 1) = \dim_I(G) - d.$$

We proved the following result.

Lemma 6.18. *Let G be a finite partial cube that is isometrically embedded into $\mathbb{Z}^d$ with nontrivial projections on each factor. Then there exists a (possibly empty) matching M in the semicube graph $\mathrm{Sc}(G)$, such that*

$$d = \dim_I(G) - |M|.$$

An isometric embedding of the graph G from Figure 6.4a into the (3×3)-grid is shown in Figure 6.7a (cf. Figure 6.4).

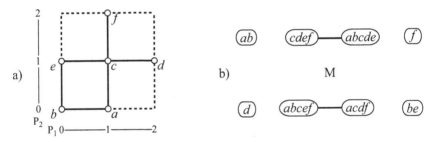

Figure 6.7. An embedding of G and the corresponding matching in $\mathrm{Sc}(G)$.

The lower and upper semicubes of G are

$$L_{1,0} = \{b, e\}, \quad U_{1,1} = \{a, c, d, f\}, \quad L_{2,0} = \{a, b\}, \quad U_{2,1} = \{c, d, e, f\},$$
$$L_{1,1} = \{a, b, c, e, f\}, \quad U_{1,2} = \{d\}, \quad L_{2,1} = \{a, b, c, d, e\}, \quad U_{2,2} = \{f\}.$$

Clearly, $M = \{L_{1,1}U_{1,1}, L_{2,1}U_{2,1}\}$. This matching is depicted in Figure 6.7b. For this embedding, $\dim_I(G) = 4$, $d = 2$, and $|M| = 2$.

In the opposite direction, suppose that a matching M in the semicube graph $\mathrm{Sc}(G)$ of a graph $G = (V, E)$ is given. We want to prove that there is an isometric embedding of G into $\mathbb{Z}^d$, where $d = \dim_I(G) - |M|$. The following construction is instrumental.

Let H be the graph obtained from M by adding edges joining each pair of opposite semicubes. M is a matching, therefore every vertex of H has a degree one or two.

Lemma 6.19. *Connected components of H are paths.*

Proof. Let H' be a component of H. If H' does not contain edges of M, then it is a path of length one connecting two opposite semicubes. Otherwise, suppose that $W_{ab}W_{cd}$ is an edge of M that belongs to H'. Then H' contains edges $W_{ba}W_{ab}$ and $W_{cd}W_{dc}$. By (6.6), $W_{dc} \subset W_{ab}$; that is, W_{dc} is a proper subset of W_{ab}. This is also true for alternating vertices in a path in H', starting at W_{ab}. Thus, H' has no cycles and is therefore a path (cf. Figure 6.8). $\qquad\square$

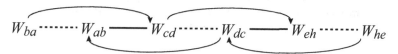

Figure 6.8. A path in the graph H. Added edges are shown by dotted lines. Arrows indicate proper inclusions of semicubes.

It follows that H is a union of disjoint paths Q_k, $1 \le k \le d$, for some positive d. There are $2 \dim_I(G)$ semicubes of G, of which $2|M|$ are matched in M. There are two ends per path Q_k, and the set of these ends consists of $2 \dim_I(G) - 2|M|$ unmatched vertices in $\mathrm{Sc}(G)$. Hence, $d = \dim_I(G) - |M|$.

Note that each path Q_k starts and ends with an edge joining a pair of opposite semicubes (cf. Figure 6.8). Accordingly, each Q_k is an M-augmented path in H (cf. Section 1.9) of odd length l_k.

Let us number vertices of each path in H, starting at one end and ending at the other one, so that $S_{k,i}$ denotes the ith vertex of Q_k, $0 \le i \le l_k$. It is convenient to define $S_{k,-1} = S_{k,l_k+1} = V$. By (6.6), there are two chains of proper set inclusions for vertices in a given path Q_k (cf. Figure 6.8):

$$S_{k,0} \subset S_{k,2} \subset \cdots \subset S_{k,2x} \subset \cdots \subset S_{k,l_k-1} \subset S_{k,l_k+1} = V \qquad (6.7)$$

and

$$V = S_{k,-1} \supset S_{k,1} \supset \cdots \supset S_{k,2x-1} \supset \cdots \supset S_{k,l_k-2} \supset S_{k,l_k}. \qquad (6.8)$$

Lemma 6.20. *For any $v \in V$, $1 \le k \le d$, there is a unique integer x such that $v \in S_{k,2x-1} \cap S_{k,2x}$.*

Proof. First we note that sets $S_{k,2y-1} \cap S_{k,2y}$ and $S_{k,2z-1} \cap S_{k,2z}$ are disjoint for $y \ne z$. Indeed, assuming that $y < z$, we have

$$(S_{k,2y-1} \cap S_{k,2y}) \cap (S_{k,2z-1} \cap S_{k,2z}) = S_{k,2z-1} \cap S_{k,2y} \subseteq S_{k,2y+1} \cap S_{k,2y} = \varnothing.$$

by (6.7), (6.8), and because the cubes $S_{k,2y}$ and $S_{k,2y+1}$ are opposite. Thus it suffices to show that there is x such that $v \in S_{k,2x-1} \cap S_{k,2x}$.

Clearly, given v and k, there is a unique x such that $S_{k,2x}$ is the first set in (6.7) containing v. If $v \in S_{k,0}$, then $v \in S_{k,-1} \cap S_{k,0} = S_{k,0}$, and we are done. Otherwise, suppose that $v \in S_{k,2x}$ and $v \notin S_{k,2x-2}$ for a positive x.

Because $S_{k,2x-2}$ and $S_{k,2x-1}$ are opposite semicubes, we have $v \in S_{k,2x-1}$. Hence, $v \in S_{k,2x-1} \cap S_{k,2x}$. □

By Lemma 6.20, the function $\lambda_k(v) = x$ is well defined for every $1 \leq k \leq d$. Therefore,

$$\lambda(v) = (\lambda_1(v), \ldots, \lambda_d(v)) \qquad (6.9)$$

defines a mapping $\lambda : V \to \mathbb{Z}^d$.

Example 6.21. Let G be the graph from Figure 6.4 and H be the graph obtained from the matching M in $\mathrm{Sc}(G)$ depicted by solid lines in Figure 6.9. The added vertices are shown by dotted lines.

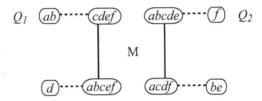

Figure 6.9. The graph H obtained from a matching M.

There are two M-augmented paths, Q_1 and Q_2, clearly visible in Figure 6.9. The sets $S_{k,2x-1} \cap S_{k,2x}$ and the values of functions λ_1 and λ_2 are entries of the two tables in Figure 6.10.

	$x = 0$	$x = 1$	$x = 2$
$k = 1$	ab	cef	d
$k = 2$	f	acd	be

	a	b	c	d	e	f
λ_1	0	0	1	2	1	1
λ_2	1	2	1	1	2	0

Figure 6.10. Sets $S_{k,2x-1} \cap S_{k,2x}$ and functions λ_1 and λ_2.

The corresponding embedding $\lambda : G \to \mathbb{Z}^2$ is depicted in Figure 6.11.

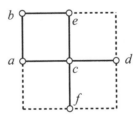

Figure 6.11. Embedding of G defined by the functions λ_1 and λ_2.

Note that this embedding is different from the one shown in Figure 6.7.

Lemma 6.22. *The mapping λ defined by (6.9) is an isometric embedding of G into $\mathcal{Z}^d$.*

Proof. Let u and v be two vertices of G. Any semicube separating u and v in G belongs to a single path Q_k. By (6.7), the number of semicubes in the form $S_{k,2x}$ that separates u and v is $|\lambda_k(u) - \lambda_k(v)|$. It is that same number for semicubes in the form $S_{k,2x-1}$ (cf. (6.8)). Therefore, the total number of semicubes separating u and v is

$$2 \sum_{k=1}^{d} |\lambda_k(u) - \lambda_k(v)| = 2d_{\mathcal{Z}^d}(\lambda(u), \lambda(v)).$$

On the other hand, if an edge ab belongs to a shortest path connecting vertices u and v, then semicubes W_{ab} and W_{ba} separate u and v, and any semicube separating u and v is in one of these forms (cf. the proof of Theorem 5.19). Hence, the number of semicubes separating u and v is twice the distance between vertices u and v in G. We proved that

$$d_{\mathcal{Z}^d}(\lambda(u), \lambda(v)) = d_G(u, v);$$

that is, λ is an isometric embedding of G into $\mathcal{Z}^d$. □

Lemmas 6.18 and 6.22 yield the following result:

Theorem 6.23. *For a finite partial cube G,*

$$\dim_{\mathcal{Z}}(G) = \dim_I(G) - |M|,$$

where M is a maximum matching in the semicube graph $\mathrm{Sc}(G)$.

Figures 6.9 and 6.11 show a maximum matching in the semicube graph $\mathrm{Sc}(G)$ in Figure 6.5, and the corresponding minimum dimension lattice embedding of the partial cube from that figure, respectively.

6.5 Lattice Dimension of Infinite Partial Cubes

The lattice dimension of infinite partial cubes may be characterized by lattice dimensions of their finite isometric subgraphs.

We begin by proving a general result about locally finite infinite graphs, that is, whose vertices have a finite number of neighbors (cf. Section 1.8).

Theorem 6.24. *Let G be a locally finite infinite graph. For any vertex v_0 of G, there is a ray in G with the initial vertex v_0.*

Proof. Let $\mathcal{L}_0$ be the set of all paths in G with an end at v_0. Because G is infinite, so is $\mathcal{L}_0$. Because G is locally finite, there is a neighbor v_1 of v_0 that belongs to infinitely many paths in $\mathcal{L}_0$. Let $\mathcal{L}_1$ be the set of all paths starting at v_0 and containing v_1. Because $\mathcal{L}_1$ is infinite and G is locally finite, there is a neighbor v_2 of v_1 different from v_0 that belongs to infinitely many paths in $\mathcal{L}_1$. By continuing this process, we obtain an infinite chain of infinite sets $\mathcal{L}_0 \supseteq \mathcal{L}_1 \supseteq \cdots$. The corresponding sequence of vertices $v_0, v_1, v_2, \ldots$ is a ray in G. $\qquad\square$

Next we establish some auxiliary results with a clear geometric flavor. Let $G = (V, E)$ be a connected graph. The *convex hull* of a set $X \subseteq V$ is the intersection of all convex subsets of V containing X (cf. Exercise 3.25). V is clearly convex, therefore this notion is well defined. Note that the convex hull of a set X is the minimum convex set containing the set X (cf. Exercise 6.7).

Theorem 6.25. *Let $G = (\mathcal{F}, E_{\mathcal{F}})$, where $\mathcal{F}$ is a wg-family of finite subsets of a set X, and let $\mathcal{X}$ be a subset of $\mathcal{F}$. Then the convex hull of $\mathcal{X}$ is the set*

$$C(\mathcal{X}) = \{R \in \mathcal{F} : \cap \mathcal{X} \subseteq R \subseteq \cup \mathcal{X}\}.$$

Proof. First we prove that $C(\mathcal{X}) \subseteq \mathcal{Y}$ for any convex set $\mathcal{Y}$ containing $\mathcal{X}$. Suppose to the contrary that there is $R \in C(\mathcal{X})$ such that $R \notin \mathcal{Y}$. Let P be an arbitrary chosen set in $\mathcal{X}$ and let $P_0 = P, P_1, \ldots, P_n = R$ be a shortest PR-path in $\mathcal{F}$. Because $P \in \mathcal{X} \subseteq \mathcal{Y}$ and $R \notin \mathcal{Y}$, there is k such that $P_k \in \mathcal{Y}$ and $P_{k+1} \notin \mathcal{Y}$. By Theorem 5.8 , there is $x \in X$ such that $P_k \triangle P_{k+1} = \{x\}$. There are two possible cases.

(i) $x \in P_{k+1}$, $x \notin P_k$. By Theorem 5.7, P_{k+1} lies between P_k and R; that is,

$$P_k \cap R \subseteq P_{k+1} \subseteq P_k \cup R.$$

It follows that $x \in R \subseteq \cup \mathcal{X}$. Therefore there is $Q \in \mathcal{X} \subseteq \mathcal{Y}$ such that $x \in Q$. We clearly have

$$P_k \cap Q \subseteq P_{k+1} \subseteq P_k \cup Q;$$

that is, P_{k+1} lies between P_k and Q. Because $P_k, Q \in \mathcal{Y}$ and $\mathcal{Y}$ is convex, we must have $P_{k+1} \in \mathcal{Y}$, a contradiction.

(ii) $x \notin P_{k+1}$, $x \in P_k$. As in the previous case, we have

$$P_k \cap R \subseteq P_{k+1} \subseteq P_k \cup R.$$

Hence, $x \notin R \supseteq \cap \mathcal{X}$. Therefore there is $Q \in \mathcal{X} \subseteq \mathcal{Y}$ such that $x \notin Q$. Then

$$P_k \cap Q \subseteq P_{k+1} \subseteq P_k \cup Q,$$

which yields a contradiction as in the previous case.

It follows that $C(\mathcal{X}) \subseteq \mathcal{Y}$ for any convex set $\mathcal{Y}$ containing $\mathcal{X}$. It remains to show that $C(\mathcal{X})$ is a convex set. Suppose that R lies between $P, Q \in C(\mathcal{X})$. Then

$$\bigcap \mathfrak{X} \subseteq P \cap Q \subseteq R \subseteq P \cup Q \subseteq \bigcup \mathfrak{X}.$$

Hence, $R \in C(\mathfrak{X})$. Therefore, $C(\mathfrak{X})$ is convex.   □

Corollary 6.26. *Let G be a partial cube. The convex hull of any finite set of vertices of G is a finite set.*

Proof. We may assume that $G = (\mathfrak{F}, E_{\mathfrak{F}})$, where $\mathfrak{F}$ is a wg-family of finite sets. Let $\mathfrak{X}$ be a finite subset of $\mathfrak{F}$. By Theorem 6.25, $C(\mathfrak{X})$ is a finite set.   □

Note that the result of Corollary 6.26 does not necessarily hold for graphs that are not partial cubes (cf. Exercise 6.14).

For $v_0 \in V$ and any integer $k \geq 0$ we set

$$B_k(v_0) = \{v \in V : d_G(v, v_0) \leq k\}$$

and call this set the *ball of radius k and center v_0*. It is clear that $B_0(v_0) = v_0$ and $B_i(v_0) \subseteq B_{i+1}(v_0)$.

Lemma 6.27. *If the ball $B_k(v_0)$ is infinite, then it contains a vertex of infinite degree.*

Proof. $B_0(v_0)$ is finite and $B_k(v_0)$ is infinite, therefore there is i such that $B_i(v_0)$ is finite and $B_{i+1}(v_0)$ is infinite. Because every vertex of $B_{i+1}(v_0)$ is adjacent to a vertex in $B_i(v_0)$, there is a vertex in $B_i(v_0)$ of infinite degree.
□

We note that some balls in a partial cube G may be nonisometric subgraphs of G (consider, for instance, a ball of radius 2 in the cycle C_6), whereas their convex hulls are isometric subgraphs.

Theorem 6.28. *For a partial cube G and an integer $k \geq 0$, $\dim_{\mathcal{Z}}(G) \leq k$ if and only if $\dim_{\mathcal{Z}}(G') \leq k$ for all finite isometric subgraphs of G' of G.*

Proof. It suffices to prove that, if all finite isometric subgraphs G' of G can be embedded into $\mathcal{Z}^d$, then so can G itself. Let v_0 be a fixed vertex of G. Suppose that, for some i, the ball $B_i(v_0)$ is infinite. By Lemma 6.27, this ball has a vertex u with infinitely many neighbors. Thus we can construct a subgraph that is the star $K_{1,2d+2}$ with the center at u. This star is clearly an isometric subgraph of G. By Lemma 6.8, the lattice dimension of this star is $d + 1$, which contradicts our assumption that all isometric subgraphs of G have dimension at most d. This contradiction proves that all balls $B_i(v_0)$ are finite. By Corollary 6.26, the convex hulls of these balls are also finite. These convex hulls form a nested sequence

$$G_0 = \{v_0\} \subseteq G_1 \subseteq \cdots \subseteq G_i \subseteq \cdots$$

of finite isometric subgraphs of G such that $\cup_i G_i = G$. Note that all graphs G_i are embeddable into $\mathcal{Z}^d$.

Let us form a graph H, the vertices of which are all embeddings of the graphs G_i into $\mathcal{Z}^d$ such that v_0 is mapped into the zero vertex of $\mathcal{Z}^d$. The edges of H are pairs of embeddings, $f : G_{i-1} \to \mathcal{Z}^d$ and $g : G_i \to \mathcal{Z}^d$, such that f is the restriction of g to G_{i-1}. The graph H is connected, inasmuch as any vertex of H is connected to the unique embedding $f_0 : G_0 \to \mathcal{Z}^d$. This graph is locally finite because there are only finitely many embeddings of any G_i that map v_0 into the zero vertex. By Lemma 6.24, there is a ray in H with the initial vertex f_0. Any two embeddings of this ray that both map the same vertex v of G at the same vertex of $\mathcal{Z}^d$ must map v at the same vertex as each other, and every vertex of G is mapped into $\mathbf{Z}^d$ by all but finitely many embeddings in the ray, so we can derive from this ray a consistent embedding of all vertices of G into $\mathbf{Z}^d$. This embedding is isometric, as any path in G belongs to some subgraph G_i for sufficiently large i. $\square$

Theorem 6.29. *If the lattice dimension of a partial cube G is infinite, then* $\dim_{\mathcal{Z}}(G) = \dim_I(G)$.

Proof. By Theorem 6.3, $\dim_I(G) \geq \dim_{\mathcal{Z}}(G)$. Consider an isometric embedding of G into $\mathcal{Z}(X)$, where $|X| = \dim_{\mathcal{Z}}(G)$. By composing this embedding with the embedding of Theorem 6.3 from $\mathcal{Z}(X)$ into $\mathcal{H}(X \times \mathbf{Z})$, we obtain

$$\dim_I(G) \leq |X \times \mathbf{Z}| = |X| = \dim_{\mathcal{Z}}(G),$$

because X is an infinite set. Hence, $\dim_{\mathcal{Z}}(G) = \dim_I(G)$. $\square$

6.6 Lattice Dimensions of Products and Pasted Graphs

Let us recall (cf. Theorem 5.40) that the isometric dimension of the Cartesian product of two partial cubes is the sum of their isometric dimensions. A similar result holds for the lattice dimension.

Theorem 6.30. *Let $G = (V, E)$ be the Cartesian product of two finite partial cubes $G_1 = (V_1, E_1)$ and $G_2 = (V_2, E_2)$. Then*

$$\dim_{\mathcal{Z}}(G) = \dim_{\mathcal{Z}}(G_1) + \dim_{\mathcal{Z}}(G_2).$$

Proof. Let $W_{(a,b)(c,d)}$ be a semicube of the graph G. Because (a, b) and (c, d) are adjacent vertices of G, there are two possible cases:

(i) $c = a$, $bd \in E_2$. Let (x, y) be a vertex of G. Then

$$d_G((x, y), (a, b)) = d_{G_1}(x, a) + d_{G_2}(y, b)$$

and

$$d_G((x, y), (c, d)) = d_{G_1}(x, c) + d_{G_2}(y, d).$$

Hence,

$$d_G((x, y), (a, b)) < d_G((x, y), (c, d)) \text{ if and only if } d_{G_2}(y, b) < d_{G_2}(y, d).$$

It follows that

$$W_{(a,b)(c,d)} = V_1 \times W_{bd}. \tag{6.10}$$

(ii) $d = b$, $ac \in E_1$. By an argument similar to (i), we have

$$W_{(a,b)(c,d)} = W_{ac} \times V_2. \tag{6.11}$$

Clearly, two semicubes given by (6.10) form an edge in the semicube graph $Sc(G)$ if and only if their second factors form an edge in the semicube graph $Sc(G_2)$. The same is true for semicubes in the form (6.11) with respect to their first factors. It is also clear that semicubes in the form (6.10) and in the form (6.11) are not joined by an edge in $Sc(G)$. Therefore the semicube graph $Sc(G)$ is isomorphic to the disjoint union of semicube graphs $Sc(G_1)$ and $Sc(G_2)$. If M_1 is a maximum matching in $Sc(G_1)$ and M_2 is a maximum matching in $Sc(G_2)$, then $M = M_1 \cup M_2$ is a maximum matching in $Sc(G)$. The result follows from Theorems 6.23 and 5.40. □

Let G be the partial cube obtained by vertex-pasting together finite partial cubes G_1 and G_2. In the case of the lattice dimension, we can claim only a weaker result than the one stated in Theorem 5.42 for the isometric dimension.

Theorem 6.31. *Let G be a partial cube obtained by vertex-pasting together finite partial cubes G_1 and G_2. Then*

$$\max\{\dim_{\mathcal{Z}}(G_1), \dim_{\mathcal{Z}}(G_2)\} \leq \dim_{\mathcal{Z}}(G) \leq \dim_{\mathcal{Z}}(G_1) + \dim_{\mathcal{Z}}(G_2).$$

Proof. The first inequality is trivial because both G_1 and G_2 are isomorphic to subgraphs of G.

Let $a = \{a_1, a_2\}$ be the vertex of G obtained by identifying vertices a_1 and a_2 of graphs G_1 and G_2, respectively, and let $d_i = \dim_{\mathcal{Z}}(G_i)$ for $i \in \{1, 2\}$. The graphs $\mathcal{Z}^{d_1}$ and $\mathcal{Z}^{d_2}$ can be identified with subgraphs of $\mathcal{Z}^{d_1+d_2} = \mathcal{Z}^{d_1} \square \mathcal{Z}^{d_2}$ with intersection at the zero vertex. We embed the graphs G_1 and G_2 into $\mathcal{Z}^{d_1}$ and $\mathcal{Z}^{d_2}$, respectively, in such a way that both a_1 and a_2 are mapped into the zero vertex of $\mathcal{Z}^{d_1+d_2}$. Clearly, we obtain an isometric embedding of G into $\mathcal{Z}^{d_1+d_2}$. Thus $\dim_{\mathcal{Z}}(G) \leq \dim_{\mathcal{Z}}(G_1) + \dim_{\mathcal{Z}}(G_2)$. □

The following example illustrates possible cases for inequalities in Theorem 6.31.

Example 6.32. The star $K_{1,6}$ can be obtained from the stars $K_{1,2}$ and $K_{1,4}$ by vertex-pasting these two stars along their centers. By Lemma 6.27,

$$\dim_{\mathcal{Z}}(K_{1,6}) = 3, \quad \dim_{\mathcal{Z}}(K_{1,2}) = 1, \quad \text{and} \quad \dim_{\mathcal{Z}}(K_{1,4}) = 2.$$

Hence,

$$\max\{\dim_{\mathcal{Z}}(K_{1,2}), \dim_{\mathcal{Z}}(K_{1,4})\} < \dim_{\mathcal{Z}}(K_{1,6}) = \dim_{\mathcal{Z}}(K_{1,2}) + \dim_{\mathcal{Z}}(K_{1,4}).$$

The same star $K_{1,6}$ is obtained from two copies of the star $K_{1,3}$ by vertex-pasting along their centers. We have $\dim_{\mathcal{Z}}(K_{1,3}) = 2$, $\dim_{\mathcal{Z}}(K_{1,6}) = 3$, so

$$\max\{\dim_{\mathcal{Z}}(K_{1,3}), \dim_{\mathcal{Z}}(K_{1,3})\} < \dim_{\mathcal{Z}}(K_{1,6}) < \dim_{\mathcal{Z}}(K_{1,3}) + \dim_{\mathcal{Z}}(K_{1,3}).$$

Let us vertex-paste two stars $K_{1,3}$ along two leaves. The resulting graph T is a tree with four leaves. Therefore,

$$\max\{\dim_{\mathcal{Z}}(K_{1,3}), \dim_{\mathcal{Z}}(K_{1,3})\} = \dim_{\mathcal{Z}}(T) < \dim_{\mathcal{Z}}(K_{1,3}) + \dim_{\mathcal{Z}}(K_{1,3}).$$

Thus we have exact lower and upper bounds for $\dim_{\mathcal{Z}}(G)$ in Theorem 6.31.

We now turn our attention to partial cubes obtained by edge-pasting together two finite partial cubes. For notations used in the rest of this section the reader is referred to Section 5.8.

We need some results about semicube graphs for our proof of an analogue of Theorem 6.31 for a partial cube obtained by edge-pasting together two finite partial cubes.

Lemma 6.33. *Let G be a partial cube and $W_{pq}W_{uv}$, $W_{qp}W_{xy}$ be two edges in the semicube graph $\mathrm{Sc}(G)$. Then $W_{xy}W_{uv}$ is an edge in $\mathrm{Sc}(G)$.*

Proof. By conditions (6.6), $W_{qp} \subset W_{uv}$ and $W_{yx} \subset W_{qp}$. Hence, $W_{yx} \subset W_{uv}$. By the same conditions, $W_{xy}W_{uv} \in \mathrm{Sc}(G)$. □

If G is obtained by edge-pasting together graphs G_1 and G_2, we identify graphs G_1 and G_2 with subgraphs of the graph G. Then $G_1 \cup G_2 = G$ and $G_1 \cap G_2 = (\{a, b\}, \{ab\}) = K_2$ (cf. Figure 5.8).

Lemma 6.34. *Let G be a partial cube obtained by edge-pasting together partial cubes G_1 and G_2. Let $W_{uv}^{(1)}W_{xy}^{(1)}$ (respectively, $W_{uv}^{(2)}W_{xy}^{(2)}$) be an edge in the semicube $\mathrm{Sc}(G_1)$ (respectively, $\mathrm{Sc}(G_2)$). Then $W_{uv}W_{xy}$ is an edge in $\mathrm{Sc}(G)$.*

Proof. It suffices to consider the case of $\mathrm{Sc}(G_1)$ (cf. Figure 6.12). By conditions (6.6), $W_{vu}^{(1)} \subset W_{xy}^{(1)}$ and $W_{yx}^{(1)} \subset W_{uv}^{(1)}$. Suppose that $a \in W_{vu}^{(1)}$ and $b \in W_{yx}^{(1)}$ (the case when $b \in W_{vu}^{(1)}$ and $a \in W_{yx}^{(1)}$ is treated similarly). Then $ab\,\theta_1 xy$ and $ab\,\theta_1 uv$. By transitivity of θ_1, we have $uv\,\theta_1 xy$, a contradiction, inasmuch as semicubes $W_{uv}^{(1)}$ and $W_{xy}^{(1)}$ are distinct. Therefore we may assume

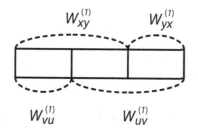

Figure 6.12. Semicubes forming an edge in $\mathrm{Sc}(G_1)$.

that, say, $a, b \in W_{uv}^{(1)}$. Then, by Lemma 5.44, $W_{vu} = W_{vu}^{(1)} \subset V_1$. Because $W_{vu}^{(1)} \subset W_{xy}^{(1)} \subseteq W_{xy}$, we have $W_{vu} \subset W_{xy}$. By conditions (6.6), $W_{uv}W_{xy}$ is an edge in $\mathrm{Sc}(G)$. $\qquad\square$

Lemma 6.35. *Let M_1 and M_2 be matchings in graphs $\mathrm{Sc}(G_1)$ and $\mathrm{Sc}(G_2)$. There is a matching M in $\mathrm{Sc}(G)$ such that*

$$|M| \geq |M_1| + |M_2| - 1.$$

Proof. By Lemma 6.34, M_1 and M_2 induce matchings in $\mathrm{Sc}(G)$ that we denote by the same symbols. The intersection $M_1 \cap M_2$ is either empty or a subgraph of the empty graph with vertices W_{ab} and W_{ba}.

If $M_1 \cap M_2$ is empty, then $M = M_1 \cup M_2$ is a matching in $\mathrm{Sc}(G)$ and the result follows.

If $M_1 \cap M_2$ is an empty graph with a single vertex, say, in M_1, we remove from M_1 the edge that has this vertex as its end vertex, resulting in the matching M_1'. Clearly, $M = M_1' \cup M_2$ is a matching in $\mathrm{Sc}(G)$ and $|M| = |M_1| + |M_2| - 1$.

Suppose now that $M_1 \cap M_2$ is the empty graph with vertices W_{ab} and W_{ba}. Let $W_{ab}W_{uv}$, $W_{ba}W_{pq}$ (respectively, $W_{ab}W_{xy}$, $W_{ba}W_{rs}$) be edges in M_1 (respectively, M_2). By Lemma 6.33, $W_{xy}W_{rs}$ is an edge in $\mathrm{Sc}(G_2)$. Let us replace edges $W_{ab}W_{xy}$ and $W_{ba}W_{rs}$ in M_2 by a single edge $W_{xy}W_{rs}$, resulting in the matching M_2'. Then $M = M_1 \cup M_2'$ is a matching in $\mathrm{Sc}(G)$ and $|M| = |M_1| + |M_2| - 1$. $\qquad\square$

Corollary 6.36. *Let M_1 and M_2 be maximum matchings in $\mathrm{Sc}(G_1)$ and $\mathrm{Sc}(G_2)$, respectively, and M be a maximum matching in $\mathrm{Sc}(G)$. Then*

$$|M| \geq |M_1| + |M_2| - 1. \tag{6.12}$$

By Theorem 6.23, we have

$$\dim_I(G_1) = \dim_{\mathcal{Z}}(G_1) + |M_1|, \quad \dim_I(G_2) = \dim_{\mathcal{Z}}(G_2) + |M_2|,$$

and

$$\dim_I(G) = \dim_{\mathcal{Z}}(G) + |M|,$$

where M_1 and M_2 are maximum matchings in $\mathrm{Sc}(G_1)$ and $\mathrm{Sc}(G_2)$, respectively, and M is a maximum matching in $\mathrm{Sc}(G)$. Therefore, by Theorem 5.46 and (6.12), we have the following result (cf. Theorem 6.31).

Theorem 6.37. *Let G be a partial cube obtained by edge-pasting together finite partial cubes G_1 and G_2. Then*

$$\max\{\dim_{\mathcal{Z}}(G_1), \dim_{\mathcal{Z}}(G_2)\} \leq \dim_{\mathcal{Z}}(G) \leq \dim_{\mathcal{Z}}(G_1) + \dim_{\mathcal{Z}}(G_2).$$

Example 6.38. Let us consider two edge-pastings together, the stars $G_1 = K_{1,3}$ and $G_2 = K_{1,3}$ of lattice dimension 2 shown in Figures 4.16 and 4.17. In the first case the resulting graph is the star $G = K_{1,5}$ of lattice dimension 3. Then we have

$$\max\{\dim_{\mathcal{Z}}(G_1), \dim_{\mathcal{Z}}(G_2)\} < \dim_{\mathcal{Z}}(G) < \dim_{\mathcal{Z}}(G_1) + \dim_{\mathcal{Z}}(G_2).$$

In the second case the resulting graph is a tree with 4 leaves. Therefore,

$$\max\{\dim_{\mathcal{Z}}(G_1), \dim_{\mathcal{Z}}(G_2)\} = \dim_{\mathcal{Z}}(G) < \dim_{\mathcal{Z}}(G_1) + \dim_{\mathcal{Z}}(G_2).$$

Let $c_1 a_1$ and $c_2 a_2$ be edges of stars $G_1 = K_{1,4}$ and $G_2 = K_{1,4}$ (each of which has lattice dimension 2), where c_1 and c_2 are centers of the respective stars. Let us edge-paste these two graphs by identifying c_1 with c_2 and a_1 with a_2, respectively. The resulting graph G is the star $K_{1,7}$ of lattice dimension 4. Thus,

$$\max\{\dim_{\mathcal{Z}}(G_1), \dim_{\mathcal{Z}}(G_2)\} \leq \dim_{\mathcal{Z}}(G) = \dim_{\mathcal{Z}}(G_1) + \dim_{\mathcal{Z}}(G_2).$$

Thus we have exact lower and upper bounds for $\dim_{\mathcal{Z}}(G)$ in Theorem 6.37.

Notes

Isometric embeddings of trees into integer lattices were studied by Hadlock and Hoffman (1978) who established the result of Theorem 6.12. Our presentation follows Ovchinnikov (2004).

The concept of the lattice dimension of a finite partial cube was introduced in Eppstein (2005) together with the result of Theorem 6.23. Theorem 6.24 is known as König's Infinity Lemma (Diestel, 2005). Most of the results presented in Section 6.5 originally appeared in Chapter 8 of the monograph by Eppstein et al. (2008).

Lattice dimensions of Cartesian products and pasted graphs were computed in Ovchinnikov (2008c).

Exercises

6.1. Show that a finite tree with exactly two leaves is a path.

6.2. Show that for any leaf v of a tree T the hanging path to v is well defined.

6.3. Prove Lemma 6.10.

6.4. Show that translations, permutations, and reflections are automorphisms of an integer lattice.

6.5. For given m, n, find the automorphism groups of

a) The path P_n
b) The grid $P_n \square P_m$

6.6. Show that

$$H_r \cap H_p = \{\iota\}, \quad H_r \cap K = \{\iota\}, \quad H_p \cap K = \{\iota\},$$

where ι is the identity element of the group $\mathrm{Aut}(\mathcal{Z}(X))$ (see Section 6.3).

6.7. Prove that the convex hull of a set U of vertices of a connected graph is the minimum convex set containing U.

6.8. Give an example of a graph in which every interval is a finite set and convex hulls of some intervals are infinite.

6.9. Let P and Q be two vertices of $\mathcal{Z}(X)$, and let

$$\{P_1, \ldots, P_n\} \quad \text{and} \quad \{Q_1, \ldots, Q_n\}$$

be sets of neighbors of P and Q, respectively. There is an automorphism γ of $\mathcal{Z}(X)$ such that $\gamma(P) = Q$ and $\gamma(P_i) = Q_i$ for all $1 \leq i \leq n$ (cf. Theorem 3.32).

6.10. Show that the three embeddings in Figure 6.1 are pairwise noncongruent.

6.11. Show that the semicube graph of a cube is empty.

6.12. Describe the semicube graph of the path P_n and show that it has a unique maximum matching M with $|M| = n - 1$.

6.13. Prove that $\dim_{\mathcal{Z}}(C_{2n}) = \dim_I(C_{2n})$, where C_{2n} is an even cycle.

6.14. Let $K_{X,Y}$ be an infinite bipartite graph with $|X| \geq 2$, $|Y| \geq 2$. Show that the result of Corollary 6.26 does not hold for $K_{X,Y}$.

6.15. Prove that the semicube graph of a finite partial cube is disconnected.

7

Hyperplane Arrangements

In this chapter, we study region graphs of hyperplane arrangements in Euclidean spaces. These graphs are highly symmetrical partial cubes and the theory has a distinctive geometric flavor. Two nongeometric applications are presented in Sections 7.5 and 7.7.

7.1 Hyperplanes

A function $L(x) = a_1 x_1 + \cdots + a_d x_d + b$, where $(a_1, \ldots, a_d)$ is a nonzero vector in $\mathbf{R}^d$ is said to be an (affine) *linear function* on the d-dimensional Euclidean space $\mathbf{R}^d$. An (affine) *hyperplane* H in $\mathbf{R}^d$ is the solution set of the equation $L(x) = 0$, that is,

$$H = \{x \in \mathbf{R}^d : L(x) = 0\}.$$

The hyperplane H is a translate of the $(d-1)$-dimensional vector subspace of $\mathbf{R}^d$ defined by

$$\{x \in \mathbf{R}^d : a_1 x_1 + \cdots + a_d x_d = 0\}$$

(cf. Exercise 7.1b). By definition, the dimension of H is also $d - 1$.

Example 7.1. In 1-dimensional space $\mathbf{R}$, a hyperplane is a singleton. Hyperplanes in the plane $\mathbf{R}^2$ are straight lines. They are the usual planes in $\mathbf{R}^3$.

The complement $\mathbf{R}^d \setminus H$ of a hyperplane H in $\mathbf{R}^d$ is the union of two open *half-spaces* (cf. Figure 7.1):

$$H^+ = \{x \in \mathbf{R}^d : a_1 x_1 + \cdots + a_d x_d + b > 0\}$$

and

$$H^- = \{x \in \mathbf{R}^d : a_1 x_1 + \cdots + a_d x_d + b < 0\}.$$

The three sets H, H^+, and H^- are convex and form a partition of $\mathbf{R}^d$ (cf. Exercise 7.2). Observe that notation H^+, H^- depends on the choice of the function L defining the hyperplane H (cf. Exercise 7.1a).

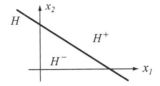

Figure 7.1. A straight line separating two open half-planes in $\mathbf{R}^2$.

Two sets A and B in $\mathbf{R}^d$ are said to be *separated* by a hyperplane H if A is contained in one of the half-spaces determined by H and B is contained in the other half-space.

A nonempty intersection of a finite family of distinct hyperplanes in $\mathbf{R}^d$ is called a *linear manifold*. Every linear manifold M is a translate of a unique vector subspace V of $\mathbf{R}^d$. By definition, the dimension $\dim(M)$ of M is the dimension of V. It is clear that a hyperplane in $\mathbf{R}^d$ is a linear manifold of dimension $d-1$, and a point in $\mathbf{R}^d$ is a linear manifold of dimension zero.

We need the result of the following lemma in Section 7.3.

Lemma 7.2. *Let $M = \bigcap_{i=1}^{n} H_i$ be a linear manifold in $\mathbf{R}^d$ of dimension less than $d-1$ and a be a point in $\mathbf{R}^d$. There exists a hyperplane H containing both the point a and the manifold M.*

Proof. The result is trivial if a belongs to one of the hyperplanes H_i. Thus we may assume that $a \notin H_i$ for all i. Because $\dim(M) < d-1$, we have $n \geq 2$. Let L_1 and L_2 be linear functions defining hyperplanes H_1 and H_2, respectively. The linear function (cf. Exercise 7.7)

$$L(x) = L_1(x) - \frac{L_1(a)}{L_2(a)} L_2(x)$$

is well defined because $a \notin H_2$. Inasmuch as $L_1(x) = L_2(x) = 0$ for $x \in M$, we have $L(x) = 0$ for all $x \in M$, so $M \subseteq H = \{x \in \mathbf{R}^d : L(x) = 0\}$. Finally, $a \in H$, because $L(a) = 0$. □

7.2 Arrangements of Hyperplanes

A family of sets $\mathcal{X} = \{X_i\}_{i \in I}$ in $\mathbf{R}^d$ is said to be *locally finite* if any ball in $\mathbf{R}^d$ intersects only a finite number of sets in $\mathcal{X}$. There are at most countably many elements in a locally finite family of sets, so the index set I is either finite or countable (cf. Exercise 7.3).

Definition 7.3. An *arrangement* $\mathcal{A}$ (of hyperplanes) is a locally finite family $\mathcal{A} = \{H_i\}_{i \in I}$ of distinct hyperplanes in $\mathbf{R}^d$. An arrangement $\mathcal{A}$ is said to be *finite* if I is a finite set. A *central arrangement* is an arrangement of hyperplanes all of which contain the origin of $\mathbf{R}^d$. A central arrangement is necessarily finite. If $\bigcap_{i \in I} H_i \neq \varnothing$, then coordinates in $\mathbf{R}^d$ may be chosen so each hyperplane contains the origin. In this case we also call $\mathcal{A}$ central.

Examples of arrangements of three lines in the plane are displayed in Figure 7.2.

Figure 7.2. A central and two noncentral arrangements of three lines in the plane.

Two examples of infinite line arrangements are shown in Figure 7.3. The left arrangement is the family of lines

$$\{(x_1, x_2) : x_i + k = 0,\ i \in \{1, 2\},\ k \in \mathbf{Z}\},$$

whereas the right arrangement is the union of three families

$$\{(x_1, x_2) : x_2 + k = 0\}_{k \in \mathbf{Z}}, \quad \{(x_1, x_2) : \sqrt{3}x_1 + x_2 + k = 0\}_{k \in \mathbf{Z}},$$
$$\{(x_1, x_2) : \sqrt{3}x_1 - x_2 + k = 0\}_{k \in \mathbf{Z}}.$$

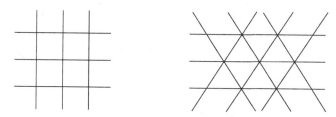

Figure 7.3. Infinite families of parallel lines forming two line arrangements.

We also note that a locally finite set of points in the real line is an arrangement in $\mathbf{R}^1$.

In what follows we assume that the reader is familiar with basic topological properties of the Euclidean space $\mathbf{R}^d$.

Definition 7.4. A *region* of an arrangement $\mathcal{A}$ in $\mathbf{R}^d$ is a connected component of the complement of the hyperplanes, $\mathbf{R}^d \setminus \bigcup_{H \in \mathcal{A}} H$.

Example 7.5. There are, respectively, six, seven, and four regions of the arrangements displayed in Figure 7.2. The regions of the left arrangement in Figure 7.3 are open unit squares; they are open triangles in the right drawing.

For $a \notin \bigcup_{H \in \mathcal{A}} H$, we denote by $D_H(a)$ the unique half-space determined by H that contains a, and let

$$D_{\mathcal{A}}(a) = \bigcap_{H \in \mathcal{A}} D_H(a).$$

The set $D_{\mathcal{A}}(a)$ is a convex (and therefore connected) subset of $\mathbf{R}^d$ (cf. Exercise 7.2).

Theorem 7.6. *Let a be a point in a region R of an arrangement $\mathcal{A} = \{H_i\}_{i \in I}$. Then $R = D_{\mathcal{A}}(a)$.*

Proof. R is a connected component of $\mathbf{R}^d \setminus \bigcup_{H \in \mathcal{A}} H$ and $a \in R \cap D_{\mathcal{A}}(a)$, thus we have $D_{\mathcal{A}}(a) \subseteq R$. Suppose that there is $x \in R$ such that $x \notin D_{\mathcal{A}}(a)$. Then there is a hyperplane in $\mathcal{A}$ separating points a and x. This hyperplane intersects a path connecting a and x in R (Exercise 7.5). However this is impossible because no hyperplane in $\mathcal{A}$ intersects a region. This contradiction proves that $R = D_{\mathcal{A}}(a)$. $\qquad \square$

We proved that regions of an arrangement are nonempty intersections of open half-spaces determined by hyperplanes in the arrangement.

Theorem 7.7. *A region of an arrangement $\mathcal{A}$ in $\mathbf{R}^d$ is an open set.*

Proof. Let $R = D_{\mathcal{A}}(a)$ be a region. Inasmuch as $\mathcal{A}$ is a locally finite family, there is an open ball with the center at a that does not intersect any of the hyperplanes in $\mathcal{A}$. Because this ball is a connected set, it is contained in R. The result follows, because a is an arbitrary point in R. $\qquad \square$

A region of a finite arrangement is an interior of a *polyhedron*. If R is a bounded region of an arbitrary arrangement $\mathcal{A}$, that is, R is a subset of an open ball in $\mathbf{R}^d$, then R is the intersection of a finite subfamily of hyperplanes in $\mathcal{A}$, because $\mathcal{A}$ is a locally finite family. In this case R is an interior of a *polytope*, that is, a bounded polyhedron. In general, a region of an infinite arrangement is not necessarily a polyhedron.

Example 7.8. Consider the arrangement $\mathcal{A}$ of tangent lines to the graph of parabola $y = x^2$ at integer points (n, n^2), $n \in \mathbf{Z}$. The region $R = D_{\mathcal{A}}((0, 1))$ is bounded by all lines in $\mathcal{A}$ (cf. Figure 7.4), so it is not a polyhedron (it has infinitely many sides).

In the conclusion of this section we consider some examples of finite arrangements in $\mathbf{R}^d$ that anticipate some later material.

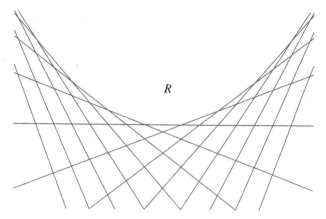

Figure 7.4. A fragment of the arrangement from Example 7.8.

Example 7.9. The *Boolean arrangement* $\mathbf{B}_d$ is the central arrangement of the coordinate hyperplanes $x_i = 0$, $1 \leq i \leq d$, in $\mathbf{R}^d$. The regions of $\mathbf{B}_d$ are in one-to-one correspondence with the vertices of the cube $[-1, 1]^d$.

Example 7.10. The *braid arrangement* $\mathcal{BA}_d$ consists of the hyperplanes

$$x_i - x_j = 0, \quad 1 \leq i < j \leq d.$$

Thus there are $\binom{d}{2}$ hyperplanes in $\mathcal{BA}_d$. Let $a = (1, 2, \ldots, d)$. For a permutation σ of the set $\{1, 2, \ldots, d\}$, we let $a_\sigma = (\sigma(1), \sigma(2), \ldots, \sigma(d))$. It is clear that any region of $\mathcal{BA}_d$ is in the form $R_\sigma = D_{\mathcal{BA}_d}(a_\sigma)$, and all these regions are distinct. Therefore there are $d!$ regions of $\mathcal{BA}_d$. Note that $\mathcal{BA}_d$ is a central arrangement.

Example 7.11. The *semiorder arrangement* $\mathcal{SO}_d$ consists of the hyperplanes

$$x_i - x_j = 1, \quad i \neq j, \quad 1 \leq i, j \leq d.$$

Example 7.12. Let $b = (b_1, \ldots, b_d)$ be a vector in $\mathbf{R}^d$ with positive coordinates. The hyperplanes

$$x_i - x_j = b_j, \quad i \neq j, \quad 1 \leq i, j \leq d$$

form the *deformation* $\mathcal{A}_{d,b}$ of the braid arrangement. Clearly, $\mathcal{A}_{d,b}$ is the semiorder arrangement for $b = (1, 1, \ldots, 1)$.

7.3 Region Graphs

Two regions of an arrangement $\mathcal{A}$ either are separated by a given hyperplane $H \in \mathcal{A}$ or "lie on the same side" of H; that is, they belong to one of the two half-spaces determined by H.

We say that two regions of an arrangement $\mathcal{A}$ are *adjacent* if they are separated by a unique hyperplane from $\mathcal{A}$.

Definition 7.13. Let $\mathcal{R}(\mathcal{A})$ be the set of regions of an arrangement $\mathcal{A}$. The *region graph* $G(\mathcal{A})$ of $\mathcal{A}$ has $\mathcal{R}(\mathcal{A})$ as the set of vertices; edges of $G(\mathcal{A})$ are pairs of adjacent regions in $\mathcal{R}$.

Region graphs of the three line arrangements in Figure 7.2 are shown in Figure 7.5. The first graph is the cycle C_6, the next graph is the cube Q_3 with a deleted vertex, and the last graph is the path P_4.

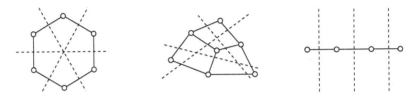

Figure 7.5. Examples of region graphs.

Infinite region graphs of arrangements in Figure 7.3 are displayed in Figure 7.6. They are square and hexagon lattices, respectively.

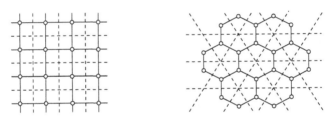

Figure 7.6. Fragments of two infinite line arrangements and their region graphs.

Note that all graphs in Figures 7.5 and 7.6 are partial cubes. Our goal is to show that this is true for any region graph.

We begin by showing that region graphs are connected and establishing a lower bound for the graph distance between two regions (Lemma 7.15). The drawing in Figure 7.7 is instructive.

Lemma 7.14. *Let P and Q be two regions of an arrangement $\mathcal{A}$ in $\mathbf{R}^d$. There are points $p \in P$ and $q \in Q$ such that each point of the line segment $[p, q]$ belongs to at most one hyperplane in $\mathcal{A}$.*

Proof. Let B be an open ball in $\mathbf{R}^d$ intersecting both P and Q, and let p be a fixed point in $P \cap B$. Because $\mathcal{A}$ is locally finite, there are finitely many linear manifolds determined by hyperplanes in $\mathcal{A}$ that intersect B. If all these manifolds are hyperplanes, the result follows for an arbitrary choice of a point $q \in Q \cap B$. Otherwise, let $\{M_1, \ldots, M_k\}$ be the set of the manifolds whose dimension is less than $d - 1$. By Lemma 7.2, there are hyperplanes H_i such that each H_i contains p and M_i for $1 \leq i \leq k$. Any region of the arrangement

$$\mathcal{A}' = \mathcal{A} \cup \{H_1, \ldots, H_k\}$$

is contained in a region of $\mathcal{A}$. Let R be a region of $\mathcal{A}'$ that is contained in Q and has a nonempty intersection with B. Then $[p, q]$ is the desired line segment for any $q \in R \cap B \subseteq Q$.   □

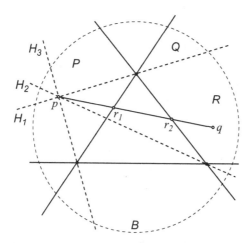

Figure 7.7. Proofs of Lemmas 7.14 and 7.15.

Lemma 7.15. *The region graph $G(\mathcal{A})$ of an arrangement $\mathcal{A}$ in $\mathbf{R}^d$ is connected. Furthermore, for any two regions P and Q, the distance $d_G(P, Q)$ is less than or equal to the number of hyperplanes in $\mathcal{A}$ separating P and Q.*

Proof. Let P and Q be two regions of $\mathcal{A}$. By Lemma 7.14, there are points $p \in P$ and $q \in Q$ such that different hyperplanes from $\mathcal{A}$ separating regions P and Q intersect the line segment $[p, q]$ at different points. Let us number these points together with points p and q in the direction from p to q as follows,

$$p = r_0, r_1, \ldots, r_{k+1} = q$$

(cf. Figure 7.7). Thus k is the number of hyperplanes in $\mathcal{A}$ separating regions P and Q. Consider the sequence of line segments

$$I_0 = [r_0, r_1), \ldots, I_j = (r_j, r_{j+1}), \ldots, I_k = (r_k, r_{k+1}].$$

Each interval I_j is the intersection of $[p, q]$ with a unique region R_j. Clearly, regions R_j and R_{j+1} are adjacent in G. Thus the sequence

$$P = R_0, R_1, \ldots, R_k = Q$$

is a PQ-path of length k in $G(\mathcal{A})$. The result follows.   □

Now we construct an isometric embedding of $G = G(\mathcal{A})$ into a cube.

Let $\mathcal{A}$ be an arrangement of hyperplanes in $\mathbf{R}^d$. For each $H \in \mathcal{A}$ we fix half-spaces H^+ and H^- by selecting a linear function determining H and define a mapping φ from the set of regions $\mathcal{R}$ into the power set $\mathcal{P}(\mathcal{A})$ by

$$\varphi(R) = \{H \in \mathcal{A} : R \subseteq H^+\}, \quad R \in \mathcal{R}.$$

Lemma 7.16. *Let P and Q be regions of $\mathcal{A}$. The set $\varphi(P) \triangle \varphi(Q)$ consists of the hyperplanes separating P and Q.*

Proof. Clearly, $H \notin \varphi(R)$ if and only if $R \subseteq H^-$. Therefore,

$$\varphi(P) \triangle \varphi(Q) = \{H \in \mathcal{A} : P \subseteq H^+, Q \subseteq H^-\} \cup \{H \in \mathcal{A} : P \subseteq H^-, Q \subseteq H^+\},$$

which proves the lemma.   □

By Lemmas 7.16 and 7.15, φ is an embedding of G into a connected component of the graph $\mathfrak{P}(\mathcal{A})$. Therefore, the distance $d_G(P, Q)$ is greater than or equal to the Hamming distance $|\varphi(P) \triangle \varphi(Q)|$. By the same lemmas,

$$d_G(P, Q) = |\varphi(P) \triangle \varphi(Q)|;$$

that is, φ is an isometric embedding.

Because connected components of the graph $\mathfrak{P}(\mathcal{A})$ are cubes (cf. Theorem 3.9), we obtained the following result.

Theorem 7.17. *The region graph G of a hyperplane arrangement $\mathcal{A}$ in $\mathbf{R}^d$ is a partial cube. Furthermore, the graph distance between two regions P and Q of $\mathcal{A}$ equals the number of hyperplanes in $\mathcal{A}$ separating P and Q in $\mathbf{R}^d$.*

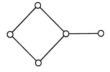

Figure 7.8. Graph B9 from Figure 4.5.

Not all partial cubes are the region graphs of hyperplane arrangements. For instance, the partial cube in Figure 7.8 is not a region graph of any arrangement of hyperplanes in any $\mathbf{R}^d$ (Exercise 7.6).

The region graph of a hyperplane arrangement in $\mathbf{R}^d$ can be "drawn" in $\mathbf{R}^d$ by selecting a "representing" point in each region and connecting selected points in adjacent regions by a line segment.

Example 7.18. The points of $\{-1,1\}^d \subseteq \mathbf{R}^d$ represent regions of the Boolean arrangement $\mathbf{B}^d$ (Example 7.9). The region graph of $\mathbf{B}^d$ is the cube Q_d. For $d = 2$, the graph is displayed in Figure 7.9.

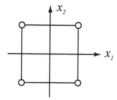

Figure 7.9. The graph of the Boolean arrangement $\mathbf{B}_2$.

Example 7.19. The regions of the braid arrangement $\mathcal{BA}_d$ (cf. Example 7.10) are represented by points $x_\sigma = (\sigma(1), \ldots, \sigma(d))$ where σ is a permutation of the set $\{1, \ldots, d\}$. Note that the intersection of all hyperplanes in $\mathcal{BA}_d$ is the one-dimensional subspace defined by

$$x_1 = x_2 = \cdots = x_d.$$

For any permutation σ,

$$\sigma(1) + \sigma(2) + \cdots + \sigma(d) = \frac{d(d+1)}{2}.$$

Therefore, the points x_σ belong to the hyperplane H defined by the equation

$$x_1 + x_2 + \cdots + x_d = \frac{d(d+1)}{2}.$$

The convex hull of all points x_σ is a $(d-1)$-dimensional polytope (it is contained in H, which is a $(d-1)$-dimensional linear manifold; cf. Exercise 7.8) in $\mathbf{R}^d$. This polytope (and its graph) is called the *permutohedron* Π_{d-1}.

Two vertices of Π_{d-1} are connected by an edge if and only if the corresponding permutations differ by an adjacent transposition. The graph Π_{d-1} is the region graph of the braid arrangement $\mathcal{BA}_d$ (cf. Exercise 7.9).

The graph of the permutohedron Π_2 is shown in Figure 7.10. The dotted lines are intersections of the planes in $\mathcal{BA}_3$ with the plane

$$x_1 + x_2 + x_3 = 6.$$

The vertices are labeled by permutations of the set $\{1, 2, 3\}$.

The permutohedron Π_3 is shown in Figure 7.11. It is the graph of the *truncated octahedron* in geometry. Because it is the graph of a convex polytope, it is planar. Its planar drawing is shown in Figure 7.12.

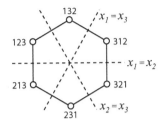

Figure 7.10. The permutohedron Π_2.

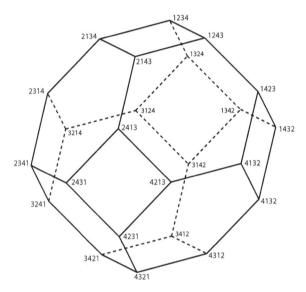

Figure 7.11. The permutohedron Π_3.

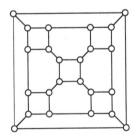

Figure 7.12. A planar drawing of Π_3.

Example 7.20. The planes of the semiorder arrangement $\mathcal{SO}_3$ are orthogonal to the plane

$$x_1 + x_2 + x_3 = 1.$$

The intersection of $\mathcal{SO}_3$ with this plane is shown by the dotted lines in Figure 7.13. The region graph of $\mathcal{SO}_3$ is displayed in the same figure.

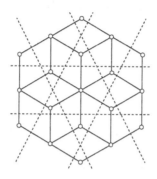

Figure 7.13. The region graph of the semiorder arrangement $\mathcal{SO}_3$.

7.4 The Lattice Dimension of a Region Graph

Region graphs have many attractive combinatorial and geometric properties that make some tasks simpler than in the general case of partial cubes. To exemplify this statement, we compute in this section the lattice dimension (cf. Definition 6.5) of the region graph of an arrangement $\mathcal{A}$.

Lemma 7.21. *Let G be the region graph of an arrangement $\mathcal{A}$. A region R belongs to the semicube Q_{PS} if and only if R and P lie on the same side of the (unique) hyperplane H separating P and S. Accordingly, the semicubes in the region graph G of an arrangement $\mathcal{A}$ are the sets of all regions that lie on one side of hyperplanes in $\mathcal{A}$.*

Proof. The result follows (cf. Exercise 7.10) from this geometric observation: let p, q, r be points in the regions P, Q, R, respectively, satisfying properties in Lemma 7.14. A hyperplane in $\mathcal{A}$ intersecting one of the sides of the triangle pqr intersects precisely one more side of the triangle.   □

It follows from Lemma 7.21 that $(PQ, RS) \in \theta$ if and only if regions P and Q are separated by the same hyperplane of the arrangement $\mathcal{A}$ as regions R and S are. Thus we have a one-to-one correspondence between the equivalence classes of the relation θ and the hyperplanes in $\mathcal{A}$. By Theorem 5.34, we have

$$\dim_I(G) = |E(G)/\theta| = |\mathcal{A}|.$$

Therefore, by Theorem 6.23, in order to evaluate $\dim_Z(G)$, we need to find the cardinality of a maximum matching M in the semicube graph $\mathrm{Sc}(G)$.

Lemma 7.22. *Let $\mathcal{A}$ be a finite arrangement of parallel hyperplanes in $\mathbf{R}^d$, G be its region graph, and M be a maximum matching in $\mathrm{Sc}(G)$. Then*

$$|M| = |\mathcal{A}| - 1.$$

Proof. As we indicated before, $\dim_I G = |\mathcal{A}|$. Clearly, G is a path. Hence, $\dim_Z(G) = 1$. We have

$$|M| = \dim_I(G) - \dim_Z(G) = |\mathcal{A}| - 1,$$

by Theorem 6.23.   □

Theorem 7.23. *Let $\mathcal{A} = \mathcal{A}_1 \cup \cdots \cup \mathcal{A}_m$ be a union of hyperplane arrangements in $\mathbf{R}^d$ such that hyperplanes in each $\mathcal{A}_i$ are parallel and hyperplanes in distinct $\mathcal{A}_i$ are not parallel. Let G be the region graph of $\mathcal{A}$. Then $\dim_Z(G) = m$.*

Proof. By Theorem 6.28, we may consider only finite isometric subgraphs of the graph G. Any such subgraph can be viewed as part of the arrangement of a finite subset of $\mathcal{A}$, so we need to consider only the case when each $\mathcal{A}_i$ is finite. By Definition 6.16, two semicubes are connected by an edge in $\mathrm{Sc}(G)$ if they have a nonempty intersection and their union is the set of all regions of the arrangement $\mathcal{A}$. By Lemma 7.21, if two hyperplanes in $\mathcal{A}$ are not parallel, the four semicubes defined by these hyperplanes are not mutually adjacent in $\mathrm{Sc}(G)$. Therefore, any matching in $\mathrm{Sc}(G)$ consists of edges connecting semicubes defined by parallel hyperplanes in $\mathcal{A}$. Let M be a maximum matching in $\mathrm{Sc}(G)$. By Lemma 7.22,

$$|M| = \sum_{i=1}^{m}(|\mathcal{A}_i| - 1) = |\mathcal{A}| - m.$$

By Theorem 6.23,

$$\dim_Z(G) = \dim_I(G) - |M| = |\mathcal{A}| - (|\mathcal{A}| - m) = m$$

and the result follows.   □

Example 7.24. Let G be the graph of the arrangement shown in Figure 7.14. This arrangement consists of three pairs of parallel planes. Thus, $\dim_I(G) = 6$ and $\dim_Z(G) = 3$. The embedding of G into $\mathbf{Z}^3$ is shown in the same figure. Clearly, G cannot be isometrically embedded into the 2-dimensional integer lattice.

As hyperplanes in a central arrangement cannot be parallel, the result can be simplified for that case.

Corollary 7.25. *If $\mathcal{A}$ is a central arrangement and G is its region graph, then*

$$\dim_I(G) = \dim_Z(G) = |\mathcal{A}|.$$

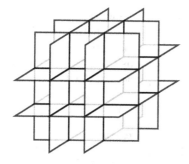 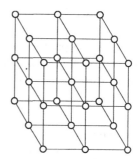

Figure 7.14. A hyperplane arrangement consisting of three pairs of parallel planes and its region graph.

7.5 Acyclic Orientations

Let us recall that an *orientation* of a graph G (cf. Section 1.7) is a digraph D obtained from G by replacing every edge by one of the two arcs with the same ends. A *directed walk* in D is a sequence

$$v_0 e_1 v_1 e_2 \cdots v_{k-1} e_k v_k,$$

where $v_0, \ldots, v_k$ are vertices of G, $e_1, \ldots, e_k$ are arcs of D, and $e_i = (v_{i-1}, v_i)$, for $1 \leq i \leq k$. The notions of a *directed path* and a *directed cycle* are defined similarly (cf. Section 1.3). An orientation of G is said to be *acyclic* if does not contain directed cycles.

In this section, G is a nonempty finite graph with vertex set $\{1, 2, \ldots, d\}$. We denote by $\mathcal{AO}(G)$ the set of all acyclic orientations of G. Two acyclic orientations are said to be *adjacent* if they differ on a single edge of G. This adjacency relation defines a graph on $\mathcal{AO}(G)$ that we denote by the same symbol $\mathcal{AO}(G)$.

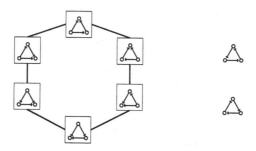

Figure 7.15. Orientations of the cycle C_3.

Example 7.26. There are 2^3 orientations of the cycle C_3; six are acyclic and two are directed cycles. The graph $\mathcal{AO}(C_3)$ and two directed cycles obtained from C_3 are shown in Figure 7.15 (cf. Exercise 7.13).

It is instructive to compare the graph $\mathcal{AO}(C_3)$ shown in Figure 7.15 with the region graph Π_2 of the braid arrangement $\mathcal{BA}_3$ (cf. Figure 7.10). These graphs are isomorphic. In fact, any graph $\mathcal{AO}(G)$ is isomorphic to the region graph of some subarrangement of $\mathcal{BA}_d$. To establish this fact, we first define an arrangement $\mathcal{A}(G)$ as the family of hyperplanes H_{ij} in $\mathbf{R}^d$ given by the equations

$$x_i - x_j = 0, \ \text{ for } ij \in E(G).$$

The arrangement $\mathcal{A}(G)$ is called a *graphic arrangement*. Clearly, the arrangement $\mathcal{AO}(G)$ is a subarrangement of the braid arrangement $\mathcal{BA}_d$; that is, $\mathcal{AO}(G) \subseteq \mathcal{BA}_d$.

To save writing, we denote by $\mathcal{AO}$ and $\mathcal{A}$ the family of all acyclic orientations of a given graph G on the set $\{1, 2, \ldots, d\}$ and the graphic arrangement associated with G, respectively, and by $\mathcal{R}$ the set of regions of $\mathcal{A}$. First we construct a bijection from $\mathcal{AO}$ onto $\mathcal{R}$.

Let $D \in \mathcal{AO}$ and $i \in V(D)$. We denote by $p_i(D)$ the number of vertices that can be reached from the vertex i along a directed nontrivial walk and define

$$p(D) = (p_1(D), \ldots, p_d(D)) \in \mathbf{R}^d.$$

Lemma 7.27. *Let $ij \in E(G)$. The pair (i, j) is an arc of D if and only if $p_i(D) > p_j(D)$.*

Proof. (Necessity.) Let (i, j) be an arc of D. Every vertex of D that can be reached from j can also be reached from i. Thus, $p_i(D) \geq p_j(D)$. The vertex j can be reached from i but cannot be reached from itself because D is acyclic. Hence, $p_i(D) > p_j(D)$.

(Sufficiency.) Follows from the necessity part because either (i, j) or (j, i) is an arc of D. □

It follows that, for any $D \in \mathcal{AO}$, the point $p(D)$ belongs to a (unique) region of $\mathcal{A}$ and hence defines a mapping

$$\psi : \mathcal{AO} \to \mathcal{R},$$

where $\psi(D) = D_A(p(D))$ (cf. Theorem 7.6).

Lemma 7.28. *The mapping ψ is one-to-one.*

Proof. Let D and D' be two distinct orientations of G. Then there is an edge ij of G such that (i, j) is an arc of D and (j, i) is an arc of D'. By Lemma 7.27, $p_i(D) > p_j(D)$ and $p_i(D') < p_j(D')$. Therefore the hyperplane H_{ij} separates points $p(D) \in \psi(D)$ and $p(D') \in \psi(D')$, so the region $\psi(D)$ is distinct from the region $\psi(D')$. □

Lemma 7.29. *ψ maps $\mathcal{AO}$ onto $\mathcal{R}$.*

Proof. Let R be a region of $\mathcal{A}$ and let $x = (x_1, \ldots, x_d)$ be a point in R. For every edge $ij \in E(G)$, we choose the arc (i, j) if $x_i > x_j$ and the arc (j, i), otherwise. It is clear that the resulting orientation D of G is acyclic. By Lemma 7.27, $x_i > x_j$ if and only if $p_i(D) > p_j(D)$, for every $ij \in E(G)$. Thus $R = D_\mathcal{A}(x) = D_\mathcal{A}(p(D)) = \psi(D)$, so ψ is onto. $\square$

By Lemmas 7.28 and 7.29, ψ is a bijection from $\mathcal{AO}$ onto $\mathcal{R}$. By Lemma 7.27, two acyclic orientations D and D' of G are adjacent if and only if there is a unique $ij \in E(G)$ such that

$$(p_i(D) - p_j(D))(p_i(D') - p_j(D')) < 0,$$

or, equivalently, there is a unique hyperplane H_{ij} separating points $p(D)$ and $p(D')$. Note that a hyperplane in $\mathcal{A}$ separates points $p(D)$ and $p(D')$ if and only if it separates the regions $\psi(D) = D_\mathcal{A}(p(D))$ and $\psi(D') = D_\mathcal{A}(p(D'))$. Hence, two acyclic orientations D and D' of G are adjacent if and only if the regions $\psi(D)$ and $\psi(D')$ are adjacent. It follows that ψ is a graph isomorphism from $\mathcal{AO}(G)$ onto $G(\mathcal{A})$.

Theorem 7.30. *The graph $\mathcal{AO}(G)$ of all acyclic orientations of a nonempty graph G is isomorphic to the region graph of the graphic arrangement $\mathcal{A}(G)$.*

Accordingly, the graph of all acyclic orientations of the complete graph K_n is isomorphic to the permutohedron Π_{n-1} (cf. Example 7.26).

7.6 Zonotopal Tilings

The region graph of a line arrangement $\mathcal{A}$ in $\mathbf{R}^2$ is of course planar. One can draw this graph by selecting a point in each region and connecting the points corresponding to adjacent regions by line segments (see drawings in Section 7.3). In this section we describe an elegant geometric method for drawing region graphs of line arrangements. The method is based on the construction of Minkowski sums of intervals in $\mathbf{R}^2$.

The *Minkowski sum* (or *vector sum*) of two sets $P, Q \subseteq \mathbf{R}^2$ is defined to be

$$P + Q = \{x + y : x \in P,\ y \in Q\},$$

where $x + y$ stands for the vector sum of x and y. The Minkowski sum of a finite family of sets is defined in a similar way.

Let us consider first a simple case of a central arrangement $\mathcal{A} = \{L_1, L_2\}$ of two lines in $\mathbf{R}^2$. Clearly, the region graph of $\mathcal{A}$ is the cycle C_4, which can be drawn in the plane as a quadrilateral. We draw this graph by using the Minkowski sum of two intervals as follows. Let $\{e_i, -e_i\}$, for $i \in \{1, 2\}$, be two normal unit vectors to the line L_i. It is not difficult to see that the Minkowski sum

$$[e_1, -e_1] + [e_2, -e_2]$$

is the rhombus shown in Figure 7.16. The boundary of this rhombus is the desired drawing of the region graph of $\mathcal{A}$.

Note that the same rhombus can be constructed by using intervals of length two, which are perpendicular to the lines of the arrangement $\mathcal{A}$.

Figure 7.16. The region graph of an arrangement of two lines.

We use the same method to draw the region graph of a central arrangement of p lines $\mathcal{A}$ in $\mathbf{R}^2$. The region graph of this arrangement is the even cycle C_{2p}. Let $\{e_i, -e_i\}$ be normal unit vectors to the lines L_i for $1 \le i \le p$, and let

$$[e_1, -e_1] + [e_2, -e_2] + \cdots + [e_p, -e_p]$$

be the Minkowski sum of the intervals $[e_i, -e_i]$, $1 \le i \le p$. We invite the reader to prove that this sum is a centrally symmetric convex $2p$-gon with sides of length two. The boundary of this polygon is a drawing of the region graph of the arrangement $\mathcal{A}$. The case of $p = 3$ is illustrated in Figure 7.17 (cf. the drawing in Figure 7.10). The rhombus in the left drawing is the Minkowski sum of two intervals. After adding the third interval, we obtain a hexagon shown as the shaded region in the right drawing. This example hints at an inductive proof in the general case.

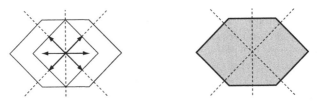

Figure 7.17. The region graph of an arrangement of three lines.

Let $\mathcal{A}$ be an arbitrary finite line arrangement in $\mathbf{R}^2$. We say that a point $a \in \mathbf{R}^2$ is a *vertex* of $\mathcal{A}$ if a belongs to at least two lines in $\mathcal{A}$. In what follows, we assume that $\mathcal{A}$ has at least one vertex. Otherwise, all lines in $\mathcal{A}$ are parallel and the region graph of $\mathcal{A}$ is a path. Each vertex of the arrangement $\mathcal{A}$ defines a central arrangement consisting of all lines in $\mathcal{A}$ containing this vertex, and the arrangement itself is the union of these central arrangements. Let us form a set of centrally symmetric convex polygons that we call *tiles*, one per vertex of the arrangement, as described in the foregoing paragraph, and glue together two tiles whenever they correspond to adjacent vertices along the same line of the arrangement (drawings in Figure 7.18 are instructive). In the constructed figure, which is called a *zonotopal tiling*, the tiles meet at interior vertices (corresponding to bounded regions of the arrangement) at angles totaling 2π, because the sides of these angles are perpendicular to the sides of a polygon. For the same reason, each boundary vertex of the figure (corresponding to an unbounded region of the arrangement) has angles totaling at most π (cf. Exercise 7.14). Thus the union of tiles of the constructed zonotopal tilings is a convex polygon in the plane $\mathbf{R}^2$. The vertices of tiles together with their edges form a drawing of the region graph of the line arrangement $\mathcal{A}$. An example of a zonotopal tiling of a hexagon by rhombi can be obtained from the drawing in Figure 7.13. If all tiles of a tiling are regular polygons, the tiling is called a *mosaic*.

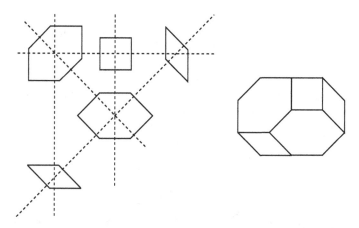

Figure 7.18. A line arrangement (left) and the corresponding zonotopal tiling (right).

Tiles corresponding to the vertices of an infinite line arrangement can be used in the same way to construct infinite tilings of convex sets in the plane. Note that the resulting sets are not necessarily polygons (cf. Exercise 7.15). However, some "regular" infinite line arrangements produce highly symmetrical zonotopal tilings of the plane by convex centrally symmetric polygons. In the rest of this section, we give some examples of these tilings.

A line arrangement consisting of k infinite families of parallel lines, equally spaced within each family, is called a *multigrid* or a *k-grid*. The zonotopal tiling formed by the region graph of a 2-grid is a tiling of a plane by edge-to-edge rhombi or squares such as the one shown in Figure 7.6 (left), which is a mosaic of squares. A mosaic of hexagons is obtained from the 3-grid in Figure 7.3 (right) (cf. the right drawing in Figure 7.6). By Theorem 7.23, the graph of this mosaic is isometrically embeddable into the integer lattice $\mathbb{Z}^3$.

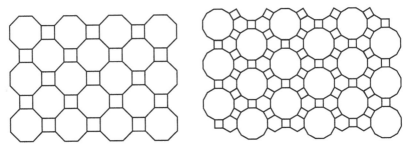

Figure 7.19. The mosaic of octagons and squares (left), and the mosaic of do-decagons, hexagons, and squares (right).

Two more mosaics of octagons and squares and of dodecagons, hexagons, and squares, are shown in Figure 7.19. These again are drawings of region graphs of line arrangements with regularly spaced lines having four and six different slopes, respectively (cf. Exercise 7.16).

A more sophisticated example of an infinite partial cube is given by the graph of a *Penrose rhombic tiling* shown in Figure 7.20. These kinds of tilings can be formed from the region graphs of line arrangements known as *penta-grids* that are 5-grids having equal angles between each pair of parallel line families, and satisfying certain additional constraints on the placement of the grids relative to each other.

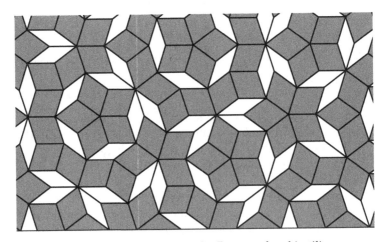

Figure 7.20. A fragment of a Penrose rhombic tiling.

Example 7.31. As in Section 7.6, one can use Minkowski sums to construct the "tile" for a central arrangement of planes in $\mathbf{R}^d$. The corresponding tile is a highly symmetrical polyhedron called a *zonotope*. For instance, the permutohedron Π_3 is a 3-dimensional zonotope defined by the braid arrangement $\mathcal{BA}_4$ (note that it "lives" in a subspace of $\mathbf{R}^4$ defined by $\sum_{i=1}^4 x_i = 0$).

Another example of a zonotope is obtained from the central arrangement of six planes in $\mathbf{R}^3$ given by equations:

$$2x_2 + x_3 = 0 \qquad\qquad x_2 + 2x_3 = 0$$
$$2x_1 \quad\;\, + x_3 = 0 \qquad\qquad x_1 + 2x_2 \quad\;\, = 0$$
$$2x_1 + x_2 \quad\;\, = 0 \qquad\qquad x_1 \quad\;\, + 2x_3 = 0$$

It is not difficult to verify that intersections of these planes with the plane given by $x_1 + x_2 + x_3 = 1$ are precisely the lines of the semiorder arrangement in Figure 7.13. For this reason, we call the resulting zonotope a *semiorder zonotope*. Two views of this zonotope are shown in Figure 7.21. The center of the left figure is the "north pole" of the zonotope, the "equatorial" view of which is shown in the right figure. It is instructive to compare the left figure with the drawing in Figure 7.13.

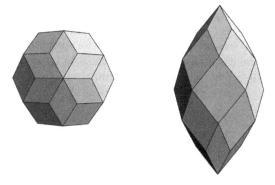

Figure 7.21. Two views of the 3-dimensional semiorder zonotope.

7.7 Families of Binary Relations

As we showed in Section 5.3, the family of partial orders on a finite set is well-graded and therefore can be modeled as a partial cube (cf. drawing in Figure 5.3). In this section, we construct partial cube models for some other classes of binary relations on a finite set.

In what follows, X is a finite set of cardinality greater than two, and letters P, Q, R, and so on, stand for binary relations on X.

First we consider linear orders on X. In Section 5.3 a linear order on X was defined as an irreflexive, transitive, and complete binary relation on X. It can be easily seen that the family of linear orders on X is not a well-graded set. Indeed, let P and Q be two distinct linear orders, and let, say, $(a, b) \in P$, $(a, b) \notin Q$. Then, $\{(a, b), (b, a)\} \subseteq P \triangle Q$, so the Hamming distance $d(P, Q) > 1$ for any two linear orders. However, we can use the following result (cf. Exercise 7.21) to cast the set $\mathcal{LO}$ of all linear orders on X as the vertex set of a partial cube (cf. Section 5.13).

Theorem 7.32. *A binary relation R on X is a linear order if and only if there exists a one-to-one function $f : X \to \mathbf{R}$ such that*

$$(a, b) \in R \text{ if and only if } f(a) > f(b), \tag{7.1}$$

for all $a, b \in X$.

Any one-to-one function f defining the relation R by (7.1) is said to be a *representing function* for R.

The set of all real functions on X is a vector space $V = \mathbf{R}^X$ that is isomorphic to $\mathbf{R}^d$, where $d = |X|$. For each pair $\{a, b\}$ of distinct elements of X, we define a subset H_{ab} of V by

$$H_{ab} = \{f \in V : f(a) = f(b)\}.$$

It is clear that these sets are hyperplanes in V forming a central arrangement $\mathcal{A}$ which is the same as the braid arrangement $\mathcal{BA}_d$ (cf. Example 7.10). Note that $V \setminus \cup \mathcal{A}$ is the set of one-to-one real-valued functions on X, that is, the set of representing functions for linear orders on X.

Theorem 7.33. *Two functions $f, g \in V \setminus \cup \mathcal{A}$ represent the same linear order if and only if they belong to the same region of $\mathcal{A}$. Accordingly, there is a one-to-one correspondence between the elements of $\mathcal{LO}$ and the regions of the arrangement $\mathcal{A}$.*

Proof. (Necessity.) Let us assume that functions f and g represent the same linear order R and belong to different regions of $\mathcal{A}$. Then there is a pair $\{a, b\}$ such that the hyperplane H_{ab} separates f and g; that is,

$$[f(a) - f(b)][g(a) - g(b)] < 0.$$

By (7.1), $(a, b) \in R$ and $(b, a) \in R$, contradicting our assumption that R is a linear order.

(Sufficiency.) Likewise, if f and g belong to the same region, then for any pair of distinct elements $\{a, b\}$, we have

$$[f(a) - f(b)][g(a) - g(b)] > 0.$$

By (7.1), f and g define the same linear order on X.   □

It follows from Theorem 7.33 that linear orders on a d-element set X can be represented by the vertices of the permutohedron Π_{d-1}. In this representation, two linear orders P and Q are adjacent if there is a pair of distinct elements $\{a, b\}$ of X such that $P \triangle Q = \{(a, b), (b, a)\}$; that is, the Hamming distance between P and Q equals two (cf. Exercise 7.22).

Now we employ the method used in the foregoing paragraphs to show that the family of labeled interval orders on a finite set is well-graded.

Definition 7.34. Let ρ be a positive function on X. A binary relation R on X is said to be a *labeled interval order* if there exists a function $f : X \to \mathbf{R}$ such that

$$(a, b) \in R \text{ if and only if } f(a) > f(b) + \rho(b), \tag{7.2}$$

for all $a, b \in X$. Any function f defining the relation R by (7.2) is said to be a *representing function* for R. For a given ρ, we denote by $\mathcal{IO}_\rho$ the family of all labeled interval orders on X. If ρ is a constant function, then (7.2) defines a *semiorder* on X. The family of all semiorders on X is denoted by $\mathcal{SO}$.

Interpretation. For a given function $f : X \to \mathbf{R}$, we "represent" elements of the set X by intervals

$$I_a = [f(a), f(a) + \rho(a)], \quad a \in X.$$

Then, by (7.2), $(a, b) \in R$ if and only if there is f such that the interval I_a lies strictly to the right of the interval I_b.

As before, let $V = \mathbf{R}^X$ be the vector space of all real-valued functions on X. Any function f in V defines a labeled interval order by (7.2). Let $J = (X \times X) \setminus \{(x, x)\}_{x \in X}$. For $(a, b) \in J$ we define hyperplanes $H_{(a,b)}$ by

$$H_{(a,b)} = \{f \in V : f(a) - f(b) - \rho(b) = 0\}.$$

The hyperplanes $H_{(a,b)}$ are distinct (Exercise 7.23) and form an arrangement $\mathcal{A}$ in V which is the deformation $\mathcal{A}_{d,\rho}$ of the braid arrangement $\mathcal{BA}_d$ (cf. Example 7.12 and note that ρ is a vector in V; here, as before, $d = |X|$.) If ρ is a constant function, then $\mathcal{A}_{d,\rho}$ is the semiorder arrangement $\mathcal{SO}_d$ in $V = \mathbf{R}^d$ (cf. Exercise 7.24).

Any two functions f and g in a region of $\mathcal{A}$ define the same labeled interval order by (7.2). Indeed, suppose to the contrary, that these functions define distinct labeled orders P and Q, respectively. We may assume that there is $(a, b) \in J$ such that $(a, b) \in P$ and $(a, b) \notin Q$. Then

$$f(a) - f(b) - \rho(b) > 0 \text{ and } g(a) - g(b) - \rho(b) \le 0;$$

that is, f and g do not lie on the same side of the hyperplane $H_{(a,b)}$. This contradicts our assumption that f and g belong to the same region.

On the other hand, if f and g belong to distinct regions, then there is a hyperplane $H_{(a,b)}$ separating f and g; that is,

$$[f(a) - f(b) - \rho(b)][g(a) - g(b) - \rho(b)] < 0.$$

Hence, f and g define distinct interval orders on X by (7.2). It follows that there is a one-to-one correspondence $\varphi : \mathcal{R} \to \mathcal{J}$ between the set $\mathcal{R}$ of regions of $\mathcal{A}$ and the set $\mathcal{J}$ of labeled interval orders representable by functions from the union of regions. We want to show that $\mathcal{J} = \mathcal{J}\mathcal{O}_\rho$.

Lemma 7.35. *Let R be a labeled interval order on X. There is a function f representing R such that $f \in V \setminus \cup_{(a,b)\in J}H_{(a,b)}$.*

Proof. First, let us note that the zero function on X represents the empty labeled interval order and does not belong to any of the hyperplanes $H_{(a,b)}$.

Let h be a representing function of a nonempty labeled interval order R on X. Because X is a finite set, and $h(x) - h(y) > \rho(y)$ for all $(x, y) \in R$, there is $\delta > 0$ such that

$$\max_{(x,y)\in R} \frac{\rho(y)}{h(x) - h(y)} < \delta < 1. \tag{7.3}$$

We define $f = \delta h$ and show first that f is a representing function for R. For an arbitrary ordered pair $(a, b) \in R$, we have

$$f(a) - f(b) = \delta[h(a) - h(b)] > \rho(b),$$

by the first inequality in (7.3). On the other hand, if for some ordered pair (a, b) we have $f(a) - f(b) > \rho(b)$, then

$$h(a) - h(b) > \frac{\rho(b)}{\delta} > \rho(b),$$

by the second inequality in (7.3). Hence, $(a, b) \in R$. This proves that f is a representing function for R.

Suppose now that f belongs to a hyperplane $H_{(c,d)}$; that is,

$$f(c) - f(d) = \rho(d).$$

Then

$$h(c) - h(d) = \frac{\rho(d)}{\delta} > \rho(d),$$

that is, $(c, d) \in R$. Because f is a representing function for R, we must have $f(c) - f(d) > \rho(d)$, a contradiction. Therefore, $f \in V \setminus \cup_{(a,b)\in J}H_{(a,b)}$. □

We proved that there is one-to-one correspondence between the set of regions of the arrangement $\mathcal{A}_{d,\rho}$ in $V = \mathbf{R}^X$ and the set of labeled interval orders $\mathcal{J}\mathcal{O}_\rho$ on X. By Theorem 7.17, the graph distance between two regions of $\mathcal{A}_{d,\rho}$ equals the number of hyperplanes in $\mathcal{A}_{d,\rho}$ separating these regions. The latter number is the Hamming distance between labeled interval orders corresponding to the regions (cf. Exercise 7.25). Thus we have the following result.

Theorem 7.36. *The family of labeled interval orders* $\mathfrak{IO}_\rho$ *on a finite set* X *is well-graded. In particular, the family of semiorders* $\mathfrak{SO}$ *is well-graded.*

The correspondence $\varphi : \mathcal{R} \to \mathfrak{IO}_\rho$ *is a graph isomorphism between the region graph of the arrangement* $\mathcal{A}_{d,\rho}$ *and the graph of the wg-family* $\mathfrak{IO}_\rho$.

In the rest of this section, we establish well-gradedness of the family of interval orders on a finite set.

Definition 7.37. A binary relation R on X is called an *interval order* if there exist real-valued functions $f < g$ such that

$$(a, b) \in R \text{ if and only if } f(a) > g(b) \qquad (7.4)$$

for all $a, b \in X$. Any pair of functions (f, g) with $f < g$ defining the relation R by (7.4) is said to be a *representing function* for R. (The pair (f, g) is a function $(f, g) : X \to \mathbf{R}^2$.) The set of interval orders on X is denoted by $\mathfrak{IO}$.

Some remarks are in order. If R is a labeled interval order in $\mathfrak{IO}_\rho$ with a representing function f, then it is also an interval order with the representing function $(f, f + \rho)$. On the other hand, an interval order represented by function (f, g) is the labeled interval order in $\mathfrak{IO}_\rho$ with $\rho = g - f$. Therefore, $\mathfrak{IO} = \cup_\rho \mathfrak{IO}_\rho$. As in the case of labeled interval orders, interval orders can be regarded as binary relations "representable" by naturally ordered intervals $[f(a), g(a)]$, $a \in X$.

We need modified versions of definitions from Sections 7.2 and 7.3. Let V be a finite-dimensional vector space and W be an open convex space in V. For a given hyperplane arrangement $\mathcal{A}$, we denote $\mathcal{A}_W$ a hyperplane arrangement consisting of those hyperplanes in $\mathcal{A}$ that have nonempty intersections with W. The region graph G_W of $\mathcal{A}_W$ is obtained from the region graph G of $\mathcal{A}$ by deleting vertices representing regions of $\mathcal{A}$ that do not intersect the set W. The following theorem is a generalization of Theorem 7.17. The proof is left to the reader (cf. Exercise 7.28).

Theorem 7.38. *The region graph* G_W *of the hyperplane arrangement* $\mathcal{A}_W$ *in* V *is a partial cube. Furthermore, the graph distance between two regions* P *and* Q *of* $\mathcal{A}_W$ *equals the number of hyperplanes in* $\mathcal{A}_W$ *separating* P *and* Q *in* V.

Let V be the vector space of all functions from X into $\mathbf{R}^2$, and let W be a subset of V defined by

$$W = \{(f, g) \in V : f(x) < g(x), \quad \text{for all } x \in X\}. \qquad (7.5)$$

It is not difficult to see that W is an open convex cone in V (cf. Exercise 7.27). This set consists of functions representing interval orders on X.

For $(a, b) \in J = (X \times X) \setminus \{x, x\}_{x \in X}$, we define hyperplanes

$$H_{(a,b)} = \{(f, g) \in V : f(a) = g(b)\}.$$

These hyperplanes form a central arrangement $\mathcal{A}$ in V which is a subset of the braid arrangement in V (cf. Exercise 7.29). Any function in the set W defined by (7.5) represents a linear order on X defined by (7.4).

Suppose that two functions (f_1, g_1) and (f_2, g_2) in a region of $\mathcal{A}_W$ define distinct linear orders P and Q, so there is a pair $(a, b) \in J$ such that $(a, b) \in P$ and $(a, b) \notin Q$. Then

$$f_1(a) > g_1(b) \quad \text{and} \quad f_2(a) \le g_2(b),$$

that is, the functions (f_1, g_1) and (f_2, g_2) do not lie on the same side of $H_{(a,b)}$. Hence, two functions in the same region of $\mathcal{A}_W$ define the same interval order on X.

On the other hand, it is clear that functions in distinct regions of $\mathcal{A}_W$ define distinct interval orders on X. As in the case of interval linear orders, it follows that there is one-to-one correspondence $\varphi : \mathcal{R}_W \to \mathcal{J}$ between the set $\mathcal{R}_W$ of regions of $\mathcal{A}_W$ and the set $\mathcal{J}$ of interval orders representable by functions from the union of regions which is the set $W \setminus \cup \mathcal{A}_W$. The next lemma shows that $\mathcal{J} = \mathcal{JO}$.

Lemma 7.39. *Let R be an interval order on X. There is a function in the set $W \setminus \cup \mathcal{A}_W$ representing R.*

Proof. First note that a function (f, g), where f and g are constant functions on X with $f < g$, represents the empty interval order on X and does not belong to any of the hyperplanes in $\mathcal{A}_W$.

Let R be a nonempty interval order on X and let (f, g) be its representing function. X is a finite set, thus there is α such that

$$0 < \alpha < \min_{(x,y)\in R}[f(x) - g(y)]. \tag{7.6}$$

Let us show that $(f - \alpha, g)$ is a representing function for R. Clearly, the function $(f - \alpha, g)$ belongs to W.

If $(a, b) \in R$, then $f(a) - \alpha > g(b)$, by (7.6). On the other hand, if

$$f(a) - \alpha > g(b)$$

for some $(a, b) \in J$, then, again by (7.6), $f(a) > g(b)$. Therefore, $(a, b) \in R$, inasmuch as (f, g) is a representing function for R.

Suppose that $(f - \alpha, g)$ belongs to a hyperplane $H_{(a,b)}$ for some $(a, b) \in J$. Then $f(a) - \alpha = g(b)$, implying $f(a) > g(b)$. Because (f, g) is a representing function for R, we have $(a, b) \in R$. Because $(f - \alpha, g)$ is also a representing function for R, we have $f(a) - \alpha > g(b)$, a contradiction.

We proved that $(f - \alpha, g)$ is a representing function for R which belongs to a region of $\mathcal{A}_W$. □

Following the arguments we used in the case of labeled interval orders (cf. Theorem 7.36), we obtain the following result.

Theorem 7.40. *The family of interval orders on a finite set is well-graded.*

Notes

Hyperplane arrangements are fundamental constructions in various areas of mathematics ranging from Lie algebras to geometry of quasi-crystals. Standard texts on finite hyperplane arrangements are the books by Zaslavsky (1975) and Orlik and Terano (1992). These geometric objects have rich algebraic and combinatorial structures. The reader is referred to Björner et al. (1999), Ziegler (2006), and Stanley (2007) for extensive studies of these structures. A rigorous treatment of infinite hyperplane arrangements is found in Bourbaki (2002, Chapter 5) and Abramenko and Brown (2008, Section 2.7).

Region graphs are also known as *chamber graphs* in the theory of reflection groups (Abramenko and Brown, 2008) and as *tope graphs* in the oriented matroids theory (Björner et al., 1999). The fact that the region graph of an arrangement is a partial cube (Theorem 7.17) justifies our interest in hyperplane arrangements. For finite arrangements, the result follows from Proposition 4.2.3 in Björner et al. (1999). Geometric proofs are found in Abramenko and Brown (2008) and Borovik and Borovik (2010). Our proof is taken from Ovchinnikov (2005) where the result is established for arbitrary arrangements. The term "permutohedron" was coined by Guilbaud and Rosenstiehl (1963). This polytope is a popular example in books on convex polyhedra (cf. Ziegler, 2006).

As we noted in Section 7.2, regions of a finite arrangement $\mathcal{A}$ are polyhedra. The topological closures of the faces of these polyhedra are naturally ordered by inclusion. The resulting poset is a *face poset* $\mathcal{F}(\mathcal{A})$ of the arrangement $\mathcal{A}$ (Björner et al., 1999, Section 2,1). Regions are maximal elements of this poset. For an example of modeling the face poset of an arrangement as a partial cube see Ovchinnikov (2008b).

According to Theorem 7.30, the number of acyclic orientations of a finite graph G equals the number of regions of the graphic arrangements $\mathcal{A}(G)$ (Greene, 1977). The latter number was determined by Zaslavsky (1975) for an arbitrary finite hyperplane arrangement. Counting the number of elements of a finite set is a typical problem in *enumerative combinatorics* (Stanley, 1997).

The reader will find more on mosaics and tilings, together with further references, in Grünbaum and Shephard (1987) and Senechal (1995). Penrose tilings were discovered by Sir Roger Penrose, who investigated these objects in the 1970s. An algebraic theory of Penrose tilings was developed by de Bruijn (1981), who demonstrated that these kinds of tilings can be formed from region graphs of pentagrids. By Theorem 7.23, the graph of a Penrose tiling is isometrically embeddable into $\mathcal{Z}^5$ (cf. Deza and Shtogrin, 2002). The drawing of the Penrose tiling in Figure 7.20 was obtained using *Mathematica*® computer algebra system. A systematic treatment of zonotopes is found in Ziegler (2006). The drawing in Figure 7.31 was created by *Mathematica* code from Eppstein (1996).

The graph of the family $\mathcal{PO}$ of partial orders on a finite set X was our first nontrivial example of a partial cube (see Section 5.3). The families of semiorders and interval orders on X are well-graded (Theorems 7.36 and 7.40, respectively) and therefore also define partial cubes. Although the family of linear orders on X is not well-graded, it can be naturally modeled by vertices of a partial cube (cf. Section 5.13 and Theorem 7.33). These representations of families of binary relations are of interest in some applications in social and cognitive sciences (cf. Eppstein et al., 2008). In our presentation we followed Ovchinnikov (2005). For purely algebraic proofs of some of these results see Doignon and Falmagne (1997).

Note that we defined partial and linear orders axiomatically, that is, by listing their defining properties. On the other hand, semiorders and interval orders were defined by means of representing functions. However, these relations can also be introduced "algebraically" as follows (cf. Fishburn, 1985).

An *interval order* on an arbitrary nonempty set X is a binary relation R on X that satisfies

$$[(a, x) \in R \text{ and } (b, y) \in R] \text{ implies } [(a, y) \in R \text{ or } (b, x) \in R] \quad (7.7)$$

for all a, b, x, and y in X. A *semiorder* on X is an interval order R on X that satisfies

$$[(a, b) \in R \text{ and } (b, c) \in R] \text{ implies } [(a, x) \in R \text{ or } (x, c) \in R] \quad (7.8)$$

for all a, b, c, and x in X. Relations satisfying (7.8) are known as "partial semiorders" (Fishburn, 1985) or "ac-orders" (Doble et al., 2001). It is not a trivial task to show that, in the case of a finite set X, the "algebraic" properties define exactly the same classes of relations as representing functions do (cf. Fishburn, 1985).

Exercises

7.1. a) Show that two linear functions

$$L(x) = a_1 x_1 + \cdots + a_d x_d + b \text{ and } L'(x) = a'_1 x_1 + \cdots + a'_d x_d + b'$$

define the same hyperplane if and only if the two vectors

$$(a_1, \ldots, a_d, b) \text{ and } (a'_1, \ldots, a'_d, b')$$

are proportional.

b) Show that two hyperplanes defined by L and L' are parallel (do not intersect) if and only if $k(a_1, \ldots, a_d) = (a'_1, \ldots, a'_d)$ for some $k \neq 0$ and $kb \neq b'$.

7.2. A subset A of the space $\mathbf{R}^d$ is said to be *convex* if for any two points $x, y \in A$ the point $tx + (1 - t)y$ with $0 \leq t \leq 1$ belongs to A. Thus a set A is convex if it contains together with any two points $x, y \in A$ the line segment $[x, y]$ connecting these points.

a) Show that H, H^+, and H^- are convex subsets of $\mathbf{R}^d$ and

$$\mathbf{R}^d = H \cup H^+ \cup H^-.$$

b) Show that the intersection of any family of convex sets in $\mathbf{R}^d$ is convex.
c) Give an example of two convex sets in $\mathbf{R}^d$ such that their union is not convex.

7.3. Show that there are at most countably many elements in a locally finite famuly of sets in $\mathbf{R}^d$.

7.4. True or false: A set $A \subseteq \mathbf{R}$ is locally finite if and only if

$$\inf\{|x - y| : x, y \in A, \ x \neq y\} > 0.$$

7.5. Let H be a hyperplane and let P be a path (a continuous curve) connecting points $x \in H^+$ and $y \in H^-$. Prove that P intersects H.

7.6. Prove that the region graph of a hyperplane arrangement contains a vertex of degree one, if and only if this graph is a finite path or a ray.

7.7. Prove that a nontrivial linear combination of linear functions is a linear function.

7.8. Prove that the permutohedron Π_{d-1} is not contained in a linear manifold of dimension less than $d - 1$.

7.9. Prove that two regions of the braid arrangement $\mathcal{BA}_d$ are adjacent if and only if their representing points differ by transposition of two consecutive coordinates.

7.10. Finish the proof of Lemma 7.21.

7.11. Describe the graph $\mathcal{AO}(G)$, where G is

a) A path P_n, $n > 1$
b) A cycle C_n, $n > 2$
c) A complete graph K_n, $n > 1$

7.12. Let G be a nonempty finite graph. Prove that the set $\mathcal{AO}(G)$ of all acyclic orientations of G is not empty.

7.13. Prove that the graph $\mathcal{AO}(C_n)$, where C_n is a cycle of length n, is isomorphic to a graph obtained from the cube Q_n by deleting two opposite vertices.

7.14. a) Prove that the sum of exterior angles of a bounded convex polygon is 2π.

b) Prove that the sum of exterior angles of an unbounded convex polygon does not exceed π.

Hint: The sum of interior angles of a convex n-gon is $(n-2)\pi$.

7.15. Construct the tiling corresponding to the line arrangement in Example 7.8. Show that the union of the tiles is a convex set which not a polygon.

7.16. Show that two planar graphs in Figure 7.19 are region graphs of line arrangements with regularly spaced lines having four and six different slopes, respectively.

7.17. Find a zonotopal tiling that is a drawing of the region graph of the line arrangement in Figure 7.22.

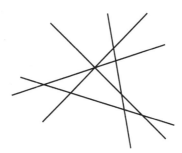

Figure 7.22. Exercise 7.17.

7.18. Find an example of a tiling of a nonsimply connected region of the plane by strictly convex centrally symmetric tiles that is not the drawing of a partial cube.

7.19. Find an example of a tiling of a convex region of the plane by convex but not necessarily strictly convex centrally symmetric tiles (i.e., some internal angles can equal π) that is not the drawing of a partial cube.

7.20. A finite *plane* graph G (i.e., a drawing of a planar graph) partitions the plane into a number of connected open sets. These sets are called the *faces* of G. Find a plane graph G such that all its bounded faces are convex quadrilaterals and G is not a partial cube.

7.21. Prove Theorem 7.32.

7.22. Show that the family of linear orders $\mathcal{LO}$ on a finite set X is 2-graded; that is, for any two distinct linear orders P and Q there is a sequence of linear orders $R_0 = P, R_1, \ldots, R_n = Q$ such that $d(R_i, R_{i+1}) = 2$ for $0 \le i \le n-1$, and $d(P, Q) = 2n$.

7.23. Show that all hyperplanes $H_{(a,b)}$ in the arrangement $\mathcal{A}_{d,\rho}$ are distinct.

7.24. Show that $\mathcal{A}_{d,\rho} = \mathcal{SO}_d$ if ρ is a constant function.

7.25. Prove that the Hamming distance $d(P,Q)$ between two labeled interval orders P and Q is equal to the number of hyperplanes in $\mathcal{A}_{d,\rho}$ separating regions corresponding to these orders.

7.26. Let $\mathcal{PO}$ be the set of all partial orders on a given finite set X of cardinality $d > 2$, and let ρ be a positive function on X. Prove the following statements.

a) $\dim_I(\mathcal{IO}_\rho) = d(d-1)$ and $\dim_Z(\mathcal{IO}_\rho) = d(d-1)/2$.
b) If $d = 3$, then $\mathcal{SO} = \mathcal{IO}_\rho = \mathcal{PO}$.
c) If $d \geq 4$, then $\mathcal{IO}_\rho$ is a proper subset of $\mathcal{PO}$.
d) If $d = 4$, then $\mathcal{SO} \subseteq \mathcal{IO}_\rho$.
e) If $d = 4$ and ρ is not a constant function on X, then $\mathcal{SO}$ is a proper subset of $\mathcal{IO}_\rho$.
f) If $d \geq 5$, then there is ρ such that

$$\mathcal{SO} \setminus \mathcal{IO}_\rho \neq \varnothing \text{ and } \mathcal{IO}_\rho \setminus \mathcal{SO} \neq \varnothing.$$

7.27. Prove that the set W defined by (7.5) is an open convex cone in the space $V = (\mathbf{R}^2)^X$.

7.28. Prove Theorem 7.38.

7.29. Let $X = \{a_1, \ldots, a_n\}$ be a finite set. Show that the vector space V of all functions $X \to \mathbf{R}^2$ is isomorphic to the space $\mathbf{R}^{2n}$ under the mapping

$$(f,g) \to (f(a_1), \ldots, f(a_n), g(a_1), \ldots, g(a_n)).$$

7.30. Prove Theorem 7.40.

7.31. Find the lattice dimension of the graph of the family of interval orders on a d-element set.

7.32. Let X and Y be two finite sets. A relation $R \subseteq X \times Y$ is called a *biorder* *from X to Y* if there are functions $f : X \to \mathbf{R}$ and $g : Y \to \mathbf{R}$ such that

$$(x,y) \in R \text{ if and only if } f(x) > g(y),$$

for all $x \in X$ and $y \in Y$. Prove that the family of all biorders from X to Y is well-graded.

8

Token Systems

The chapter deals with algebraic and stochastic structures of token systems. Cubical token systems and media are defined as systems satisfying two sets of axioms, and elements of formal theories of these systems are introduced. Cubical systems and media serve as algebraic components of random walk models introduced in the last section of the chapter.

8.1 Algebraic Preliminaries

A *semigroup* is a nonempty set S that is closed under associative binary operation $(x, y) \mapsto xy$. Thus

$$xy \in S \text{ and } (xy)z = x(yz), \text{ for all } x, y, z \in S.$$

The element xy is the *product* of elements x and y. The associativity property permits us to write products such as xyz without parentheses.

A *monoid* is a semigroup M with a unit element $1 \in M$ such that

$$1m = m1 = m, \text{ for all } m \in M.$$

It can be easily shown that the unit element 1 is unique (cf. Exercise 8.1).

We introduce below two classes of monoids that are used in our constructions.

Let Σ be a set. The *free monoid* Σ^* with *base* Σ is defined as follows. The elements of Σ^* are n-tuples

$$s = (\sigma_1, \ldots, \sigma_n), \quad (n \geq 0)$$

of elements of Σ. By definition, the only 0-tuple is the empty tuple ( ). If $t = (\tau_1, \ldots, \tau_m)$ is another element of Σ^*, the product st is defined by concatenation; that is,

$$st = (\sigma_1, \ldots, \sigma_n, \tau_1, \ldots, \tau_m).$$

This operation defines a monoid with the unit $1 = (\ )$. It is customary to write σ instead of 1-tuple (σ). Then, for $n > 0$, an n-tuple $s = (\sigma_1, \ldots, \sigma_n)$ can be written as a string $s = \sigma_1 \cdots \sigma_n$. For this reason, the element s is called a *word* of length n and the set Σ itself is called an *alphabet*. We call the unit element 1 of Σ^* the *empty word*. The convention $\sigma = (\sigma)$ allows for treating Σ as a subset of Σ^*. Note that the set of nonempty words in Σ^* is a semigroup, provided that $\Sigma \neq \varnothing$.

Let $s \in \Sigma^*$. An element $t \in \Sigma^*$ is a *segment* of s if $s = utv$ for some $u, v \in \Sigma^*$. If $u = 1$, then t is an *initial segment* of s; if $v = 1$, then t is a *terminal segment* of s. Note that $t = 1$ or $t = s$ are not excluded.

Now let X be a nonempty set and let FX be the set of all functions $f : X \to X$. This set is a monoid under the operation of composition of two functions. The unit element of FX is the identity function $1_X : x \mapsto x, x \in X$. Note that the monoid FX and the free monoid FX^* are different algebraic structures. For instance, the identity function 1_X (the unit element of FX) is not the unit element of the monoid FX^*.

Example 8.1. Let $X = \{a\}$ be a singleton. Then the monoid FX consists of the unit element $e = 1_X$. On the other hand, the elements of the free monoid FX^* are words $e \cdots e$ (n letters e, $n > 0$) plus the unit element (the empty word).

8.2 Automata and Token Systems

The principal aim of this section is to introduce a language for "systems" based on cubical graphs that are studied in subsequent sections. First, we define *automata*, a wide class of systems used in theoretical studies and applications.

Let Σ be an alphabet. An *automation* $\mathfrak{A}$ over Σ consists of the following four sets:

1. *States*: a nonempty set Q of elements called *states*.

2. *Initial states*: a subset I of Q; the states in I are called *initial*.

3. *Terminal states*: a subset T of Q; the states in T are called *terminal*.

4. *Edges*: a subset E of $Q \times \Sigma \times Q$. A triple (p, σ, q) in E is called an *edge* of the automation $\mathfrak{A}$. The edge (p, σ, q) begins at p, ends at q, and carries the *label* σ. Notations $\sigma : p \to q$ and $p \xrightarrow{\sigma} q$ are frequently used to indicate edges.

It is quite clear that the underlying structure of the automation $\mathfrak{A}$ is a labeled digraph with two distinguished subsets of states, I and T (see Figure 8.1).

A *path* c in $\mathfrak{A}$ is a sequence

$$c = (q_0, \sigma_1, q_1)(q_1, \sigma_2, q_2) \cdots (q_{k-1}, \sigma_k, q_k),$$

which we can graphically write as

$$q_0 \xrightarrow{\sigma_1} q_1 \xrightarrow{\sigma_2} \cdots \xrightarrow{\sigma_k} q_k.$$

The integer $k > 0$ is the *length* of the path. It is convenient to define each state q as a path of length zero. The element $s = \sigma_1 \cdots \sigma_k \in \Sigma^*$ is called the *label* of c.

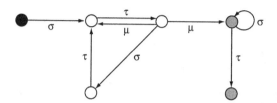

Figure 8.1. In this automation, the initial and terminal states are marked as black and grey circles, respectively. The states are not labeled.

We now introduce an instance of automation that we call a "token system".

A *token system* is an ordered pair $(\mathcal{S}, \mathcal{T})$ where $\mathcal{S}$ is a nonempty set and $\mathcal{T}$ is a nonempty set of functions $\mathcal{S} \to \mathcal{S}$ different from the identity function. The elements of $\mathcal{S}$ are the *states* of the token system $(\mathcal{S}, \mathcal{T})$ and the elements of $\mathcal{T}$ are its *tokens*.

In what follows, we write the function on the right; that is, $S\tau = \tau(S)$ for $S \in \mathcal{S}$ and $\tau \in \mathcal{T}$, and denote the identity function on $\mathcal{S}$ (which is not a token!) by τ_0.

A token system $(\mathcal{S}, \mathcal{T})$ can be cast as an automation. Indeed, let us set $\Sigma = \mathcal{T}$, $Q = \mathcal{S}$, $I = T = \varnothing$, and define

$$E = \{(P, \tau, Q) : Q = P\tau \text{ for } P, Q \in \mathcal{S}, \tau \in \mathcal{T}\}.$$

Clearly, we obtained an automation over $\mathcal{T}$ without initial and terminal states. One distinguished feature of this automation is that for any $P \in \mathcal{S}$ and $\tau \in \mathcal{T}$ there is a unique $Q \in \mathcal{S}$ such that $P \xrightarrow{\tau} Q$.

In the rest of this section, we introduce concepts and terminology that are specific to token systems. To avoid trivialities we assume that $|\mathcal{S}| \geq 2$.

Let P and Q be two states of a token system $(\mathcal{S}, \mathcal{T})$. We say that Q is *adjacent* to P if $Q \neq P$ and $Q = P\tau$ for some token $\tau \in \mathcal{T}$. Two distinct states P and Q are *adjacent* in $(\mathcal{S}, \mathcal{T})$ if P is adjacent to Q and Q is adjacent to P. A token $\tilde{\tau}$ is a *reverse* of a token τ if, for all distinct states $P, Q \in \mathcal{S}$, we have

$$P\tau = Q \text{ if and only if } Q\tilde{\tau} = P.$$

Example 8.2. Let $\mathcal{S} = \{P, Q\}$ and $\mathcal{T} = \{\tau\}$, where $P\tau = Q$ and $Q\tau = Q$ (see Figure 8.2). In the token system $(\mathcal{S}, \mathcal{T})$, the state Q is adjacent to the state P, whereas P is not adjacent to Q. Hence, the states P and Q are not adjacent in $(\mathcal{S}, \mathcal{T})$. The token τ does not have a reverse.

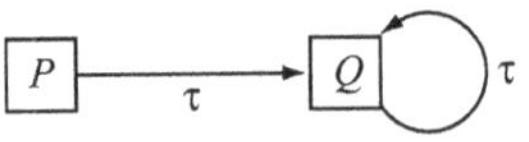

Figure 8.2. Token system $(S, \mathcal{T})$ from Example 8.2.

Suppose that tokens μ and ν are reverses of a token τ. Then, for $P \neq Q$,

$$Q\mu = P \text{ if and only if } P\tau = Q \text{ if and only if } Q\nu = P.$$

It follows that $\mu = \nu$. Therefore, if a reverse of a token exists, then it is unique. It is clear that $\tilde{\tilde{\tau}} = \tau$, provided that $\tilde{\tau}$ exists.

Example 8.3. Let $S = \{P, Q\}$ and $\mathcal{T} = \{\tau\}$ where τ is the nontrivial bijection of S onto itself (see Figure 8.3):

$$P \xrightarrow{\tau} Q, \quad Q \xrightarrow{\tau} P.$$

In the token system $(S, \mathcal{T})$, the states P and Q are adjacent and $\tilde{\tau} = \tau^{-1} = \tau$.

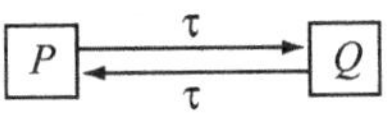

Figure 8.3. Token system $(S, \mathcal{T})$ from Example 8.3.

Example 8.4. Let $S = \{P, Q, R\}$, $\mathcal{T} = \{\tau, \tilde{\tau}\}$ with tokens defined by

$$P \xrightarrow{\tau} Q, \quad Q \xrightarrow{\tau} R, \quad R \xrightarrow{\tau} R,$$

and

$$P \xrightarrow{\tilde{\tau}} P, \quad Q \xrightarrow{\tilde{\tau}} P, \quad R \xrightarrow{\tilde{\tau}} Q.$$

(see Figure 8.4). It can be easily seen that $\tilde{\tau}$ is indeed the reverse of τ.

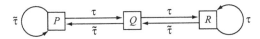

Figure 8.4. Token system $(S, \mathcal{T})$ from Example 8.4.

However, unlike in the previous example, τ^{-1} does not exist. This is typical for token systems considered later in this chapter.

A *message* of a token system $(S, \mathcal{T})$ is a word $\boldsymbol{m} = \tau_1 \cdots \tau_n$ in the monoid $\mathcal{T}^*$. If a token τ occurs in the string $\tau_1 \tau_2 \ldots \tau_n$, we say that the message $\boldsymbol{m} = \tau_1 \cdots \tau_n$ contains τ.

A message $m = \tau_1\tau_2\cdots\tau_n$ defines a function

$$S \mapsto Sm = ((\cdots((S\tau_1)\tau_2)\cdots)\tau_n)$$

on the set of states S. By definition, the empty message defines the identity transformation τ_0 of S. The family of functions

$$\mathcal{T}^\dagger = \{S \mapsto Sm : S \in S, \, m \in \mathcal{T}^*\}$$

is a monoid with the unit τ_0. Note that this monoid is different from the monoid $\mathcal{T}^*$ (cf. Example 8.1).

If $Q = Pm$ for some message m and states $P, Q \in S$, then we say that m *produces* Q *from* P or, equivalently, that m *transforms* P *into* Q. More generally, if $m = \tau_1\cdots\tau_n$, then we say that m *produces a sequence of states* (P_i), where $P_i = P\tau_0\tau_1\cdots\tau_i$ for $0 \le i \le n$.

We denote by $\widetilde{m} = \tilde{\tau}_n\cdots\tilde{\tau}_1$ the *reverse* of the message $m = \tau_1\cdots\tau_n$, provided that the tokens in $\widetilde{m}$ exist.

A message $m = \tau_1\cdots\tau_n$ is *vacuous* if the set of indices $\{1,\ldots,n\}$ can be partitioned into pairs i, j with $i \ne j$, such that τ_i and τ_j are mutual reverses. A message m is *effective* (respectively, *ineffective*) for a state P if $Pm \ne P$ (respectively, $Pm = P$) for the corresponding function m. A message $m = \tau_1\cdots\tau_n$ is *stepwise effective* for P if $P_i \ne P_{i-1}$, $1 \le i \le n$, in the sequence of states (P_i) produced by m from P. A message is *closed* for a state P if it is stepwise effective and ineffective for P.

Two token systems $(S, \mathcal{T})$ and $(S', \mathcal{T}')$ are said to be *isomorphic* if there is a pair (α, β) of bijections $\alpha : S \to S'$ and $\beta : \mathcal{T} \to \mathcal{T}'$ such that

$$P\tau = Q \text{ if and only if } \alpha(P)\beta(\tau) = \alpha(Q),$$

for all $P, Q \in S$ and $\tau \in \mathcal{T}$.

Example 8.5. Here are some examples of messages of the token system $(S, \mathcal{T})$ from Example 8.4:

(i) $\tau\tilde{\tau}\tilde{\tau}\tau$ — vacuous and closed for Q.
(ii) $\tau\tau\tilde{\tau}\tilde{\tau}$ — vacuous, not closed for Q.
(iii) $\tau\tau\tilde{\tau}$ — closed, not vacuous for Q.
(iv) $\tau\tau$ — effective, not stepwise effective for Q.
(v) $\tau\tilde{\tau}\tilde{\tau}$ — effective and stepwise effective for Q. Note that a stepwise effective message for a state may not be effective for that state; it may be closed for the state.

8.3 Cubical Token Systems

Definition 8.6. A token system $(S, \mathcal{T})$ is called a *cubical token system* (on S) if the following axioms are satisfied.

[C1] Every token $\tau \in \mathcal{T}$ has a reverse $\tilde{\tau} \neq \tau$ in $\mathcal{T}$.

[C2] For any two distinct states S and T there is a stepwise effective message producing T from S.

[C3] A message that is stepwise effective for some state is closed for that state if and only if it is vacuous.

[C4] If $m = \tau_1 \ldots \tau_n$ is a stepwise effective message for some state, then occurrences of a token and its reverse alternate in m. More specifically, if $\tau_i = \tau_j = \tau$ for $i < j$ and some $\tau \in \mathcal{T}$, then $\tau_k = \tilde{\tau}$ for some $i < k < j$.

In the rest of the section, we leave out the word "token" in "cubical token system".

Theorem 8.7. *Axioms [C1]–[C4] are independent.*

Proof. Each diagram in Figure 8.5 shows a token system satisfying exactly three of the four axioms defining cubical systems. Each drawing is labeled by the failing axiom. We omit the proofs (cf. Exercise 8.4).   □

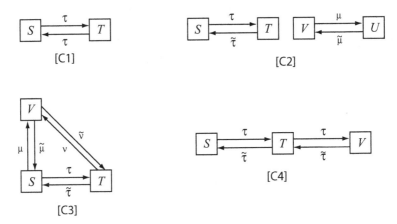

Figure 8.5. Digraphs of four token systems. Each system is labeled by the unique failing axiom.

Note that loops in diagrams in Figure 8.5 are omitted. We follow this practice in our drawings below.

Example 8.8. A unique (up to isomorphism) cubical system on a 2-element set $\mathcal{S} = \{S, T\}$ is shown in Figure 8.6. Indeed, there are precisely four functions $\mathcal{S} \to \mathcal{S}$. Two of these functions are bijections that cannot be tokens by Axiom [C1]. The remaining two functions are mutual reverses, τ and $\tilde{\tau}$. The reader is encouraged to verify the four axioms for a cubical system.

Example 8.9. A unique cubical system on a 3-element set $\mathcal{S} = \{S, T, V\}$ is shown in Figure 8.7 (cf. Exercise 8.6).

Example 8.10. An example of a cubical system on a 4-element set is displayed in Figure 8.8 (cf. Exercise 8.7).

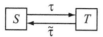

Figure 8.6. Cubical system on $\{S, T\}$.

Figure 8.7. Cubical system on $\{S, T, V\}$.

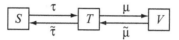

Figure 8.8. Cubical system on $\{S, T, Q, P\}$. Reverses of tokens are not shown.

A "canonical" example of a cubical system is constructed from a cubical graph. (This is why we call these systems "cubical".)

Definition 8.11. Let G be a connected cubical graph. We may assume that G is a subgraph of the cube $\mathcal{H}(X)$ on some set X. Let $\mathcal{F} = V(G)$ and $\mathcal{E} = E(G)$ be the vertex and edge sets of G, respectively. A *G-system on* $\mathcal{F}$ is a pair $(\mathcal{F}, \mathcal{T}_G)$ where $\mathcal{T}_G$ is a family of functions defined by

$$\gamma_x : S \mapsto S\gamma_x = \begin{cases} S \cup \{x\}, & \text{if } \{S, S \cup \{x\}\} \in \mathcal{E}, \\ S, & \text{otherwise,} \end{cases} \tag{8.1}$$

$$\tilde{\gamma}_x : S \mapsto S\tilde{\gamma}_x = \begin{cases} S \setminus \{x\}, & \text{if } \{S, S \setminus \{x\}\} \in \mathcal{E}, \\ S, & \text{otherwise,} \end{cases} \tag{8.2}$$

for all $x \in \cup \mathcal{F} \setminus \cap \mathcal{F}$.

Example 8.12. Let $X = \{x, y\}$ and $G = (\mathcal{F}, \mathcal{E})$, where $\mathcal{F} = 2^X$ and

$$\mathcal{E} = \{\{\varnothing, \{x\}\}, \{\{x\}, \{x, y\}\}, \{\{y\}, \{x, y\}\}\}.$$

This G-system on $\mathcal{F}$ is isomorphic to the token system displayed in Figure 8.8 under isomorphism (α, β) defined by

$$\alpha(S) = \varnothing, \quad \alpha(T) = \{x\}, \quad \alpha(Q) = \{x,y\}, \quad \alpha(P) = \{y\},$$
$$\beta(\tau) = \gamma_x, \quad \beta(\tilde{\tau}) = \tilde{\gamma}_x, \quad \beta(\mu) = \gamma_y, \quad \beta(\tilde{\mu}) = \tilde{\gamma}_y.$$

Note that $\varnothing \gamma_y = \varnothing \neq \{y\}$ (see Figure 8.9).

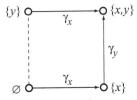

Figure 8.9. The cube on $X = \{x,y\}$ and a G-system on $\mathcal{H}(X)$.

Theorem 8.13. *A G-system on $\mathcal{F}$ is a token system and, for any $x \in \cup \mathcal{F} \setminus \cap \mathcal{F}$, the tokens γ_x and $\tilde{\gamma}_x$ are mutual reverses.*

Proof. We show first that the functions defined by (8.1) and (8.2) are tokens; that is, they are distinct from the identity transformation τ_0. Clearly, for any $x \in \cup \mathcal{F} \setminus \cap \mathcal{F}$ there are two sets $S, T \in \mathcal{F}$ such that $x \notin S$ and $x \in T$. Because G is a connected graph, there is a sequence $S_0 = S, S_1, \ldots, S_n = T$ of sets in $\mathcal{F}$ such that $\{S_i, S_{i+1}\} \in \mathcal{E}$ for $0 \le i < n$. Thus, $|S_i \triangle S_{i+1}| = 1$. Because $x \notin S$ and $x \in T$, there is j such that $x \notin S_j$ and $x \in S_{j+1}$, so, by (8.1),

$$S_{j+1} = S_j \cup \{x\} = S_i \gamma_x.$$

It follows that $\gamma_x \neq \tau_0$ for any $x \in \cup \mathcal{F} \setminus \cap \mathcal{F}$. The case of functions defined by (8.2) is treated similarly. It is clear that the tokens γ_x and $\tilde{\gamma}_x$ are mutual reverses. $\qquad \square$

To show that G-systems are cubical, we begin with a simple observation. Let $S_0 = S, S_1, \ldots, S_n = T$ be an ST-walk in G. For an edge $\{S_{i-1}, S_i\}$, we denote $\{x_i\} = S_{i-1} \triangle S_i$, $\tau_i = \gamma_{x_i}$ if $S_i = S_{i-1} \cup \{x_i\}$, and $\tau_i = \tilde{\gamma}_{x_i}$ otherwise. Then $\boldsymbol{m} = \tau_1 \cdots \tau_n$ is a stepwise effective message for S of the G-system $(\mathcal{F}, \mathcal{T}_G)$. Conversely, a stepwise effective message $\boldsymbol{m} = \tau_1 \cdots \tau_n$ of $(\mathcal{F}, \mathcal{T}_G)$ producing a state T from a state S defines an ST-walk $W_{\boldsymbol{m}}$ in G with vertices $S_i = S\tau_0\tau_1 \cdots \tau_n$. Thus there is a one-to-one correspondence between the stepwise effective messages for a state S of a G-system and walks in G originated in S.

Theorem 8.14. *A G-system on $\mathcal{F}$ is a cubical system on the set of states $\mathcal{F}$.*

Proof. Let $(\mathcal{F}, \mathcal{J}_G)$ be a G-system on $\mathcal{F}$. Axiom [C1] holds trivially and [C2] holds because G is a connected graph.

Let $m = \tau_1 \cdots \tau_n$ be a stepwise effective message for a state S. Suppose that there are two consecutive occurrences of γ_x in m, say, $\tau_i = \gamma_x$ and $\tau_j = \gamma_x$ with $i < j$, such that there is no occurrence of $\tilde{\gamma}_x$ between τ_i and τ_j. Then $x \in S_i = S\tau_0\tau_1 \cdots \tau_i$ which implies $x \in S_{j-1}$, because $\tilde{\gamma}_x$ does not occur between τ_i and τ_j. It follows that $\gamma_x = \tau_j$ is not effective for the state S_{j-1}, a contradiction. Thus occurrences of a token and its reverse must alternate in m, so [C4] holds for $(\mathcal{F}, \mathcal{J}_G)$. A minor modification of this argument shows that [C3] also holds for $(\mathcal{F}, \mathcal{J}_G)$ (cf. Exercise 8.8). □

8.4 Properties of Tokens and Contents of Messages

Lemma 8.15. *The following statements hold for a cubical system* $(\mathcal{S}, \mathcal{J})$.

(i) $\tilde{\tilde{\tau}} = \tau$ *for any* $\tau \in \mathcal{J}$.
(ii) *For any two adjacent states S and T there is a unique token producing T from S.*
(iii) *If S, T, and P are three distinct states such that $S\tau = T$ and $T\mu = P$, for some tokens τ and μ, then $\mu \neq \tau$ and $\mu \neq \tilde{\tau}$.*
(iv) *No token can be a one-to-one function.*

Proof. (i) By [C1], $\tilde{\tau}$ exists, so $\tilde{\tilde{\tau}} = \tau$.

(ii) Suppose that $S\tau = S\mu = T$. By [C1] and [C3], the message $\tau\tilde{\mu}$ is well defined and vacuous, so $\tau = \tilde{\tilde{\mu}} = \mu$.

(iii) $\tau\mu$ is a stepwise effective message for S, therefore we have $\mu \neq \tau$, by [C4]. If $\mu = \tilde{\tau}$, then $S = T\tilde{\tau} = P$, a contradiction, because $S \neq P$ and $\tilde{\tau}$ is a function.

(iv) τ is not the identity transformation, thus there are states S and T such that $S\tau = T$. By (iii), $S\tau = T = T\tau$, so τ is not a one-to-one function. □

Property (ii) of Lemma 8.15 is a very strong property of tokens of a cubical system. It asserts that two tokens τ and μ transforming some state S into a different state T are equal transformations; that is, $V\tau = V\mu$ for all $V \in \mathcal{S}$.

Definition 8.16. Let $(\mathcal{S}, \mathcal{J})$ be a cubical system. For any token τ and any message m, we define $\#(\tau, m)$ as the number of occurrences of τ in m. For any message m, the *content* of m is the set $\mathcal{C}(m)$ defined by

$$\mathcal{C}(m) = \{\tau \in \mathcal{J} \mid \#(\tau, m) > \#(\tilde{\tau}, m)\}.$$

For any state S, the *content* $\hat{S}$ of S is the union $\cup_m \mathcal{C}(m)$ taken over the set of all stepwise effective messages producing the state S.

Example 8.17. For the cubical system from Example 8.10 we have

$$\widehat{S} = \{\tilde{\tau}, \tilde{\mu}\}, \quad \widehat{T} = \{\tau, \tilde{\mu}\}, \quad \widehat{Q} = \{\tau, \mu\}, \quad \widehat{P} = \{\tilde{\tau}\mu\}.$$

In the rest of the section, we assume that a cubical system $(\mathcal{S}, \mathcal{T})$ is given.

The following properties of the functions $\#$ and $\mathcal{C}$ are immediate and are used implicitly in this section (cf. Exercise 8.11).

$$\#(\tilde{\tau}, \widetilde{m}) = \#(\tau, m), \quad \#(\tilde{\tau}, m) = \#(\tau, \widetilde{m})$$
$$\tau \in \mathcal{C}(m) \quad \text{if and only if} \quad \tilde{\tau} \in \mathcal{C}(\widetilde{m}) \tag{8.3}$$
$$\tau \in \mathcal{C}(m) \quad \text{implies} \quad \tilde{\tau} \notin \mathcal{C}(m)$$

Lemma 8.18. *If m is a stepwise effective message for some state, then*

$$\tau \in \mathcal{C}(m) \quad \text{if and only if} \quad \#(\tau, m) = \#(\tilde{\tau}, m) + 1. \tag{8.4}$$

Accordingly, for any $\tau \in \mathcal{T}$,

$$\#(\tau, m) - \#(\tilde{\tau}, m) \in \{-1, 0, 1\}. \tag{8.5}$$

Proof. By Axiom [C4] the occurrences of τ and $\tilde{\tau}$ in m alternate. Therefore,

$$\tau \in \mathcal{C}(m) \quad \text{implies} \quad \#(\tau, m) > \#(\tilde{\tau}, m) \quad \text{implies} \quad \#(\tau, m) = \#(\tilde{\tau}, m) + 1.$$

The converse implication in (8.4) is trivial. It is clear, that (8.5) follows from the equivalence in (8.4). □

Lemma 8.19. *The content of a state cannot contain both a token and its reverse.*

Proof. Suppose that $\tau, \tilde{\tau} \in \widehat{S}$ for some token τ and some state S. Then there are two stepwise effective messages m and n both producing S and such that $\tau \in \mathcal{C}(m)$ and $\tilde{\tau} \in \mathcal{C}(n)$. Therefore, by (8.4),

$$\#(\tau, m) = \#(\tilde{\tau}, m) + 1 \quad \text{and} \quad \#(\tilde{\tau}, n) = \#(\tau, n) + 1.$$

It follows that

$$\#(\tau, m\tilde{n}) = \#(\tau, m) + \#(\tau, \tilde{n}) = \#(\tilde{\tau}, m) + 1 + \#(\tilde{\tau}, \tilde{n}) + 1$$
$$= \#(\tilde{\tau}, m\tilde{n}) + 2,$$

which contradicts Axiom [C4], inasmuch as $m\tilde{n}$ is a stepwise effective message for some state. Therefore $\widehat{S}$ cannot contain both τ and $\tilde{\tau}$. □

Theorem 8.20. *For any token τ and any state S, we have either $\tau \in \widehat{S}$ or $\tilde{\tau} \in \widehat{S}$ (but not both).*

Proof. τ is a token, thus there are distinct states V and W such that $V\tau = W$. By Axiom [C2], there are stepwise effective messages $\boldsymbol{m}$ and $\boldsymbol{n}$ such that $V\boldsymbol{m} = S$ and $W\boldsymbol{n} = S$. (If S equals either V or W, the corresponding message is empty.) By Axiom [C3], the message $\tau\boldsymbol{n}\widetilde{\boldsymbol{m}}$ is vacuous. Therefore,

$$\#(\tau, \tau\boldsymbol{n}\widetilde{\boldsymbol{m}}) = \#(\tilde{\tau}, \tau\boldsymbol{n}\widetilde{\boldsymbol{m}}).$$

We have

$$\#(\tau, \tau\boldsymbol{n}\widetilde{\boldsymbol{m}}) = 1 + \#(\tau, \boldsymbol{n}) + \#(\tilde{\tau}, \boldsymbol{m})$$

and

$$\#(\tilde{\tau}, \tau\boldsymbol{n}\widetilde{\boldsymbol{m}}) = \#(\tilde{\tau}, \boldsymbol{n}) + \#(\tau, \boldsymbol{m}).$$

From the last three displayed equations we obtain

$$[\#(\tau, \boldsymbol{m}) - \#(\tilde{\tau}, \boldsymbol{m})] + [\#(\tilde{\tau}, \boldsymbol{n}) - \#(\tau, \boldsymbol{n})] = 1.$$

By (8.5), we must have either $\#(\tau, \boldsymbol{m}) - \#(\tilde{\tau}, \boldsymbol{m}) = 1$ or $\#(\tilde{\tau}, \boldsymbol{n}) - \#(\tau, \boldsymbol{n}) = 1$. Therefore, by (8.4), either $\tau \in \mathcal{C}(\boldsymbol{m})$ or $\tilde{\tau} \in \mathcal{C}(\boldsymbol{n})$. By Lemma 8.19, either $\tau \in \widehat{S}$ or $\tilde{\tau} \in \widehat{S}$. $\qquad\square$

Theorem 8.21. *If S and V are two distinct states, with $S\boldsymbol{m} = V$ for some stepwise effective message $\boldsymbol{m}$, then $\widehat{V} \setminus \widehat{S} = \mathcal{C}(\boldsymbol{m})$. Therefore,*

$$\widehat{S} \triangle \widehat{V} = \mathcal{C}(\boldsymbol{m}) + \mathcal{C}(\widetilde{\boldsymbol{m}})$$

(recall that $+$ stands for the disjoint union of two sets). Accordingly,

$$\widehat{S} \triangle \widehat{V} = \{\tau, \tilde{\tau}\},$$

if $S\tau = V$.

Proof. Let τ be a token in $\mathcal{C}(\boldsymbol{m})$. Then $\tau \in \widehat{V}$ and $\tilde{\tau} \in \mathcal{C}(\widetilde{\boldsymbol{m}})$ implying that $\tilde{\tau} \in \widehat{S}$. By Lemma 8.19, $\tau \notin \widehat{S}$. It follows that $\tau \in \widehat{V} \setminus \widehat{S}$. Thus $\mathcal{C}(\boldsymbol{m}) \subseteq \widehat{V} \setminus \widehat{S}$.

Suppose now that $\tau \in \widehat{V} \setminus \widehat{S}$, so $\tau \in \widehat{V}$ and $\tau \notin \widehat{S}$. There is a stepwise effective message $\boldsymbol{n}$ producing V and such that $\tau \in \mathcal{C}(\boldsymbol{n})$. By (8.4),

$$\#(\tau, \boldsymbol{n}) - \#(\tilde{\tau}, \boldsymbol{n}) = 1. \tag{8.6}$$

$\tau \notin \widehat{S}$, therefore we have $\tau \notin \mathcal{C}(\widetilde{\boldsymbol{m}})$ which implies, by (8.4) and (8.5),

$$\#(\tau, \widetilde{\boldsymbol{m}}) - \#(\tilde{\tau}, \widetilde{\boldsymbol{m}}) \in \{-1, 0\}. \tag{8.7}$$

The message $\boldsymbol{n}\widetilde{\boldsymbol{m}}$ is stepwise effective and produces the state S. We have $\tau \notin \mathcal{C}(\boldsymbol{n}\widetilde{\boldsymbol{m}})$, inasmuch as $\tau \notin \widehat{S}$. Therefore, by (8.4), (8.5), and (8.6),

$$1 > \#(\tau, n\widetilde{m}) - \#(\tilde{\tau}, n\widetilde{m})$$
$$= [\#(\tau, n) + \#(\tau, \widetilde{m})] - [\#(\tilde{\tau}, n) + \#(\tilde{\tau}(\widetilde{m})]$$
$$= [\#(\tau(n) - \#(\tilde{\tau}, n)] + [\#(\tau, \widetilde{m}) - \#(\tilde{\tau}, \widetilde{m})]$$
$$= 1 + [\#(\tau, \widetilde{m}) - \#(\tilde{\tau}, \widetilde{m})].$$

By (8.7), $\#(\tilde{\tau}, \widetilde{m}) - \#(\tau, \widetilde{m}) = 1$, or, equivalently, $\#(\tau, m) - \#(\tilde{\tau}, m) = 1$. By (8.4), $\tau \in \mathcal{C}(m)$. Hence, $\widehat{V} \setminus \widehat{S} \subseteq \mathcal{C}(m)$. The result follows.   □

Lemma 8.22. *A stepwise effective message m is closed if and only if*

$$\mathcal{C}(m) = \varnothing.$$

Proof. A closed stepwise effective message m is vacuous by Axiom [C3]. By (8.4), $\mathcal{C}(m) = \varnothing$.

Conversely, if $\mathcal{C}(m) = \varnothing$ for some stepwise effective message m, then, by Axiom [C4] and (8.4), m is vacuous. By Axiom [C3], m is closed.   □

Theorem 8.23. *For any two states S and V,*

$$\widehat{S} = \widehat{V} \text{ if and only if } S = T.$$

Proof. Suppose that $\widehat{S} = \widehat{V}$ and let m be a stepwise effective message producing V from S. By Theorem 8.21, $\mathcal{C}(m) = \varnothing$. By Lemma 8.22, m is a closed message. Thus, $S = V$. The converse implication is trivial.   □

Theorem 8.24. *Let m and n be two stepwise effective messages transforming some state S. Then*

$$Sm = Sn \text{ if and only if } \mathcal{C}(m) = \mathcal{C}(n).$$

Proof. Suppose that $Sm = Sn = V$. By Theorem 8.21,

$$\mathcal{C}(m) = \widehat{V} \setminus \widehat{S} = \mathcal{C}(n).$$

Conversely, suppose that $\mathcal{C}(m) = \mathcal{C}(n)$ and let $V = Sm$ and $W = Sn$. By Theorem 8.21,

$$\widehat{V} \triangle \widehat{S} = \mathcal{C}(m) + \mathcal{C}(\widetilde{m}) = \mathcal{C}(n) + \mathcal{C}(\widetilde{n}) = \widehat{W} \triangle \widehat{S},$$

implying $\widehat{V} = \widehat{W}$. By Theorem 8.23, $V = W$.   □

8.5 Graphs of Cubical Systems

Definition 8.25. The *graph* G *of a cubical system* $(S, \mathcal{T})$ has S as the set of its vertices; two vertices are adjacent in G if the corresponding states are adjacent in $(S, \mathcal{T})$.

Theorem 8.26. *Let* $(S, \mathcal{T})$ *be a cubical system. There exists a connected subgraph* $G = (\mathcal{F}, \mathcal{E})$ *of some cube* $\mathcal{H}(X)$ *such that* $(S, \mathcal{T})$ *is isomorphic to the* G-*system* $(\mathcal{F}, \mathcal{T}_G)$ *on the family* $\mathcal{F}$.

The claim of this theorem gives a reason for calling G-systems "canonical" examples of cubical token systems.

Proof. By Axiom [C2], the graph G of the cubical system $(S, \mathcal{T})$ is connected. Let $J = \{\{\tau, \tilde{\tau}\}\}_{\tau \in \mathcal{T}}$ and call elements of J *labels*. By Lemma 8.15(ii), a unique label is assigned to each edge of G.

We begin by constructing the family $\mathcal{F}$.

Let S_0 be a fixed state of the cubical system $(S, \mathcal{T})$. By [C2], for any state $T \neq S_0$, there is a stepwise effective message m such that $S_0 m = T$. We denote by W_m the walk in G produced by the message m and define a set J_T by

$$J_T = \{j \in J \mid \text{the number of occurrences of } j \text{ in } W_m \text{ is odd}\}.$$

In addition, we set $J_{S_0} = \varnothing$.

We need to show that the sets J_T are well defined. Suppose that n is another stepwise effective message producing T from S_0. By Axiom [C3], the number of occurrences of $j \in J$ in the closed walk $W_m W_{\tilde{n}}$ is even. Hence, the number of occurrences of j in W_m is odd if and only if the number of its occurrences in W_n is odd. Thus the set J_T is well defined and the assignment $T \mapsto J_T$ defines a mapping $\alpha : S \to \mathcal{P}_f(J)$, where $\mathcal{P}_f(J)$ stands for the family of finite subsets of J.

Let us prove that α is a one-to-one mapping. Let $J_S = J_T$ for some states S and T. By Axiom [C2] there are stepwise effective messages m, n, and p such that $S_0 m = S$, $Sn = T$, and $Tp = S_0$, so $W_m W_n W_p$ is a closed walk in G. By Axiom [C3], any label $j \in J$ occurs an even number of times in this walk. If $j \in J_S = J_T$, then j occurs an odd number of times in each walk W_m and W_p. Hence, j occurs an even number of times in W_n. If $j \notin J_S = J_T$, then j occurs an even number of times in each walk W_m and W_p. Hence, j occurs an even number of times in W_n. Thus any label occurs an even number of times in W_n. By Axiom [C4], the message n is vacuous, and, by Axiom [C3], $S = T$. Hence, α is a one-to-one mapping.

We now show that α is an embedding of G into the cube $\mathcal{H}(J)$. The sets J_S are vertices of the cube $\mathcal{H}(J)$. Let P and Q be two adjacent states of the cubical system $(S, \mathcal{T})$, so $P\tau = Q$ for some $\tau \in \mathcal{T}$, and let $j = \{\tau, \tilde{\tau}\}$ be the label of the edge $\{P, Q\}$ in the graph G. By Axiom [C2], there are stepwise effective messages p and q producing states P and Q, respectively, from S_0. By Axiom [C3], j occurs an even number of times in the closed walk $W_p W_\tau W_{\tilde{q}}$.

It follows that the label j occurs an odd number of times either in W_p or in W_q, so $j \in J_P \triangle J_Q$. Any other label k occurs an even number of times in the walk $W_{\bar{p}}W_q$, so $k \notin J_P \triangle J_Q$. Thus, $J_P \triangle J_Q = \{j\}$, so $\{\alpha(P), \alpha(Q)\}$ is an edge of $\mathcal{H}(J)$. It follows that α defines an embedding of the graph G into the cube $\mathcal{H}(J)$. In the rest of the proof we identify $\alpha(G)$ with G.

Let $\mathcal{F} = \{J_S\}_{S \in \mathcal{S}}$ and $(\mathcal{F}, \mathcal{T}_G)$ be the corresponding G-system on $\mathcal{F}$. (Clearly, $\cap \mathcal{F} = \varnothing$ and $\cup \mathcal{F} = J$.) Let us prove that the cubical system $(\mathcal{S}, \mathcal{T})$ is isomorphic to $(\mathcal{F}, \mathcal{T}_G)$.

Let P and Q be two adjacent states of the cubical system $(\mathcal{S}, \mathcal{T})$. Because $\alpha(P) = J_P$ and $\alpha(Q) = J_Q$ are adjacent in the graph G, we have $J_P \triangle J_Q = \{j\}$ for some $j \in J$, so we may assume that $J_Q = J_P + \{j\}$. Inasmuch as P and Q are adjacent states, we have $P\tau = Q$ for some token τ. Note that $j = \{\tau, \tilde{\tau}\}$. We define $\beta(\tau) = \gamma_j$, $\beta(\tilde{\tau}) = \tilde{\gamma}_j$ and show that these assignments do not depend on a particular choice of P and Q with $P\tau = Q$. Let S and T be another pair of adjacent states such that $S\tau = T$, and let $\boldsymbol{m}$ and $\boldsymbol{n}$ be stepwise effective messages producing Q from S_0 and T from Q, respectively. By Axiom [C4], there are an even number of occurrences of the label j in the walk $W_\tau W_n W_{\tilde{\tau}}$ connecting P with S, so there are an even number of occurrences of j in W_n. Because $j \in J_Q$, there are an odd number of occurrences of j in W_m. Therefore, there is an odd number of occurrences of j in the walk $W_m W_n$ connecting S_0 with T, and an even number of occurrences of j in the walk $W_m W_n W_{\tilde{\tau}}$ connecting S_0 with S. It follows that $j \in J_T \setminus J_S$. Thus, $J_T = \gamma_j(J_S) = J_S + \{j\}$, so β is well-defined. Moreover, the above arguments show that

$$P\tau = Q \text{ if and only if } \alpha(P)\beta(\tau) = \alpha(Q),$$

for any $P, Q \in \mathcal{S}$ and $\tau \in \mathcal{T}$. It is clear that α and β are bijections, so (α, β) is an isomorphism from $(\mathcal{S}, \mathcal{T})$ onto $(\mathcal{F}, \mathcal{T}_G)$. $\square$

Cubical systems $(\mathcal{S}, \mathcal{T})$ and $(\mathcal{F}, \mathcal{T}_G)$ of Theorem 8.26 are isomorphic, thus their graphs are isomorphic to the graph G. The next result is obvious.

Theorem 8.27. *The graph of a cubical system is cubical. Conversely, any cubical graph G defines a cubical system (a G-system).*

8.6 Examples of Cubical Systems

We begin by introducing a class of finite G-systems that serves as a source of our examples.

Definition 8.28. Let $\mathcal{F}$ be a family of subsets of a finite set X with $|\mathcal{F}| \geq 2$. A set $S \in \mathcal{F}$ is said to be *downgradable* if there exists $x \in S$ such that $S \setminus \{x\} \in \mathcal{F}$. The family $\mathcal{F}$ itself is *downgradable* if all its nonminimal sets are downgradable. Likewise, a set $S \in \mathcal{F}$ is said to be *upgradable* if there exists $x \in X \setminus S$ such that $S \cup \{x\} \in \mathcal{F}$. The family $\mathcal{F}$ itself is *upgradable* if all but its maximal sets are upgradable.

It is clear that any downgradable family $\mathcal{F}$ of sets containing the empty set is connected; that is, for any two distinct sets $S, V \in \mathcal{F}$ there is a sequence of sets $S_0 = S, S_1, \ldots, S_n = V$ such that $|S_i \triangle S_{i+1}| = 1$. Likewise, any upgradable family of subsets of X containing the set X itself is connected. Let $\mathcal{F}$ be any of such families. Then the subgraph F of $\mathcal{H}(X)$ induced by $\mathcal{F}$ is connected and therefore defines a cubical system (an F-system).

Example 8.29. A finite graph $G = (V, E)$ is said to be a *comparability graph* if there exists a partial order P on the vertex set V such that

$$uv \in E \text{ if and only if either } (u, v) \in P \text{ or } (v, u) \in P \qquad (8.8)$$

We denote $\mathcal{CG}$ the family of all comparability graphs on a fixed set V and identify this family with the family of the edge sets of these graphs. Clearly, $\mathcal{CG}$ contains the empty graph on V. The family $\mathcal{PO}$ of all partial orders on V is well-graded (cf. Theorem 5.13). It can be easily seen that this fact implies the family $\mathcal{CG}$ is downgradable (cf. Exercise 8.12) and therefore defines a cubical system. Note that the well-gradedness property of the family $\mathcal{PO}$ does not imply that $\mathcal{CG}$ is well-graded (see the graphs in Figure 8.10).

Figure 8.10. Two comparability graphs with 6 and 8 edges, respectively. The distance between the two edge sets is 2. There is no comparability graph on distance 1 from each of these two graphs (cf. Exercise 8.13). Thus the family $\mathcal{CG}$ is not well-graded.

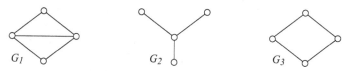

Figure 8.11. Indifference graph (G_1); interval graph that is not an indifference graph (G_2); not an interval graph (G_3).

Example 8.30. *Interval* and *indifference graphs* (cf. Figure 8.11) are complements of comparability graphs arising from interval orders and semiorders on a given vertex set V, respectively, via relation (8.8). (Thus P in (8.8) is either interval order or semiorder.) As the families of all interval orders and all semiorders are well-graded (cf. Section 7.7) and both contain the empty relation, the respective families of interval and indifference graphs are upgradable and both contain the complete graph on V. Therefore we can cast each of these two families as a cubical system.

Example 8.31. Any connected subgraph G of the integer lattice $\mathcal{Z}^d$ is cubical and therefore gives rise to a cubical system. An example can be obtained from Figure 8.9.

8.7 Media

We need the concept of a "concise message" for the definition of a medium.

Definition 8.32. A message m of a token system $(\mathcal{S}, \mathcal{T})$ is said to be *concise* for a state S if: (i) m is stepwise effective for S, (ii) no token occurs twice in m, and (iii) m does not contain a token and its reverse.

Definition 8.33. A token system $(\mathcal{S}, \mathcal{T})$ is called a *medium* (on $\mathcal{S}$) if the following axioms are satisfied.

[Ma] For any two distinct states S and V in $\mathcal{S}$ there is a concise message transforming S into V.

[Mb] A message that is closed for some state is vacuous.

Theorem 8.34. *A medium is a cubical system.*

Proof. Let $(\mathcal{S}, \mathcal{T})$ be a medium. We need to verify Axioms [C1]–[C4].

Axioms [C1] and [C2]. Let τ be a token in $\mathcal{T}$. Because τ is not the identity function, there are distinct states S and V in $\mathcal{S}$ such that $S\tau = V$. By Axiom [Ma], there is a concise message m producing S from V. The message τm is stepwise effective for S and ineffective for that state, that is, it is closed for S. By Axiom [Mb], this message is vacuous. Hence, m contains the reverse of τ, so $\tilde{\tau}$ exists. m is a concise message, therefore it cannot contain both $\tilde{\tau}$ and τ. Thus, $\tilde{\tau} \neq \tau$.

Axiom [C3]. If a message m is closed for some state S, then it is vacuous by Axiom [Mb]. Conversely, let m be a vacuous and stepwise effective message for a state S. Suppose that $Sm \neq S$, and let n be a concise message producing S from Sm. By Axiom [Mb], the message mn is vacuous, so n must contain a pair of mutually reverse tokens, a contradiction. Hence, $Sm = S$.

Axiom [C4]. Let $m = \tau_1 \cdots \tau_n$ be a stepwise effective message for a state S and (S_i) be a sequence of states produced by m. Suppose that τ_i and τ_j $(i < j)$ are two consecutive occurrences of a token τ in m such that there is no occurrence of $\tilde{\tau}$ between τ_i and τ_j. By [Ma], there is a concise message n producing S_{i-1} from S_j. By [Mb], we must have two occurrences of $\tilde{\tau}$ in the concise message n, a contradiction. Thus [C4] holds for a medium. $\square$

As the following example demonstrates, the media class is a proper subclass of cubical systems.

Example 8.35. Let $(\mathcal{S}, \mathcal{T})$ be a cubical system shown in Figure 8.8. There is no concise message producing P from S, so this cubical system is not a medium.

Let $\mathcal{F}$ be a wg-family of subsets of a set X, and let $G = (\mathcal{F}, \mathcal{E}_{\mathcal{F}})$ be a partial cube on X. The G-system $(\mathcal{F}, \mathcal{T}_{\mathcal{F}})$ (cf. Definition 8.11) is an example of a medium. We prove an even stronger statement.

Theorem 8.36. $(\mathcal{F}, \mathcal{T}_{\mathcal{F}})$ *is a medium if and only if $\mathcal{F}$ is a wg-family.*

Proof. (Necessity.) Let S and T be two distinct sets in $\mathcal{F}$. By [Ma], there is a concise message $\boldsymbol{m} = \tau_1 \ldots \tau_n$ transforming S into T. Let (S_i) be a sequence of sets produced by $\boldsymbol{m}$ from S, so $S_0 = S$ and $S_n = T$. Each τ_i is either γ_{x_i} or $\tilde{\gamma}_{x_i}$ for some x_i. Because $\boldsymbol{m}$ is a concise message, all elements x_i are distinct. Suppose first that $\tau_i = \gamma_{x_i}$ for some i. Then $S_i = S_{i-1} + \{x_i\}$. Inasmuch as $\boldsymbol{m}$ is a concise message, we must have $x_i \in S_j$ for all $j \geq i$ and $x_i \notin S_j$ for all $j < i$. Hence, $x_i \in T \setminus S$. Suppose now that $\tau_i = \tilde{\gamma}_{x_i}$ for some i. Then $S_i = S_{i-1} \setminus \{x_i\}$. Arguing as in the previous case, we obtain $x_i \in S \setminus T$. Therefore, $x_i \in S \triangle T$ for any i. On the other hand, it is clear that any element of $S \triangle T$ is one of the x_i. Thus $S \triangle T = \cup_i \{x_i\}$, so $d(S, T) = n$. Clearly, we have $d(S_{i-1}, S_i) = 1$, for all i. It follows that $\mathcal{F}$ is a wg-family.

(Sufficiency.) Let $\mathcal{F}$ be a well-graded family of subsets of some set X. By Theorem 8.13, $(\mathcal{F}, \mathcal{T}_{\mathcal{F}})$ is a token system. It is clear that the tokens γ_x and $\tilde{\gamma}_x$ are mutual reverses for any $x \in \cup \mathcal{F} \setminus \cap \mathcal{F}$. We need to show that Axioms [Ma] and [Mb] are satisfied for $(\mathcal{F}, \mathcal{T}_{\mathcal{F}})$.

Axiom [Ma]. Let S and T be two distinct states in the wg-family $\mathcal{F}$, and let (S_i) be a sequence of states in $\mathcal{F}$ such that $S_0 = S$, $S_n = T$, $d(S, T) = n$, and $d(S_{i-1}, S_i) = 1$. By the last equation, for any i, there is x_i such that $S_{i-1} \triangle S_i = \{x_i\}$. Suppose that $x_i = x_j$ for some $i < j$. We have

$$(S_{i-1} \triangle S_j) \triangle (S_i \triangle S_{j-1}) = (S_{i-1} \triangle S_i) \triangle (S_{j-1} \triangle S_j) = \{x_i\} \triangle \{x_j\} = \varnothing.$$

Hence, $S_{i-1} \triangle S_j = S_i \triangle S_{j-1}$, so, by Theorem 5.7(i),

$$j - (i - 1) = d(S_{i-1}, S_j) = d(S_i, S_{j-1}) = (j - 1) - i,$$

a contradiction. Thus, all x_i are distinct. $S_{i-1} \triangle S_i = \{x_i\}$, thus we have $S_{i-1}\tau_i = S_i$, where τ_i is either γ_{x_i} or $\tilde{\gamma}_{x_i}$. Clearly, the message $\tau_1 \cdots \tau_n$ is concise and produces T from S.

Axiom [Mb]. Let $\boldsymbol{m} = \tau_1 \cdots \tau_n$ be a stepwise effective message for a state S that is ineffective for S. As before, (S_i) stands for the sequence of states produced by $\boldsymbol{m}$ from S, so $S_0 = S_n = S$. Because $S\boldsymbol{m} = S$, for any occurrence of τ in $\boldsymbol{m}$ there must be an occurrence of $\tilde{\tau}$ in $\boldsymbol{m}$. Suppose that we have two consecutive occurrences of a token $\tau = \tau_i = \tau_j = \gamma_x$ in $\boldsymbol{m}$. Then $x \in S_i$ and $x \notin S_{j-1}$. Therefore we must have an occurrence of $\tilde{\tau} = \tilde{\gamma}_x$ between this two occurrences of τ. A similar argument shows that there is an occurrence of a token between any two consecutive occurrences of its reverse, so occurrences of a token and its reverse alternate in $\boldsymbol{m}$. Finally, let τ_i be the first occurrence of τ in m. We may assume that there is more than one occurrence of τ in m. The message $n = \tau_{i+1} \cdots \tau_n \tau_1 \cdots \tau_i$ is stepwise effective and ineffective for S_i. By the previous argument, occurrences of τ and its reverse alternate in n. It follows that the number of occurrences of both τ and $\tilde{\tau}$ in m is even, so m is vacuous.   □

Theorem 8.36 justifies the following definition.

Definition 8.37. Let $\mathcal{F}$ be a wg-family of subsets of a set X. The medium $(\mathcal{F}, \mathcal{J}_{\mathcal{F}})$ with tokens defined by

$$\tau_x : S \mapsto S\tau_x = \begin{cases} S \cup \{x\}, & \text{if } S \cup \{x\} \in \mathcal{S}, \\ S, & \text{otherwise,} \end{cases}$$

and

$$\tilde{\tau}_x : S \mapsto S\tilde{\tau}_x = \begin{cases} S \setminus \{x\}, & \text{if } S \setminus \{x\} \in \mathcal{F}, \\ S, & \text{otherwise,} \end{cases}$$

for $x \in \cup \mathcal{F} \setminus \cap \mathcal{F}$ and $S \in \mathcal{S}$, is said to be the *representing medium* of $\mathcal{F}$.

8.8 Contents in Media Theory

Definition 8.38. Let $(\mathcal{S}, \mathcal{J})$ be a medium. We write $\ell(m) = n$ to denote the *length* of a message $m = \tau_1 \cdots \tau_n$. The *content* of a message m is the set $\mathcal{C}(m)$ of its distinct tokens. It is clear that $|\mathcal{C}(m)| \leq \ell(m)$. For any state $S \in \mathcal{S}$, the content $\widehat{S}$ of S is the union $\cup_m \mathcal{C}(m)$ taken over the set of all concise messages producing S.

The two concepts of "content" are different from their counterparts in the context of cubical systems. For instance, the content of a vacuous message of a cubical system is empty, whereas it is not empty in media theory. However, the main results of Section 8.4 concerning these concepts are valid for media. We establish these results in a series of theorems in the rest of this section where we assume that a medium $(\mathcal{S}, \mathcal{J})$ is given. Note that the results of Theorems 8.21 and 8.24 are especially useful in the stochastic part of token systems theory (cf. Section 8.12).

Lemma 8.39. *Let S, V, and W be three states of the medium $(\mathcal{S}, \mathcal{J})$ and suppose that $V = Sm$, $W = Vn$ for some concise messages m and n, and $S = Wp$ where p is either a concise message or empty (see the diagram in Figure 8.12). There is at most one occurrence of each pair of mutually reverse tokens in the closed message mnp.*

Proof. Let τ be a token in $\mathcal{C}(m)$. Because m is a concise message, there is only one occurrence of τ in m and $\tilde{\tau} \notin \mathcal{C}(m)$. By Axiom [Mb], the message

mnp is vacuous, so we must have $\tilde{\tau} \in \mathcal{C}(n) \cup \mathcal{C}(p)$. Suppose that $\tilde{\tau} \in \mathcal{C}(n)$ (the case when $\tilde{\tau} \in \mathcal{C}(p) \neq \varnothing$ is treated similarly). **n** is a concise message, therefore there are no more occurrences of $\tilde{\tau}$ in **n** and $\tau \notin \mathcal{C}(n)$. Thus there is only one occurrence of the pair $\{\tau, \tilde{\tau}\}$ in the message **mn**. The pair $\{\tau, \tilde{\tau}\}$ cannot occur in **p**, inasmuch as **p** is a concise message. The result follows. $\square$

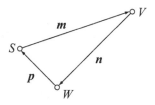

Figure 8.12. Diagram for Lemma 8.39.

One can say more in the special case when **p** is a single token.

Lemma 8.40. *Let S, V, and W be distinct states of a medium and suppose that*

$$V = Sm, \quad W = Vn, \quad S = W\tau$$

for some concise messages **m** *and* **n** *and a token* τ *(see Figure 8.13). Then*

$$\tau \notin \mathcal{C}(n), \quad \tau \notin \mathcal{C}(m),$$

and either

$$\tilde{\tau} \in \mathcal{C}(m), \quad n\tau \text{ is a concise message,} \quad \mathcal{C}(n\tau) = \mathcal{C}(\widetilde{m}), \quad \ell(m) = \ell(n) + 1,$$

or

$$\tilde{\tau} \in \mathcal{C}(n), \quad \tau m \text{ is a concise message,} \quad \mathcal{C}(\tau m) = \mathcal{C}(\widetilde{n}), \quad \ell(n) = \ell(m) + 1.$$

Accordingly,

$$|\ell(m) - \ell(n)| = 1. \tag{8.9}$$

Proof. By Lemma 8.39, $\tau \notin \mathcal{C}(n)$, $\tau \notin \mathcal{C}(m)$, and $\tilde{\tau}$ occurs either in **m** or in **n**. Suppose that $\tilde{\tau} \in \mathcal{C}(m)$. By the same lemma, neither τ nor $\tilde{\tau}$ occurs in **n**. Therefore, $n\tau$ is a concise message. The equality $\mathcal{C}(n\tau) = \mathcal{C}(\widetilde{m})$ also follows from Lemma 8.39. **m** is a concise message, therefore we have

$$\ell(m) = |\mathcal{C}(m)| = |\mathcal{C}(\widetilde{m})| = |\mathcal{C}(n\tau)| = \ell(n) + 1.$$

The case when $\tilde{\tau} \in \mathcal{C}(n)$ is treated similarly. $\square$

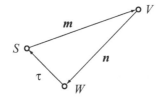

Figure 8.13. Diagram for Lemma 8.40.

Lemma 8.41. *The content of a state cannot contain both a token and its reverse.*

Proof. Suppose that $Sm = Wn = V$ for two concise messages m and n and let p be a concise message producing S from W, if $W \neq S$, and empty, if $W = S$. By Lemma 8.39, there is at most one occurrence of any token τ in the message $m\widetilde{n}p$. Hence we cannot have both $\tau \in \mathcal{C}(m)$ and $\widetilde{\tau} \in \mathcal{C}(n)$. $\square$

Theorem 8.42. *For any token τ and any state S, we have either $\tau \in \widehat{S}$ or $\widetilde{\tau} \in \widehat{S}$. Consequently, $|\widehat{S}| = |\widehat{V}|$ for any two states S and V.*

Proof. Inasmuch as τ is a token, there are two states V and W such that $W = V\tau$. By Axiom [Ma], there are concise messages m and n such that $S = Vm$ and $S = Wn$. By Lemmas 8.40 and 8.41, there are two mutually exclusive options: either $\widetilde{\tau} \in \mathcal{C}(n)$ or $\tau \in \mathcal{C}(m)$. $\square$

Theorem 8.43. *If S and V are two distinct states, with $Sm = V$ for some concise message m, then $\widehat{V} \setminus \widehat{S} = \mathcal{C}(m)$.*

Proof. Let τ be a token in $\mathcal{C}(m)$, so $\widetilde{\tau} \in \mathcal{C}(\widetilde{m})$. Thus, $\tau \in \widehat{V}$ and $\widetilde{\tau} \in \widehat{S}$. By Theorem 8.42, $\tau \notin \widehat{S}$. It follows that $\tau \in \widehat{V} \setminus \widehat{S}$; that is, $\mathcal{C}(m) \subseteq \widehat{V} \setminus \widehat{S}$.

If $\tau \in \widehat{V} \setminus \widehat{S}$, then $\tau \in \widehat{V}$ and $\tau \notin \widehat{S}$, so, by Theorem 8.42, $\widetilde{\tau} \in \widehat{S}$. Because $\tau \in \widehat{V}$, there is a concise message n producing the state V from some state W such that $\tau \in \mathcal{C}(n)$, so $\widetilde{\tau} \in \mathcal{C}(\widetilde{n})$. Let p be a concise message producing S from W (or empty if $S = W$). By Lemma 8.39, there is exactly one occurrence of the pair $\{\tau, \widetilde{\tau}\}$ in the message $m\widetilde{n}p$. Because $\widetilde{\tau} \in \widehat{S}$, we have $\tau \notin \mathcal{C}(p)$. Hence, $\tau \in \mathcal{C}(m)$. In both cases we have $\widehat{V} \setminus \widehat{S} \subseteq \mathcal{C}(m)$. The result follows. $\square$

Theorem 8.44. *For any two states S and V we have*

$$S = V \quad \text{if and only if} \quad \widehat{S} = \widehat{V}.$$

Proof. (Necessity.) Clearly, $\widehat{S} = \widehat{V}$ if $S = V$.

(Sufficiency.) Suppose that $\widehat{S} = \widehat{V}$, $S \neq V$, and let m be a concise message producing V from S. By Theorem 8.43,

$$\varnothing = \widehat{V} \setminus \widehat{S} = \mathcal{C}(m),$$

a contradiction. Hence, $S = V$. □

Theorem 8.45. *Let m and n be two concise messages transforming some state S. Then $Sm = Sn$ if and only if $\mathcal{C}(m) = \mathcal{C}(n)$.*

Proof. (Necessity.) Suppose that $V = Sm = Sn$. By Theorem 8.43,

$$\mathcal{C}(m) = \widehat{V} \setminus \widehat{S} = \mathcal{C}(n).$$

(Sufficiency.) Suppose that $\mathcal{C}(m) = \mathcal{C}(n)$ and let $V = Sm$ and $W = Sn$. By Theorem 8.43,

$$\widehat{V} \bigtriangleup \widehat{S} = \mathcal{C}(m) \cup \mathcal{C}(\widetilde{m}) = \mathcal{C}(n) \cup \mathcal{C}(\widetilde{n}) = \widehat{W} \bigtriangleup \widehat{S},$$

which implies $\widehat{V} = \widehat{W}$. By Theorem 8.44, $V = W$. □

8.9 Graphs of Media

Let $(\mathcal{S}, \mathcal{T})$ be a medium. By definition, the *graph* of $(\mathcal{S}, \mathcal{T})$ is the graph of the cubical system $(\mathcal{S}, \mathcal{T})$ (cf. Definition 8.25).

By Lemma 8.15(ii), for any two adjacent states S and T of a medium $(\mathcal{S}, \mathcal{T})$ there is a unique token τ such that $S\tau = T$ and $T\widetilde{\tau} = S$. Thus, a unique pair of mutually reversed tokens $\{\tau, \widetilde{\tau}\}$ is assigned to each edge ST of the graph of $(\mathcal{S}, \mathcal{T})$.

Let $(\mathcal{S}, \mathcal{T})$ be a medium and G be its graph. If $m = \tau_1 \cdots \tau_m$ is a stepwise effective message for a state S producing a state T, then the sequence of vertices (S_i) of G produced by m, is a walk in G; the vertex $S_0 = S$ is a tail of this walk and the vertex $S_m = T$ is its head. On the other hand, if the sequence of vertices $S_0 = S, S_1, \cdots, S_m = T$ is a walk in G, then edges $\{S_{i-1}, S_i\}$ define unique tokens τ_i such that $S_{i-1}\tau_i = S_i$. Then $m = \tau_1 \cdots \tau_m$ is a stepwise effective message for the state S producing the state T. Thus we have a one-to-one correspondence between stepwise effective messages of the medium and walks in its graph. In particular, a closed message for some state produces a closed walk in G.

A deeper connection between media and their graphs is the result of the following theorem.

Theorem 8.46. *Let* (S, T) *be a medium and* G *be its graph. If* $m = \tau_1 \cdots \tau_m$ *is a concise message producing a state* T *from a state* S, *then the sequence of vertices* (S_i) *produced by* m *forms a shortest path connecting* S *and* T *in the graph* G. *Conversely, if* $S_0 = S, S_1, \ldots, S_m = T$ *is a shortest path in* G, *then the corresponding message is a concise message of* G.

Proof. (Necessity.) Let $P_0 = S, P_1, \ldots, P_n = T$ be a path in G from S to T and $n = \mu_1 \ldots \mu_n$ be the stepwise effective message of the medium corresponding to this path. By Axiom [Mb], the message $m\tilde{n}$ is vacuous. It follows that $\ell(m) \leq \ell(\tilde{n}) = \ell(n)$, because m is a concise message for S. Thus the sequence (S_i) is a shortest path in G.

(Sufficiency.) Let $S_0 = S, S_1, \ldots, S_m = T$ be a shortest path in G and let $m = \tau_1 \cdots \tau_m$ be the corresponding stepwise effective message of the medium. By Axiom [Ma], there is a concise message n producing T from S. By the necessity part of the proof, the walk defined by n is a shortest path from S to T, so $\ell(n) = \ell(m)$. By Axiom [Mb], the message $m\tilde{n}$ must be vacuous. The message n is concise and $\ell(n) = \ell(m)$, therefore the message m must be concise. □

Theorem 8.47. *The graph* G *of a medium* (S, T) *is a partial cube.*

Proof. Let us colour the edges of G by elements of the set $J = \{\{\tau, \tilde{\tau}\}\}_{\tau \in T}$. By Theorem 8.46, shortest paths of G correspond to concise messages of (S, T). Therefore, condition (i) of Theorem 5.27 is satisfied. A closed walk W in G defines a closed message m for a vertex of W. By Axiom [Mb], the message m is vacuous. Thus every colour appears an even number of times in the walk W. The result follows from Theorem 5.27. □

Note that the converse of Theorem 8.47 does not hold: the graph of the cubical system from Example 8.12 is a partial cube, whereas the system itself is not a medium.

Let (S, T) be a medium and G be its graph. By Theorem 8.47, G is a partial cube, so there is an isometric embedding α of G into the cube $\mathcal{H}(X)$ on some set X. The set $\alpha(S)$ is a wg-family $\mathcal{F}$ of finite subsets of X. Let $(\mathcal{F}, T_{\mathcal{F}})$ be the representing medium of this wg-family. These objects are schematically shown in the diagram below, where $G_{\mathcal{F}}$ is the isometric subgraph of $\mathcal{H}(X)$ induced by the family $\mathcal{F}$.

$$(S, T) \xrightarrow[\text{of } (S, T)]{\text{graph}} G \xrightarrow{\alpha} G_{\mathcal{F}} \xrightarrow[\text{medium}]{\text{representing}} (\mathcal{F}, T_{\mathcal{F}}) \qquad (8.10)$$

We leave the proof of the last theorem in this section to the reader (cf. Exercise 8.21).

Theorem 8.48. *The media* (S, T) *and* $(\mathcal{F}, T_{\mathcal{F}})$ *in (8.10) are isomorphic.*

8.10 Examples of Media

Our examples deal with empirical applications, although we do not discuss these applications here. First, we consider media whose states are binary relations on a finite set.

Let $\mathcal{F}$ be a wg-family of partial orders on a finite set X such that

$$\cap \mathcal{F} = \varnothing \text{ and } \cup \mathcal{F} = (X \times X) \setminus \{x, x\}_{x \in X}.$$

Examples of such families include, for instance, the family $\mathcal{PO}$ of all partial orders on X and various families of interval orders introduced in Section 7.7. Tokens of the representing medium $(\mathcal{F}, \mathcal{T}_{\mathcal{F}})$ are labeled by ordered pairs (x, y) of distinct elements of X as follows (cf. Definition 8.37).

$$\tau_{(x,y)} : R \mapsto R\tau_{(x,y)} = \begin{cases} R \cup \{(x, y)\}, & \text{if } R \cup \{(x, y)\} \in \mathcal{F}, \\ R, & \text{otherwise,} \end{cases}$$

and

$$\tilde{\tau}_{(x,y)} : R \mapsto R\tilde{\tau}_{(x,y)} = \begin{cases} R \setminus \{(x, y)\}, & \text{if } R \setminus \{(x, y)\} \in \mathcal{F}, \\ R, & \text{otherwise.} \end{cases}$$

Thus we consider each partial order in $\mathcal{F}$ as a state, with transformations consisting in adding or removal of some ordered pair.

The set $\mathcal{LO}$ of linear orders on a finite set X is not well-graded (cf. Section 5.13). However, we can construct a medium on $\mathcal{LO}$ by using the result of Theorem 5.82 as follows. According to this theorem, the family

$$\mathcal{F} = \{L \cap L_0 : L \in \mathcal{LO}\},$$

where L_0 is a fixed element of $\mathcal{LO}$, is well-graded. Tokens of the representing medium $(\mathcal{F}, \mathcal{T}_{\mathcal{F}})$ are given by

$$\rho_{(x,y)} : P \mapsto P\rho_{(x,y)} = \begin{cases} P \cup \{(x, y)\}, & \text{if } P \cup \{(x, y)\} \in \mathcal{F}, \\ P, & \text{otherwise,} \end{cases}$$

and

$$\tilde{\rho}_{(x,y)} : P \mapsto P\tilde{\rho}_{(x,y)} = \begin{cases} P \setminus \{(x, y)\}, & \text{if } P \setminus \{(x, y)\} \in \mathcal{F}, \\ P, & \text{otherwise,} \end{cases}$$

for $(x, y) \in L_0$ and $P \in \mathcal{F}$. For $P = L \cap L_0$, we have

$$(L \cap L_0)\rho_{(x,y)} = \begin{cases} (L \cap L_0) \cup \{(x, y)\}, & \text{if } (L \cap L_0) \cup \{(x, y)\} \in \mathcal{F}, \\ L \cap L_0, & \text{otherwise,} \end{cases}$$

$$= \begin{cases} (L \cup \{(x, y)\}) \cap L_0, & \text{if } (L \cup \{(x, y)\}) \cap L_0 = L' \cap L_0, \\ L \cap L_0, & \text{otherwise,} \end{cases}$$

where L' is some linear order. $(y, x) \notin L_0$, thus we have

$$(L \cup \{(x, y)\}) \cap L_0 = [(L \setminus \{(y, x)\}) \cup \{(x, y)\}] \cap L_0 = L' \cap L_0.$$

By Lemmas 5.79 and 5.81, $(L \setminus \{(y, x)\}) \cup \{(x, y)\}$ is a linear order with x covering y. Let us define

$$\tau_{(x,y)} : L \mapsto L\tau_{(x,y)} = \begin{cases} (L \setminus \{(y, x)\}) \cup \{(x, y)\}, & \text{if } y \text{ covers } x \text{ in } L, \\ L, & \text{otherwise.} \end{cases}$$

Then

$$(L \cap L_0)\rho_{(x,y)} = L\tau_{(x,y)} \cap L_0.$$

A similar argument shows that, for $(x, y) \in L_0$,

$$(L \cap L_0)\tilde{\rho}_{(x,y)} = L\tau_{(y,x)} \cap L_0.$$

Let $(\mathcal{L}\mathcal{O}, \mathcal{T})$ be a token system with $\mathcal{T} = \{\tau_{(x,y)}\}_{(x,y)\in L_0}$. The tokens of this system are obtained by "pulling back" tokens from the medium $(\mathcal{F}, \mathcal{T}_{\mathcal{F}})$. Note that $\tilde{\tau}_{(x,y)} = \tau_{(y,x)}$. By Lemma 5.79, the mapping $L \mapsto L \cap L_0$ is a bijection from $\mathcal{L}\mathcal{O}$ onto $\mathcal{F}$. Clearly, the token systems $(\mathcal{L}\mathcal{O}, \mathcal{T})$ and $(\mathcal{F}, \mathcal{T}_{\mathcal{F}})$ are isomorphic. Therefore, $(\mathcal{L}\mathcal{O}, \mathcal{T})$ is a medium.

Let $X = \{1, 2, 3, 4\}$. In this case, the set of states $\mathcal{L}\mathcal{O}$ can be identified with permutations of X (the vertices of the permutohedron Π_3 shown in Figure 7.11). The token $\tau_{ij} = \tau_{(i,j)}$ replaces an adjacent pair ji by the pair ij, or does nothing if ji does not form an adjacent pair in the state. For instance, there are three tokens that are effective for the state 3142, namely:

$$3142 \xmapsto{\tau_{13}} 1342, \quad 3142 \xmapsto{\tau_{41}} 3412, \quad 3142 \xmapsto{\tau_{24}} 3124$$

(cf. Figure 7.11).

The steps used above to construct the medium $(\mathcal{L}\mathcal{O}, \mathcal{T})$ from the medium $(\mathcal{F}, \mathcal{T}_{\mathcal{F}})$ exemplify a general procedure, based on the concept of token systems isomorphism. Suppose that X is a set endowed with some structure which is not that of a wg-family of sets. Suppose also that a medium $(\mathcal{S}, \mathcal{T})$ models the set X and its structure in the sense that there is a bijection $\alpha : X \to \mathcal{S}$ and that the tokens from $\mathcal{T}$ may be regarded as representing, in some natural way, transformations occurring in the structured set X. We define a family $\mathcal{T}'$ of functions $\tau' : X \to X$ by "pulling back" (raising) tokens from $\mathcal{T}$:

$$
\begin{array}{ccc}
X & \xrightarrow{\ \tau'\ } & X \\
\alpha \downarrow & & \uparrow \alpha^{-1} \\
\mathcal{S} & \xrightarrow{\ \tau\ } & \mathcal{S}
\end{array}
$$

where $\tau' = \alpha \circ \tau \circ \alpha^{-1}$, and let $\beta : \tau' \mapsto \tau$. Then the pair (α, β) is an isomorphism from the token system $(X, \mathcal{T}')$ onto the medium $(\mathcal{S}, \mathcal{T})$. Therefore,

$(\mathfrak{X}, \mathcal{T}')$ is a medium. This method can be used, for instance, to construct a medium on the set of weak orders $\mathcal{WO}$ (cf. Exercise 8.22).

Our last example has its roots in a different application area. Let Q be a finite set of items of knowledge and $\mathfrak{K}$ be a family of subsets of Q. Elements of $\mathfrak{K}$ are regarded as *knowledge states*, representing the competence of individuals in the population of reference. The family $\mathfrak{K}$ itself is the *knowledge structure* (with respect to Q). It is assumed that $\varnothing, Q \in \mathfrak{K}$. A knowledge structure $\mathfrak{K}$ is called a *learning space* if the following axioms are satisfied.

[L1] If $K \subset L$ are two states, with $|L \setminus K| = n$, then there is a chain of states

$$K_0 = K \subset K_1 \subset \cdots \subset K_n = L$$

such that $K_i = K_{i-1} + \{q_i\}$ with $q_i \in Q$ for $1 \leq i \leq n$. In words: *If the state K of a learner is included in some other state L, then the learner can reach state L by learning one item at a time.*

[L2] If $K \subset L$ are two states, with $K \cup \{q\} \in \mathfrak{K}$ with $q \in Q$, then $L \cup \{q\} \in \mathfrak{K}$. In words: *If item q is learnable from state K, then it is also learnable from any state L that can be reached from K by learning more items.*

To cast a learning space as a medium, we take any knowledge state to be a state of the medium. The tokens consist in adding (removing) an item $q \in Q$ to (from) a state:

$$\tau_q : K \mapsto K\tau_q = \begin{cases} K \cup \{q\}, & \text{if } K \cup \{q\} \in \mathfrak{K}, \\ K, & \text{otherwise}, \end{cases}$$

and

$$\tilde{\tau}_q : K \mapsto K\tilde{\tau}_q = \begin{cases} K \setminus \{q\}, & \text{if } K \setminus \{q\} \in \mathfrak{K}, \\ K, & \text{otherwise}. \end{cases}$$

In the next section, we show that a learning space is a wg-family of subsets of Q. Thus the token system from the foregoing paragraph is indeed a medium.

8.11 Oriented and Closed Media

The representing medium $(\mathfrak{F}, \mathcal{T}_{\mathfrak{F}})$ of a wg-family of sets $\mathfrak{F}$ is endowed with an "orientation" dictated by the nature of the tokens: the set $\mathcal{T}_{\mathfrak{F}}$ is partitioned into two subsets, $\mathcal{T}_{\mathfrak{F}}^+ = \{\tau_x\}$ and $\mathcal{T}_{\mathfrak{F}}^- = \{\tilde{\tau}_x\}$ (cf. Definition 8.37). This observation motivates the following definition.

Definition 8.49. An *orientation* of a medium $(\mathcal{S}, \mathcal{T})$ is a partition of its set of tokens into two classes $\mathcal{T}^+$ and $\mathcal{T}^-$, respectively called *positive* and *negative* such that for any $\tau \in \mathcal{T}$, we have

$$\tau \in \mathcal{T}^+ \text{ if and only if } \tilde{\tau} \in \mathcal{T}^-.$$

A medium $(S, \mathcal{T})$ equipped with an orientation $\{\mathcal{T}^+, \mathcal{T}^-\}$ is said to be *oriented* by $\{\mathcal{T}^+, \mathcal{T}^-\}$ and tokens from $\mathcal{T}^+$ (respectively, $\mathcal{T}^-$) are called *positive* (respectively, *negative*). Any message containing only positive (respectively, negative) tokens is called *positive* (respectively, *negative*).

According to Theorem 8.48, any medium is isomorphic to some representing medium and therefore can be oriented. We can also orient a medium $(S, \mathcal{T})$ by selecting an arbitrary state $R \in S$ and defining $\mathcal{T}^- = \widehat{R}$ and $\mathcal{T}^+ = \mathcal{T} \setminus \widehat{R}$. By Theorem 8.42, $\{\mathcal{T}^+, \mathcal{T}^-\}$ is an orientation of $(S, \mathcal{T})$. Note that for any other state S, any concise message m producing S from R is positive. Indeed, by Theorem 8.43, $\mathcal{C}(m) = (\widehat{S} \setminus \widehat{R}) \subseteq \mathcal{T}^+$. This property of the orientation $\{\mathcal{T}^+, \mathcal{T}^-\}$ justifies the following definition.

Definition 8.50. The *root* of an oriented medium is a state R such that, for any other state S, any concise message producing S from R is positive. An oriented medium having a root is said to be *rooted*.

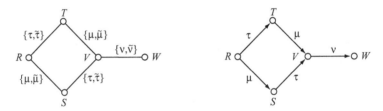

Figure 8.14. A medium and its orientation with $\mathcal{T}^+ = \{\tau, \mu, \nu\}$.

Example 8.51. Let $(S, \mathcal{T})$ be a medium with $S = \{R, S, T, V, W\}$ and six tokens $\{\tau, \tilde{\tau}, \mu, \tilde{\mu}, \nu, \tilde{\nu}\}$ with effective actions of τ, μ, and ν given by

$$R\tau = T, \quad S\tau = V, \quad R\mu = S, \quad T\mu = V, \quad V\nu = W,$$

and actions of the reversed tokens $\tilde{\tau}$, $\tilde{\mu}$, and $\tilde{\nu}$ defined accordingly. The graph of this medium is shown in Figure 8.14, left, with edges labeled by pairs of mutually reversed tokens (cf. Section 8.9). We define an orientation of this medium by choosing $\mathcal{T}^+ = \{\tau, \mu, \nu\}$ and represent this oriented medium as a digraph depicted in Figure 8.14, right. Clearly, R is the root of this oriented medium. Note that an oriented medium with orientation defined by $\mathcal{T}^+ = \{\tilde{\tau}, \mu, \tilde{\nu}\}$ is not rooted.

An oriented medium has at most one root. Indeed, suppose that R and R' are two distinct roots of the medium. By Axiom [Ma], there is a concise message m producing R' from R. Because R is a root, m is positive. Hence, the message $\widetilde{m}$ is negative and produces R from R'. This contradicts our assumption that R' is a root. An example of an oriented medium without a root is found in Example 8.51.

We consider now a class of oriented media that is of importance in the theory of learning spaces (cf. Section 8.10).

Definition 8.52. An oriented medium $(\mathcal{S}, \mathcal{T})$ is *closed* if for any state S and any two distinct positive tokens τ, μ both effective for S, we have

$$S\tau = V, \ S\mu = W \ \text{imply} \ V\mu = W\tau$$

(cf. Figure 8.15).

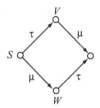

Figure 8.15. Three states and two tokens from Definition 8.52.

A typical example of a closed oriented medium is the representing medium enjoying the ∪-closedness property introduced below.

A family of sets $\mathcal{F}$ is said to be *closed under finite unions*, or ∪-*closed*, if for any finite subset $\mathcal{G}$ of $\mathcal{F}$ we have $\cup \mathcal{G} \in \mathcal{F}$.

Theorem 8.53. *Let $\mathcal{F}$ be a wg-family of finite subsets of a set X such that $\cap \mathcal{F} = \varnothing$ and $\cup \mathcal{F} = X$, and let $(\mathcal{F}, \mathcal{T}_{\mathcal{F}})$ be its representing medium endowed with orientation $\mathcal{T}_{\mathcal{F}}^+ = \{\tau_x\}_{x \in X}$, $\mathcal{T}_{\mathcal{F}}^- = \{\tilde{\tau}_x\}_{x \in X}$ (cf. Definition 8.37). The oriented medium $(\mathcal{F}, \mathcal{T}_{\mathcal{F}})$ is closed if and only if the family $\mathcal{F}$ is ∪-closed.*

Proof. (Necessity.) Let $(\mathcal{F}, \mathcal{T}_{\mathcal{F}})$ be a closed oriented medium. It suffices to show that $S \cup T \in \mathcal{F}$ for any two sets $S, T \in \mathcal{F}$. By Axiom [Ma], there is a concise message $\boldsymbol{m}$ producing T from S. There are four mutually exclusive cases:

(i) $\boldsymbol{m}$ is a positive message. Actions of positive tokens consist of adding elements of X to states, thus $S\boldsymbol{m} = T$ implies $S \subseteq T$. Hence, $S \cup T = T \in \mathcal{F}$.

(ii) $\boldsymbol{m}$ is a negative message. Then $\widetilde{\boldsymbol{m}}$ is a positive message producing S from T. The argument from (i) shows that $T \subseteq S$ implying $S \cup T \in \mathcal{F}$.

(iii) $\boldsymbol{m} = \boldsymbol{n}\boldsymbol{n}'$, with $\boldsymbol{n}$ a positive message and $\boldsymbol{n}'$ a negative message. Tokens in $\boldsymbol{n}$ add elements from $T \setminus S$ to S, therefore we have $S \cup T = S\boldsymbol{n} \in \mathcal{F}$.

(iv) $\boldsymbol{m} = \tau_1 \ldots \tau_k \tau_{k+1} \ldots \tau_n$, with τ_k negative and τ_{k+1} positive. Let (S_i) be a sequence of states produced by $\boldsymbol{m}$ from S. Both $\tilde{\tau}_k$ and τ_{k+1} are positive and the medium is closed, thus we have $S_{k-1}\tau_{k+1} = S_{k+1}\tilde{\tau}_k = V$ (cf. the diagram below).

$$S = S_0 \xrightarrow{\tau_1} S_1 \longrightarrow \cdots \longrightarrow S_{k-1} \xrightarrow{\tau_k} S_k$$

$$\tau_{k+1} \downarrow \qquad\qquad \downarrow \tau_{k+1}$$

$$V \xrightarrow{\tau_k} S_{k+1} \longrightarrow \cdots \xrightarrow{\tau_n} S_n = T$$

Thus the message $m' = \tau_1 \ldots \tau_{k+1}\tau_k \ldots \tau_n$ is concise and produces T from S. In other words , tokens τ_k and τ_{k+1} can be transposed without changing the state produced. By repeating this procedure we reduce this case to the one in item (iii).

(Sufficiency.) Suppose that $\mathcal{F}$ is $\cup$-closed and let τ_x and τ_y be two positive tokens that are effective for a state S. We have $S\tau_x = S + \{x\}$ and $S\tau_y = S + \{y\}$. Because $\mathcal{F}$ is $\cup$-closed, the set $S + \{x, y\}$ belongs to $\mathcal{F}$. Clearly,

$$(S + \{x\})\tau_y = S + \{x, y\} = (S + \{y\})\tau_x,$$

so the medium $(\mathcal{F}, \mathcal{T}_{\mathcal{F}})$ is closed.   □

Learning spaces (cf. Section 8.10) form a special class of closed oriented media as the following theorem asserts.

Theorem 8.54. *Let $\mathcal{F}$ be a finite family of sets. The following conditions are equivalent.*

(i) *$\mathcal{F}$ is a learning space.*
(ii) *$\mathcal{F}$ contains the empty set and is $\cup$-closed and well-graded.*

Proof. (i)$\Rightarrow$(ii). $\mathcal{F}$ is a learning space, therefore it contains the empty set $\varnothing$. Let us prove that $\mathcal{F}$ is $\cup$-closed. Let P and Q be two distinct nonempty sets in $\mathcal{F}$. By Axiom [L1], there is a sequence $q_1, \ldots, q_n$ in P such that

$$\varnothing \subset \{q_1\} \subset \{q_1, q_2\} \subset \cdots \subset \{q_1, \ldots, q_n\} = P$$

with $\{q_1, \ldots, q_i\} \in \mathcal{F}$ for $1 \leq i \leq n$. By Axiom [L2], $\varnothing \subset Q$, $\{q_1\} \in \mathcal{F}$ implies $Q \cup \{q_1\} \in \mathcal{F}$. By the same axiom, $\{q_1\} \subset Q \cup \{q_1\}$, $\{q_1, q_2\} \in \mathcal{F}$ implies $Q \cup \{q_1, q_2\} \in \mathcal{F}$. By repeating this argument we obtain

$$P \cup Q = \{q_1, \ldots, q_n\} \cup Q \in \mathcal{F}.$$

Turning to the well-gradedness, we take any two distinct sets P and Q in $\mathcal{F}$. If one of these sets is a subset of the other set, then well-gradedness follows from Axiom [L1]. Thus we may assume that $P \setminus Q \neq \varnothing$ and $Q \setminus P \neq \varnothing$. As shown above, we have $P \cup Q \in \mathcal{F}$. By Axiom [L1], there are two chains of subsets in $\mathcal{F}$:

$$P \subset P \cup \{q_1\} \subset \cdots \subset P \cup \{q_1, \ldots, q_n\} = P \cup Q,$$
$$Q \subset Q \cup \{q_1'\} \subset \cdots \subset Q \cup \{q_1', \ldots, q_m'\} = P \cup Q,$$

with $n = |Q \setminus P|$, $m = |P \setminus Q|$, yielding $n + m = |P \triangle Q| = d(P, Q)$. The result follows from defining the sequence:

$$R_0 = P, \ R_1 = P \cup \{q_1\}, \ldots, R_n = P \cup \{q_1, \ldots, q_n\} = P \cup Q,$$
$$R_{n+1} = (P \cup Q) \setminus \{q'_m\}, \ldots, R_{n+m} = (P \cup Q) \setminus \{q'_1, \ldots, q'_m\} = Q.$$

(ii)$\Rightarrow$(i). Let $\mathcal{F}$ be a $\cup$-closed wg-family containing $\varnothing$. It is clear that Axiom [L1] is satisfied. Suppose that $K \subset L$, with K, $K \cup \{q\}$, and L in $\mathcal{F}$. Because $\mathcal{F}$ is $\cup$-closed, we have $L \cup \{q\} = (K \cup \{q\}) \cup L \in \mathcal{F}$. Hence, Axiom [L2] holds. $\qquad\square$

8.12 Random Walks on Token Systems

In this section, we assume that $(\mathcal{S}, \mathcal{T})$ is a cubical system or a medium on an at most countable set of states $\mathcal{S}$. Our goal is to investigate "evolution" of this system under random occurrence of tokens. Let us choose an "initial state" S_0 and assume that tokens $\tau_1, \ldots, \tau_n, \ldots$ of the system occur randomly at arbitrary chosen times $t_1, \ldots, t_n, \ldots$. The sequence of tokens (τ_n) produces a sequence of states (S_n), $n \geq 0$, where $S_n = S_{n-1}\tau_n$. To describe precisely these "random walks" on the system $(\mathcal{S}, \mathcal{T})$, we first assume the existence of two probability distributions on sets $\mathcal{S}$ and $\mathcal{T}$, respectively.

Definition 8.55. A quadruple $(\mathcal{S}, \mathcal{T}, \xi, \theta)$ is a *probabilistic token system* if the following three conditions hold.

(i) $(\mathcal{S}, \mathcal{T})$ is a cubical system or a medium.
(ii) $\xi : S \mapsto \xi(S)$ is a probability distribution (the *initial distribution*) on $\mathcal{S}$; thus $\xi(S) \geq 0$ for all $S \in \mathcal{S}$, and $\sum_{S \in \mathcal{S}} \xi(S) = 1$.
(iii) $\theta : \tau \mapsto \theta_\tau$ is a probability distribution on $\mathcal{T}$ with $\theta_\tau > 0$ for all tokens τ in $\mathcal{T}$. Thus, $\sum_{\tau \in \mathcal{T}} \theta_\tau = 1$.

By selecting an initial state according to the distribution ξ, and applying occurring tokens first to the initial state and then to its images under successive tokens, we obtain a sequence of random variables $(\mathbf{S}_n)$ taking there values in the set of states. Thus $\mathbf{S}_n = S$ signifies that S is the state on trial n. We define 1-step transition probabilities, that is, the *transition matrix* $\mathbf{P}$, by

$$p(S, V) = \begin{cases} \theta_\tau & \text{if } S\tau = V, \\ 0 & \text{otherwise,} \end{cases} \quad \text{for } V \neq S,$$

and

$$p(S, S) = 1 - \sum_{V \in \mathcal{S} \setminus \{S\}} p(S, V).$$

Note that $0 < p(S, S) < 1$. Indeed, by Axiom [C2], for any state S of the token system $(\mathcal{S}, \mathcal{T})$, there is a token τ that is effective for S. Thus, $p(S, S) < 1$, because $\theta_\tau > 0$, and $p(S, S) > 0$, because $\tilde{\tau}$ is ineffective for S (Exercise 8.5). It is clear that either $p(S, V) = p(V, S) = 0$ or $p(S, V)p(V, S) > 0$ for all $S, V \in \mathcal{S}$.

Furthermore, we assume that the probability of the event

$$\mathbf{S}_0 = S_0 \text{ and } \mathbf{S}_1 = S_1 \text{ and } \cdots \text{ and } \mathbf{S}_n = S_n$$

is given by

$$\Pr(\mathbf{S}_0 = S_0, \mathbf{S}_1 = S_1, \dots, \mathbf{S}_n = S_n) = \xi(S_0)p(S_0, S_1) \cdots p(S_{n-1}, S_n)$$

for any sequence $S_0, S_1, \dots, S_n$ of states in $\mathcal{S}$. In other words, we assume that $(\mathbf{S}_n)$ is a *Markov chain* with initial distribution ξ and transition matrix $\mathbf{P}$.

Because $(\mathbf{S}_n)$ is a Markov chain, the n-step transition probabilities are given by

$$p^{(n)}(S, V) = \sum_{(S_i)} p(S_0, S_1)p(S_1, S_2) \cdots p(S_{n-1}, S_n),$$

where the sum is taken over all n-tuples of states $(S_i) = (S_0, S_1, \dots, S_n)$ with $S_0 = S$ and $S_n = V$.

By Axiom [C2], for any two distinct states S and V there is a stepwise effective message $\boldsymbol{m} = \tau_1 \cdots \tau_n$ producing V from S. Let (S_i) be a sequence of states produced by $\boldsymbol{m}$. Then

$$p(S_0, S_1)p(S_1, S_2) \cdots p(S_{n-1}, S_n) = \theta_{\tau_1}\theta_{\tau_2} \cdots \theta_{\tau_n} > 0.$$

It follows that there exists $n \geq 1$ such that $p^{(n)}(S, V) > 0$. Markov chains with this property are called *irreducible*. One can say that in an irreducible Markov chain "every state can be reached from every other state". For the n-tuple $(S, S, \dots, S)$, the product $p(S_0, S_1)p(S_1, S_2) \cdots p(S_{n-1}, S_n)$ is positive because $p(S, S) > 0$. Hence, $p^{(n)}(S, S) > 0$. Markov chains satisfying this condition are known as *aperiodic* chains. Thus, the Markov chain $(\mathbf{S}_n)$ is irreducible and aperiodic.

Now we establish another important property of the chain $(\mathbf{S}_n)$. Let S_0 be a fixed state in $\mathcal{S}$ and let v be a function on $\mathcal{S}$ given by

$$v(S) = \prod_{\tau \in \widehat{S} \setminus \widehat{S}_0} \frac{\theta_\tau}{\theta_{\tilde{\tau}}}. \tag{8.11}$$

Note that we have different definitions of content depending on whether the system under consideration is cubical or a medium. However, the properties of the contents are the same. We use these properties in the proof of the following lemma.

Lemma 8.56. *For any two distinct states S and V,*

$$v(V) = v(S) \prod_{\tau \in \widehat{V} \setminus \widehat{S}} \frac{\theta_\tau}{\theta_{\tilde{\tau}}}. \tag{8.12}$$

Proof. By Theorems 8.21 and 8.43, $\widehat{V} \setminus \widehat{S} \neq \varnothing$. We need to show that

$$\prod_{\tau \in \widehat{V} \setminus \widehat{S}_0} \frac{\theta_\tau}{\theta_{\tilde{\tau}}} = \prod_{\tau \in \widehat{S} \setminus \widehat{S}_0} \frac{\theta_\tau}{\theta_{\tilde{\tau}}} \cdot \prod_{\tau \in \widehat{V} \setminus \widehat{S}} \frac{\theta_\tau}{\theta_{\tilde{\tau}}}. \tag{8.13}$$

We use the identity

$$(\widehat{S} \setminus \widehat{S}_0) \cup (\widehat{V} \setminus \widehat{S}) = (\widehat{V} \setminus \widehat{S}_0) \cup [\widehat{S} \setminus (\widehat{S}_0 \cup \widehat{V})] \cup [(\widehat{S}_0 \cap \widehat{V}) \setminus \widehat{S}],$$

where sets $\widehat{S} \setminus \widehat{S}_0$ and $\widehat{V} \setminus \widehat{S}$ are disjoint, and the three sets $\widehat{V} \setminus \widehat{S}_0$, $\widehat{S} \setminus (\widehat{S}_0 \cup \widehat{V})$, and $(\widehat{S}_0 \cap \widehat{V}) \setminus \widehat{S}$ on the right side are pairwise disjoint (cf. Exercise 8.23). Therefore,

$$\prod_{\tau \in \widehat{S} \setminus \widehat{S}_0} \frac{\theta_\tau}{\theta_{\tilde{\tau}}} \cdot \prod_{\tau \in \widehat{V} \setminus \widehat{S}} \frac{\theta_\tau}{\theta_{\tilde{\tau}}} = \prod_{\tau \in \widehat{V} \setminus \widehat{S}_0} \frac{\theta_\tau}{\theta_{\tilde{\tau}}} \cdot \prod_{\tau \in P} \frac{\theta_\tau}{\theta_{\tilde{\tau}}} \cdot \prod_{\tau \in Q} \frac{\theta_\tau}{\theta_{\tilde{\tau}}}, \tag{8.14}$$

where $P = \widehat{S} \setminus (\widehat{S}_0 \cup \widehat{V})$ and $Q = (\widehat{S}_0 \cap \widehat{V}) \setminus \widehat{S}$. If $\mu \in P$, then $\mu \in \widehat{S}$, $\mu \notin \widehat{S}_0$, and $\mu \notin \widehat{V}$. By Theorems 8.20 and 8.42, $\tilde{\mu} \notin \widehat{S}$, $\tilde{\mu} \in \widehat{S}_0$, and $\tilde{\mu} \in \widehat{V}$, implying $\tilde{\mu} \in Q$. Hence, for any factor $\theta_\mu / \theta_{\tilde{\mu}}$ of the second product in (8.14), there is a factor $\theta_{\tilde{\mu}} / \theta_\mu$ of the third product. Likewise, for any factor $\theta_\mu / \theta_{\tilde{\mu}}$ of the third product in (8.14), there is a factor $\theta_{\tilde{\mu}} / \theta_\mu$ of the second product. Thus,

$$\prod_{\tau \in P} \frac{\theta_\tau}{\theta_{\tilde{\tau}}} \cdot \prod_{\tau \in Q} \frac{\theta_\tau}{\theta_{\tilde{\tau}}} = 1,$$

which proves (8.13). $\qquad\square$

Theorem 8.57. *Let v be a function defined by (8.11). Then*

$$v(S)p(S, V) = v(V)p(V, S), \tag{8.15}$$

for all $S, V \in \mathcal{S}$.

Proof. Equality (8.15) holds trivially if $p(S, V) = p(V, S) = 0$ or $S = V$. Otherwise, $V = S\tau$ for some token $\tau \in \mathcal{T}$, so $\widehat{V} \setminus \widehat{S} = \{\tau\}$. By (8.13),

$$v(V) = v(S) \frac{\theta_\tau}{\theta_{\tilde{\tau}}} = v(S) \frac{p(S, V)}{p(V, S)}$$

proving (8.15). $\qquad\square$

The transition matrix $\mathbf{P}$ and a positive function v satisfying equation (8.57) are said to be in *detailed balance* and the equation itself is called the *detailed balance equation*. Clearly, $\sum_{S \in \mathcal{S}} p(V, S) = 1$ for any state V, therefore we have

$$v(V) = \sum_{S \in \mathcal{S}} v(V)p(V, S) = \sum_{S \in \mathcal{S}} v(S)p(S, V),$$

or, in matrix notation, $v = v\mathbf{P}$. A positive function v satisfying this equation is called an *invariant measure* for the transition matrix $\mathbf{P}$. Thus v defined by (8.11) is an invariant measure for $\mathbf{P}$. If $\sum_{S \in \mathcal{S}} v(S) < \infty$, then we can normalize measure v and obtain a function that is called an *invariant distribution*:

$$\pi(S) = \frac{1}{\sum_{S \in \mathcal{S}} v(S)} v(S).$$

Let us define a function m on $\mathcal{S}$ by

$$m(S) = \sum_{V \in \mathcal{S}} \prod_{\tau \in \widehat{V} \setminus \widehat{S}} \frac{\theta_\tau}{\theta_{\bar{\tau}}}, \tag{8.16}$$

assuming that $\prod_{\tau \in \varnothing}(\theta_\tau / \theta_{\bar{\tau}}) = 1$ and that $m(S) = \infty$, if the series in (8.16) diverges. Clearly, $\sum_{S \in \mathcal{S}} v(S) = m(S_0)$.

Lemma 8.58. *If $m(S_0) < \infty$, then $m(S) < \infty$ for all $S \in \mathcal{S}$. Moreover, $\pi(S) = (m(S))^{-1}$ for all $S \in \mathcal{S}$.*

Proof. By Lemma 8.56,

$$m(S_0) = \sum_{V \in \mathcal{S}} v(V) = \sum_{V \in \mathcal{S}} \left(v(S) \prod_{\tau \in \widehat{V} \setminus \widehat{S}} \frac{\theta_\tau}{\theta_{\bar{\tau}}} \right) = v(S)m(S),$$

which implies

$$\pi(S) = \frac{1}{\sum_{S \in \mathcal{S}} v(S)} v(S) = \frac{1}{m(S_0)} v(S) = \frac{1}{m(S)}.$$

$\square$

The following theorem summarizes properties of the Markov chain $(\mathbf{S}_n)$.

Theorem 8.59. *The Markov chain $(\mathbf{S}_n)$ is irreducible and aperiodic. If the series*

$$m(S) = \sum_{V \in \mathcal{S}} \prod_{\tau \in \widehat{V} \setminus \widehat{S}} \frac{\theta_\tau}{\theta_{\bar{\tau}}}$$

converges for some state S, it converges for all states. In this case, $(\mathbf{S}_n)$ has an invariant distribution

$$\pi(S) = \frac{1}{m(S)} = \left[\sum_{V \in \mathcal{S}} \prod_{\tau \in \hat{V} \setminus \hat{S}} \frac{\theta_\tau}{\theta_{\tilde{\tau}}} \right]^{-1}. \tag{8.17}$$

The distribution π satisfies the detailed balance equation

$$\pi(S)p(S, V) = \pi(V)p(V, S).$$

If $\mathcal{S}$ is a finite set, then $m(S)$ is a finite function. Then (8.17) can be written in the form (cf. Exercise 8.24)

$$\pi(S) = \frac{\prod_{\tau \in \hat{S}} \theta_\tau}{\sum_{V \in \mathcal{S}} \prod_{\tau \in \hat{V}} \theta_\tau}. \tag{8.18}$$

Example 8.60. Let $(\mathcal{S}, \mathcal{T})$ be a token system with the set of states $\mathcal{S} = \{1, 2, \ldots\}$ and tokens defined by

$$k\tau_i = \begin{cases} k, & \text{if } k \neq i - 1, \\ i, & \text{if } k = i - 1, \end{cases} \quad k\tilde{\tau}_i = \begin{cases} k, & \text{if } k \neq i, \\ i - 1, & \text{if } k = i, \end{cases} \quad \text{for } i \geq 1.$$

It is easy to verify that $(\mathcal{S}, \mathcal{T})$ is a medium and that tokens τ_i and $\tilde{\tau}_i$ are indeed reverses of each other. Let $(\mathcal{S}, \mathcal{T}, \xi, \theta)$ be a probabilistic token system. We denote $\theta_i = \theta_{\tau_i}$ and $\tilde{\theta}_i = \theta_{\tilde{\tau}_i}$. Suppose that π is an invariant distribution for the Markov chain $(\mathbf{S}_n)$, that is, $\pi \mathbf{P} = \pi$. A direct calculation shows that

$$\pi(n) = \pi(0) \prod_{i=1}^{n} \frac{\theta_i}{\tilde{\theta}_i}.$$

Thus $(\mathbf{S}_n)$ has a stationary distribution if and only if

$$\sum_{n > 0} \left(\prod_{i=1}^{n} \frac{\theta_i}{\tilde{\theta}_i} \right) < \infty,$$

where the sum on the left side is $m(0)$ (cf. (8.16)). The above series converges, for instance, if

$$\limsup \frac{\theta_n}{\tilde{\theta}_n} < 1.$$

Notes

In theoretical computer science, automata theory (see Eilenberg, 1974) is the study of abstract machines (also known as state transition machines) and the problems they are able to solve. We use main concepts of automata theory in this chapter for a single purpose: to present the token systems theory in a more general context.

The term "token system" was coined by Jean-Claude Falmagne in his original paper (Falmagne, 1997), where a "stochastic token theory" was proposed. The main goal of this chapter was to introduce two instances of token systems, cubical token systems and media, and give the reader a taste of probabilistic models build upon these algebraic concepts.

Cubical token systems were introduced in Ovchinnikov (2008a) as a generalization of media. Examples in Section 8.6 demonstrate that there are cubical systems different from media that are of importance in applications. Downgradable families (cf. Section 8.6) are also known as *accessible set systems* in matroid theory (see, for instance, Björner and Ziegler, 1992).

Only very basic elements of media theory are presented in Sections 8.7–8.11. For a full account of the theory and its applications, the reader is referred to the book by Eppstein et al. (2008).

For the theory and applications of learning spaces, the reader is referred to the book by Falmagne and Doignon (2011). Note that a learning space is known in the combinatorics literature as an "antimatroid". Dilworth (1940) was the first to study antimatroids and they have been frequently rediscovered in other contexts; see Korte et al. (1991) for a comprehensive survey of antimatroid theory with many additional references.

A cubical token system (or a medium) can serve as the algebraic component of a random walk model whose states are those of the system. Material in Section 8.12 lays down a probabilistic foundation for random walk models. Applications of these models are found in Eppstein et al. (2008).

Exercises

8.1. Prove that the unit element 1 of a monoid is unique.

8.2. Let $s = u_1 v_1 = u_2 v_2$ be two factorizations of $s \in \Sigma^*$. Show that there exists a unique factorization $s = xyz$ such that either

$$u_1 = x, \quad v_1 = yz, \quad u_2 = xy, \quad v_2 = z$$

or

$$u_1 = xy, \quad v_1 = z, \quad u_2 = x, \quad v_2 = yz$$

8.3. Prove that $\mathfrak{T}^\dagger$ is a monoid with the unit τ_0.

8.4. Prove Theorem 8.7.

8.5. Let $(\mathcal{S}, \mathfrak{T})$ be a cubical system. Show that a token and its reverse cannot be both effective for a state.

8.6. Show that the cubical system on three states is unique up to isomorphism (cf. Example 8.9).

8.7. Show that the token system in Figure 8.8 is cubical. Find another example of a cubical system on four states.

8.8. Finish the proof of Theorem 8.14.

8.9. Let $(S, \mathcal{J})$ be a cubical systems. For any given $\tau \in \mathcal{J}$ we define

$$\mathcal{U}_\tau = \{S \in S \mid S\tau \neq S\}.$$

Show that

a) $\mathcal{U}_\tau \neq \varnothing$.
b) $(\mathcal{U}_\tau)\tau = \mathcal{U}_{\tilde{\tau}}$.
c) $\mathcal{U}_\tau \cap \mathcal{U}_{\tilde{\tau}} = \varnothing$.
d) The restriction $\tau|_{\mathcal{U}_\tau}$ is a bijection from $\mathcal{U}_\tau$ onto $\mathcal{U}_{\tilde{\tau}}$ with $\tau|_{\mathcal{U}_\tau}^{-1} = \tilde{\tau}|_{\mathcal{U}_{\tilde{\tau}}}$.

8.10. Find contents of the states in Examples 8.8 and 8.9.

8.11. Establish the four properties in (8.3).

8.12. Prove that the family of comparability graphs $\mathcal{CG}$ is downgradable.

8.13. Prove that graphs displayed in Figure 8.10 are comparability graphs. Show that there is no comparablity graph on the six vertices that lie between these two graphs.

8.14. a) Show that every indifference graph is an interval graph.
b) Give an example of an interval graph that is not an indifference graph.

8.15. Let $(S, \mathcal{J})$ be a finite medium with n states and $2m$ tokens.

a) Prove that $m + 1 \leq n \leq 2^m$.
b) Show that $m + 1$ and 2^m are exact bounds for n.
c) Prove that $n = m + 1$ if and only if the graph of $(S, \mathcal{J})$ is a tree.

8.16. Let $(S, \mathcal{J})$ be a medium and G be its graph. Show that the sets

$$S_\tau = \{S \in S : \tau \in \widehat{S}\}$$

are the semicubes of G.

8.17. Prove that the graph of a medium $(S, \mathcal{J})$ is a tree if and only if for any $\tau \in \mathcal{J}$ there is a unique $S \in S$ such that τ is effective for S.

8.18. Let $\mathcal{F}$ be a wg-family of subsets of a set X containing $\varnothing$ with $\cup \mathcal{F} = X$.

a) Describe the contents of the states of the representing medium $(\mathcal{F}, \mathcal{J}_\mathcal{F})$.
b) Describe the contents of the states of the medium on the set $\mathcal{PO}$ of partial orders on a finite set (cf. Section 8.10). Compare this description with the standard definition of a binary relation.
c) Describe the contents of the states of a learning space (cf. Section 8.10).

8.19. Describe the contents of the states of the medium on the set $\mathcal{LO}$ of linear orders on a finite set (cf. Section 8.10).

8.20. A medium $(\mathcal{S}, \mathcal{T})$ is said to be *complete* if for any state S and any token τ, either τ or $\tilde{\tau}$ is effective for S.

a) Prove that a medium is complete if and only if it is closed under any orientation (Eppstein et al., 2008, Section 4.3).
b) Prove that a medium is complete if and only if it is isomorphic to the representing medium on $\mathcal{P}_f(X)$ for some set X.

8.21. Prove Theorem 8.48.

8.22. Use the result of Theorem 5.91 and the "pulling back" method from Section 8.10 to construct a medium on the family $\mathcal{WO}$ of weak orders on a finite set.

8.23. For any three sets A, B, and C, show that

a) Sets $A \setminus C$ and $B \setminus A$ are disjoint.
b) Sets $B \setminus C$, $A \setminus (B \cup C)$, and $(B \cap C) \setminus A$ are disjoint.
c) $(A \setminus C) \cup (B \setminus A) = (B \setminus C) \cup [A \setminus (B \cup C)] \cup [(B \cap C) \setminus A]$.

8.24. Prove (8.18).

Notation

References

Abramenko, P. and Brown, K. S. (2008). *Buildings.* Springer Science + Business Media, LLC, New York.

Afrati, F., Papadimitriou, C. H., and Papageorgiou, G. (1985). The complexity of cubical graphs. *Information and Control,* 66:53–60.

Asratian, A. S., Denley, T. M. J., and Häggkvist, R. (1998). *Bipartite Graphs and their Applications.* Cambridge University Press, Cambridge, London, and New Haven.

Avis, D. (1981). Hypermetric spaces and the Hamming cone. *Canadian Journal of Mathematics,* 33:795–802.

Bandelt, H.-J. and Chepoi, V. (2008). Metric graph theory and geometry: a survey. In *Surveys on discrete and computational geometry, twenty years later, AMS-IMS-SIAM Joint Summer Research Conference, June 18-22, 2006, Snowbird, Utah,* volume 453 of *Contemporary Mathematics,* pages 49–86, Providence, RI. American Mathematical Society.

Bandelt, H.-J. and Mulder, H. M. (1983). Infinite median graphs, $(0, 2)$-graphs, and hypercubes. *Journal of Graph Theory,* 7:487–497.

Behzad, M. and Chartrand, G. (1971). *Introduction to the Theory of Graphs.* Allyn and Bacon, Boston.

Berge, C. (1957). Two theorems in graph theory. *Proceedings of the National Academy of Sciences of the United States of America,* 43:842–844.

Biggs, N. L. (1993). *Algebraic Graph Theory.* Cambridge University Press, Cambridge, 2nd edition.

Biggs, N. L., Lloyd, E. K., and Wilson, R. J. (1986). *Graph Theory 1736–1936.* Clarendon Press, Oxford, 2nd edition.

Birkhoff, G. (1967). *Lattice Theory.* American Mathematical Society, Providence, R.I., 3rd edition.

Björner, A., Las Vergnas, M., Sturmfels, B., White, N., and Ziegler, G. (1999). *Oriented Matroids.* Cambridge University Press, Cambridge, London, and New Haven, second edition.

Björner, A. and Ziegler, G. (1992). Introduction to greedoids. In White, N., editor, *Matroid Applications,* pages 284–357. Cambridge University Press.

Bogart, K. (1973). Preference structures. I. Distances between transitive preference relations. *Journal of Mathematical Sociology*, 3:49–67.

Bondy, J. A. and Murty, U. S. R. (1976). *Graph Theory with Applications*. North-Holland Publishing Co., Amsterdam and New York.

Bondy, J. A. and Murty, U. S. R. (2008). *Graph Theory*. Springer, New York.

Borovik, A. V. and Borovik, A. (2010). *Mirrors and Reflections*. Springer, New York.

Bourbaki, N. (1966). *General Topology*. Addison-Wesley Publishing Company, Reading, Massachusetts - Palo Alto - London - Don Mills, Ontario.

Bourbaki, N. (2002). *Lie Groups and Lie Algebras*. Springer-Verlag, Berlin, Heidelberg, and New York.

Brešar, B., Imrich, W., and Klavžar, S. (2005). Reconstructing subgraph-counting graph polynomials of increasing families of graphs. *Discrete Mathematics*, 297:159–166.

Chepoi, V. (1988). Isometric subgraphs of Hamming graphs and d-convexity. *Control and Cybernetics*, 24:6–11.

Chepoi, V. (1994). Separation of two convex sets in convexity structures. *Journal of Geometry*, 50:30–51.

Coxeter, H. S. M. (1973). *Regular Polytopes*. Dover Publications, Inc., New York.

de Bruijn, N. (1981). Algebraic theory of Penrose's non-periodic tilings of the plane. *Indagationes Mathematicae*, 43:38–66.

Deza, M. and Laurent, M. (1997). *Geometry of Cuts and Metrics*. Springer-Verlag, Berlin, Heidelberg, and New York.

Deza, M. and Shtogrin, M. (2002). Mosaics and their isometric embeddings. *Izvestia: Mathematics*, 66:443–462.

Diestel, R. (2005). *Graph Theory*. Springer-Verlag, Berlin, Heidelberg, and New York, 3rd edition.

Dilworth, R. P. (1940). Lattices with unique irreducible decomposition. *Annals of Mathematics*, 41:771–777.

Djoković, D. Ž. (1973). Distance preserving subgraphs of hypercubes. *Journal of Combinatorial Theory, Ser. B*, 14:263–267.

Doble, C., Doignon, J.-P., Falmagne, J.-C., and Fishburn, P. (2001). Almost connected orders. *Order*, 18(4):295–311.

Doignon, J.-P. and Falmagne, J.-C. (1997). Well-graded families of relations. *Discrete Mathematics*, 173:35–44.

Eilenberg, S. (1974). *Automata, Languages, and Machines*. Academic Press.

Eppstein, D. (1996). Zonohedra and zonotopes. *Mathematica in Education and Research*, 5(4):15–21.

Eppstein, D. (2005). The lattice dimension of a graph. *European Journal of Combinatorics*, 26(6):585–592.

Eppstein, D. (2008). Recognizing partial cubes in quadratic time. In *Proceedings of the nineteenth annual ACM-SIAM symposium on Discrete algorithms*, pages 1258–1266, Philadelphia, PA, USA. SIAM.

Eppstein, D., Falmagne, J.-C., and Ovchinnikov, S. (2008). *Media Theory*. Springer-Verlag, Berlin.

Falmagne, J.-C. (1997). Stochastic token theory. *Journal of Mathematical Psychology*, 41(2):129–143.

Falmagne, J.-C. and Doignon, J.-P. (1997). Stochastic evolution of rationality. *Theory and Decision*, 43:107–138.

Falmagne, J.-C. and Doignon, J.-P. (2011). *Learning Spaces*. Springer, Heidelberg.

Firsov, V. V. (1965). On isometric embedding of a graph into a Boolean cube. *Cybernetics*, 1:112–113.

Fishburn, P. (1985). *Interval orders and interval graphs*. John Wiley & Sons, Chichester.

Foldes, S. (1977). A characterization of hypercubes. *Discrete Mathematics*, 17:155–159.

Foulds, L. R. (1992). *Graph Theory Applications*. Springer-Verlag, New York.

Fukuda, K. and Handa, K. (1993). Antipodal graphs and oriented matroids. *Discrete Mathematics*, 111:245–256.

Garey, M. R. and Graham, R. L. (1975). On cubical graphs. *Journal of Combinatorial Theory (B)*, 18:84–95.

Golomb, S. W. (1994). *Polyominoes: Puzzles, Patterns, Problems, and Packings*. Princeton Science Library, Princeton University Press, NJ, 2nd edition.

Gorbatov, V. A. and Kazanskiy, A. A. (1983). Characterization of graphs embedded in n-cubes. *Engineering Cybernetics*, 20:96–102.

Graham, R. L., Grötschel, M., and Lovász, L., editors (1995). *Handbook of Combinatorics*. The M.I.T. Press, Cambridge, MA.

Graham, R. L. and Winkler, P. M. (1985). On isometric embeddings of graphs. *Transactions of the American Mathematical Society*, 288(2):527–536.

Grätzer, G. (2003). *General Lattice Theory*. Birkhäuser, Boston, Basel, Berlin, 2nd edition.

Greene, C. (1977). Acyclic orientations (note). In Aigner, M., editor, *Higher Combinatorics*, pages 65–68. Reidel, Dordrecht.

Grünbaum, B. (2003). *Convex Polytopes*. Springer, New York, Berlin, Heidelberg, 2nd edition.

Grünbaum, B. and Shephard, G. (1987). *Tilings and Patterns*. W.H. Freeman, New York.

Guilbaud, G. T. and Rosenstiehl, P. (1963). Analyse algébrique d'un scrutin. *Mathématiques et Sciences Humaines*, 4:9–33.

Gutman, I. and Cyvin, S. (1989). *Introduction to the Theory of Benzenoid Hydrocarbons*. Springer-Verlag, Berlin, Heidelberg, and New York.

Gutman, I. and Klavžar, S. (1996). A method for calculatin Wiener numbers of benzenoid hydrocarbons. *ACH – Models in Chemistry*, 133:389–399.

Hadlock, F. and Hoffman, F. (1978). Manhattan trees. *Utilitas Mathematica*, 13:55–67.

Hales, T. C. (2007). Jordan's proof of the Jordan curve theorem. *Studies in Logic, Grammar and Rhetoric*, 10(23):45–60.

Hall, P. (1935). On representatives of subsets. *Journal of London Mathematical Society*, 10:26–30.

Hamming, R. (1950). Error Detecting and Error Correcting Codes. *Bell System Technical Journal*, 26(2):147–160.

Harary, F. (1969). *Graph Theory*. Addison-Wesley, Reading, Mass.

Harary, F. (1988). Cubical graphs and cubical dimensions. *Computers & Mathematics with Applications*, 15:271–275.

Harary, F., Hayes, J. P., and Wu, H.-J. (1988). A survey of the theory of hypercube graphs. *Computers & Mathematics with Applications*, 15:277–289.

Havel, I. (1983). Embedding graphs in undirected and directed graphs. In *Graph Theory, Proceedings of a Conference held in Łagów, Poland, February 10–13, 1981*, volume 1018 of *Lecture Notes in Mathematics*, pages 60–68, Berlin, Heidelberg, New York, Tokyo. Springer-Verlag.

Havel, I. and Liebl, P. (1972a). Embedding the dichotomic tree into the n-cube. *Časopis pro pěstování matematiky*, 97:201–205.

Havel, I. and Liebl, P. (1972b). Embedding the polytomic tree into the n-cube. *Časopis pro pěstování matematiky*, 98:307–314.

Havel, I. and Morávek, R. L. (1972). B-valuations of graphs. *Czechoslovak Mathematical Journal*, 22:338–351.

Holton, D. A. and Sheehan, J. (1993). *The Petersen Graph*. Cambridge University Press, Cambridge.

Hsu, L.-H. and Lin, C.-H. (2009). *Graph Theory and Interconnection Networks*. CRC Press, Boca Raton, London, New York.

Imrich, W. and Klavžar, S. (1998). A convexity lemma and expansion procedure for bipartite graphs. *European Journal of Combinatorics*, 19:677–685.

Imrich, W. and Klavžar, S. (2000). *Product Graphs*. John Wiley & Sons, London and New York.

Kay, D. C. and Chartrand, G. (1965). A characterization of certain ptolemaic graphs. *Canadian Journal of Mathematics*, 17:342–346.

Kőnig, D. (1989). *Theory of Finite and Infinite Graphs*. Birkhäuser, Boston, Basel, Berlin.

Klavžar, S., Gutman, I., and Mohar, B. (1995). Labeling of benzenoid systems which reflects the vertex-distance relations. *Journal of Chemical Information and Computer Sciences*, 35:590–593.

Korte, B., Lovász, L., and Schrader, R. (1991). *Greedoids*. Number 4 in Algorithms and Combinatorics. Springer-Verlag.

Krumme, D., Venkataraman, K., and Cybenko, G. (1986). Hypercube embedding is NP-complete. In Heath, M., editor, *Hypercube Microprocessors 1986*. SIAM, Philadelphia.

Kuzmin, V. and Ovchinnikov, S. (1975). Geometry of preference spaces I. *Automation and Remote Control*, 36:2059–2063.

Lovász, L. and Plummer, M. D. (1986). *Matching Theory*. Annals of Discrete Mathematics, Vol. 29. North-Holland, Amsterdam.

Mirkin, B. (1979). *Group Choice*. Winston, Washington, D.C.

Mohar, B. and Thomassen, C. (2001). *Graphs on Surfaces*. The John Hopkins University Press, Baltimore and London.

Mulder, H. M. (1980). *The Interval Function of a Graph*. Mathematical Centre Tracts 132. Mathematisch Centrum, Amsterdam.

Munkres, J. R. (2000). *Topology*. Prentice Hall, Upper Saddle River, NJ 07458, 2nd edition.

Nebeský, L. (1971). Median graphs. *Commentationes Mathematicae Universitatis Carolinae*, 12:317–325.

Nebeský, L. (1974). On cubes and dichotomic trees. *Časopis pro pěstovaáni matematiky*, 99:164–167.

Nebeský, L. (2008). On the distance function of a connected graph. *Czechoslovak Mathematical Journal*, 58:1101–1106.

Orlik, P. and Terano, H. (1992). *Arrangements of Hyperplanes*. Springer-Verlag, Berlin, Heidelberg, and New York.

O'Rourke, J. (1987). *Art Gallery Theorems and Algorithms*. Oxford University Press, New York, Oxford.

Ovchinnikov, S. (1980). Convexity in subsets of lattices. *Stochastica*, IV:129–140.

Ovchinnikov, S. (2004). The lattice dimension of a tree. Electronic preprint math.CO/0402246, arXiv.org.

Ovchinnikov, S. (2005). Hyperplane arrangements in preference modeling. *Journal of Mathematical Psychology*, 49:481–488.

Ovchinnikov, S. (2008a). Cubical token systems. *Mathematical Social Sciences*, 56:149–165.

Ovchinnikov, S. (2008b). Geometric representations of weak orders. In Bouchon-Meunier, B., Yager, R., Marsala, C., and Rifqi, M., editors, *Uncertainty and Intelligent Information Systems*, pages 447–456. World Scientific Publishing, Singapore.

Ovchinnikov, S. (2008c). Partial cubes: structures, characterizations, and constructions. *Discrete Mathematics*, 308:5597–5621.

Reed, R. C. and Wilson, R. J. (1998). *An Atlas of Graphs*. Clarendon Press, Oxford.

Roberts, F. (1979). *Measurement Theory, with Applications to Decisionmaking, Utility, and the Social Sciences*. Addison-Wesley, Reading, Mass.

Roberts, F. R. (1984). *Applied Combinatorics*. Prentice-Hall, Englewood Cliffs.

Roth, R. I. and Winkler, P. M. (1986). Collapse of the metric hierarchy for bipartite graphs. *European Journal of Combinatorics*, 7:179–197.

Sabidussi, G. (1960). Graph multiplication. *Mathematische Zeitschrift*, 72:446–457.

Scapellato, R. (1990). On *F*-geodetic graphs. *Discrete Mathematics*, 17:155–159.

Senechal, M. (1995). *Quasicrystals and Geometry*. Cambridge University Press, Cambridge, London, and New Haven.

Stanley, R. P. (1997). *Enumerative Combinatorics*. Cambridge University Press, Cambridge and New York.

Stanley, R. P. (2007). An introduction to hyperplane arrangements. In Miller, E., Reiner, V., and Sturmfels, B., editors, *Geometric Combinatorics*, pages 389–496. American Mathematical Society, Institute for Advanced Study.

Szpilrajn, E. (1930). Sur l'extension de l'ordre partiel. *Fundamenta Mathematicae*, 16:386–389.

Trotter, W. (1992). *Combinatorics and Partially Ordered Sets: Dimension Theory*. The Johns Hopkins University Press, Baltimore and London.

West, D. B. (2001). *Introduction to Graph Theory*. Prentice Hall, Upper Saddle River, NJ 07458.

Winkler, P. (1984). Isometric embedding in products of complete graphs. *Discrete Applied Mathematics*, 7:221–225.

Xu, J. (2001). *Topological Structure and Analysis of Interconnection Networks*. Kluwer Academic Press, Dordrecht, Boston, London.

Zaslavsky, T. (1975). *Facing up to arrangements: face count formulas for partitions of space by hyperplanes*, volume 154 of *Memoirs of the AMS*. American Mathematical Society, Providence, R.I.

Ziegler, G. (2006). *Lectures on Polytopes*. Springer, Berlin, Heidelberg, and New York.

Index